SIMPLIFIED DESIGN OF STEEL STRUCTURES

SIMPLIFIED DESIGN OF STEEL STRUCTURES

Eighth Edition

JAMES AMBROSE

and

PATRICK TRIPENY

JOHN WILEY & SONS, INC.

Published by John Wiley & Sons, Inc., Hoboken, New Jersey

Published simultaneously in Canada

Wiley Bicentennial Logo: Richard J. Pacifico

For general information about our other products and services, please contact our Customer Care Department within the United States at (800) 762–2974, outside the United States at (317) 572–3993 or fax (317) 572–4002.

Wiley also publishes its books in a variety of electronic formats. Some content that appears in print may not be available in electronic books. For more information about Wiley products, visit our web site at www.wiley.com.

Library of Congress Cataloging-in-Publication Data:

Ambrose, James E.
Simplified design of steel structures / James Ambrose and Patrick Tripeny. — 8th ed.
p. cm.
Includes bibliographical references and index.
ISBN: 978-0-470-08631-5 (cloth)
1. Building, Iron and steel. I. Tripeny, Patrick. II. Title.
TA684.A343 2007
624.1′821—dc22 2007002363

10 9 8 7 6 5 4

CONTENTS

Preface to the Eighth Edition ix

Preface to the First Edition xiii

Introduction 1

1 Considerations for Use of Steel 12

1.1 Properties of Steel / 12
1.2 Types of Steel Products / 16
1.3 Development of Structural Systems / 24
1.4 Connection Methods / 25
1.5 Data for Steel Products / 25
1.6 Usage Considerations for Steel Structures / 26
1.7 Choice and Planning of Steel Structural Systems / 33

2 Structural Investigation and Design 45

2.1 Situations for Investigation and Design / 45
2.2 Methods of Investigation and Design / 50

2.3 Investigation of Columns and Beams / 52
2.4 Investigation of Column and Beam Frames / 60
2.5 Approximate Investigation of Indeterminate Structures / 68
2.6 Service Conditions / 72
2.7 Limit States Versus Service Conditions / 72
2.8 Resistance Factors / 73
2.9 Choice of Design Method / 73

3 Horizontal-Span Framing Systems 75

3.1 Factors in Beam Design / 75
3.2 Inelastic Versus Elastic Behavior / 77
3.3 Nominal Moment Capacity of Steel Beams / 84
3.4 Design for Bending / 89
3.5 Design of Beams for Buckling Failure / 99
3.6 Shear in Steel Beams / 103
3.7 Deflection of Beams / 108
3.8 Safe Load Tables / 118
3.9 Manufactured Trusses for Flat Spans / 129
3.10 Decks with Steel Framing / 137
3.11 Concentrated Load Effects in Beams / 144
3.12 Torsional Effects / 148
3.13 Buckling of Beams / 151
3.14 Beam Bearing Plates / 156

4 Steel Columns 161

4.1 Column Shapes / 161
4.2 Slenderness and End Conditions / 163
4.3 Safe Axial Loads for Steel Columns / 164
4.4 Design of Steel Columns / 170
4.5 Columns with Bending / 180
4.6 Column Framing and Connections / 188
4.7 Column Bases / 192

5 Frame Bents 196

5.1 Development of Bents / 196
5.2 Multiunit Rigid Frames / 197
5.3 Three-Dimensional Frames / 199

5.4 Mixed Frame and Wall Systems / 201
5.5 Special Problems of Steel Rigid Frame Bents / 204
5.6 Trussed Frames / 207

6 Miscellaneous Steel Components and Systems 213

6.1 Manufactured Systems / 213
6.2 Composite Structural Elements / 214
6.3 Tension Elements and Systems / 220

7 Horizontal-Span Steel Trusses 234

7.1 General Considerations / 234
7.2 Bracing for Trusses / 238
7.3 Loads on Trusses / 239
7.4 Investigation for Internal Forces in Planar Trusses / 242
7.5 Design Forces for Truss Members / 257
7.6 Combined Actions in Truss Members / 260
7.7 Design Considerations for Steel Trusses / 260
7.8 Two-Way Trusses / 262

8 Steel Connections 269

8.1 Basic Considerations / 269
8.2 Bolted Connections / 272
8.3 Design of a Bolted Connection / 284
8.4 Bolted Framing Connections / 290
8.5 Bolted Truss Connections / 295
8.6 Welding / 298
8.7 Design of Welded Connections / 303
8.8 Welded Steel Frames / 307
8.9 Control Joints: Design for Selected Behavior / 307

9 General Considerations for Building Structures 312

9.1 Choice of Building Construction / 312
9.2 Structural Design Standards / 313
9.3 Loads for Structural Design / 314
9.4 Dead Loads / 315
9.5 Building Code Requirements for Structures / 315
9.6 Live Loads / 318
9.7 Lateral Loads (Wind and Earthquake) / 322

9.8 Load Combinations and Factors / 326
9.9 Determination of Design Loads / 326
9.10 Structural Planning / 328
9.11 Building Systems Integration / 329
9.12 Economics / 329

10 Building Structures: Design Examples **332**

10.1 Building One / 332
10.2 Building Two / 350
10.3 Building Three / 354
10.4 Building Four / 384
10.5 Building Five / 394
10.6 Building Six / 396

Appendix A: Properties of Structural Sections **413**

Appendix B: Beam Design Aids **439**

Appendix C: Study Aids **445**

Terms / 445
Questions / 450
Answers to Questions / 452

Appendix D: Answers to Problems in Chapters **455**

References **459**

Index **461**

PREFACE TO THE EIGHTH EDITION

Publication of this book presents an opportunity for yet another generation of students of building design to access the subject of steel structures. The particular focus of this work is a concentration on widely used, simple, and ordinary methods of construction. In addition, the effort has been made to keep mathematical work at a low level, in order to emphasize the accessibility of the work to untrained persons.

The basic purpose of this "simplified" work is well expressed in the preface to the first edition by the originator of the simplified series of books, the late Professor Harry Parker of the University of Pennsylvania. Excerpts from Professor Parker's preface to the first edition follow this preface. To the extent possible, we have adhered to the spirit expressed by Professor Parker.

Of course, structural engineering is no longer simple, really. Utilization of all of the available resources for investigation and design of structures is a complex and exhaustive task. Professional engineers must climb this notable learning curve in pursuit of credibility as professionals. We do not mean to belittle the work of serious engineers by our simplified approach. Still, practical means for construction and the use of very ordinary structural products provides the resource for the vast amount of building construction—even in the computer age. It is possible, therefore, to present real structural solutions for ordinary structural tasks with a minimum of complexity. And it is also possible to present the design process for these solutions in relatively elementary form.

Readers of this book should obtain a useful overview of the field of steel structures and of the means for their design. Those wishing to pursue the study to more advanced levels can find many publications and educational opportunities for their study. For many, seeking mostly only a general view of the topic and an understanding of basic design processes, this book may suffice well.

This topic is supported by an amazing archive of reference materials, including publications and computer programs. We have used a few essential sources for this work, a primary one being the AISC Specification, published by the American Institute of Steel Construction and presented with other exhaustive information in the *Manual of Steel Construction* (Ref. 3), known more familiarly as the AISC Handbook To the extent possible, the work presented here conforms to the specifications of the latest edition of that work. We are grateful to the AISC for permission to use some materials from the AISC Handbook for this work.

Preparation of instructional material needs to include the testing of the materials in classroom situations. The authors of this book have had extensive opportunities to utilize classroom experience for development of the materials in this book. We are considerably in debt to the many students who have sat in our classes, and from whom we have undoubtedly learned more than they have from our efforts. We are also indebted to the schools that have provided our teaching opportunities, especially the University of Utah and the University of Southern California. We are very grateful for the support and encouragement provided by these schools to classroom teachers.

We are very appreciative for the support of our publishers, John Wiley & Sons. This publisher has a long history of maintaining a strong catalog of publications in the fields of architecture and construction, and we thank them for that continuing effort. We are especially grateful to our publisher, Amanda Miller; our editor, Paul Drougas; and our production liaison, Nancy Cintron. This book would indeed not have been produced without their significant contributions and support.

Finally, we need to express the gratitude we have to our families. Writing work, especially when added to an already full-time occupation, is very time-consuming. We thank our spouses and children for their patience, endurance, support, and encouragement in permitting us to achieve this work.

James Ambrose

Patrick Tripeny

PREFACE TO THE FIRST EDITION

(The following is an excerpt from Professor Parker's preface to the first edition.)

Simplified Design of Structural Steel is the fourth of a series of elementary books dealing with the design of structural members used in the construction of buildings. The present volume treats of the most common structural steel members that occur in building construction. The solution of many structural problems is difficult and involved but it is surprising, on investigation, how readily many of the seemingly difficult problems may be solved. The author has endeavored to show how the application of the basic principles of mechanics simplifies the problems and leads directly to a solution. Using tables and formulas blindly is a dangerous procedure; they can only be used safely when there is a clear understanding of the underlying principles upon which the tables or formulas are based. This book deals principally in the practical application of engineering principles and formulas in the design of structural members.

In preparing material for this book the author has assumed that the reader is unfamiliar with the subject. Consequently the discussions advance by easy stages, beginning with problems relating to simple direct stresses and continuing to the more involved examples. Most of the fundamental principles of mechanics are reviewed and, in general, the only preparation needed is a knowledge of arithmetic and high school algebra.

In addition to discussions and explanations of design procedure, it has been found that the solution of practical examples adds greatly to the value of a book of this character. Consequently, a great portion of the text consists of the solution of illustrative examples. The examples are followed by problems to be solved by the student.

The author proposes no new methods of design nor short cuts of questionable value. Instead, he has endeavored to present concise and clear explanations of the present-day design methods with the hope that the reader may obtain a foundation of sound principles of structural engineering.

HARRY PARKER
High Hollow, Southampton, Pa.
March 1945

INTRODUCTION

Design for the use of steel structures in buildings involves a broad range of considerations. These include basic properties of the materials, forms of common industrialized products, and common usages for typical building construction. However, the topic also embraces concerns for building design in general, regulatory codes, and commonly used methods of professional designers. This introduction briefly considers some of these very basic issues.

USE OF STEEL FOR BUILDING STRUCTURES

This book deals in general with the common uses of steel for the structures of ordinary buildings. If fully considered, this involves a considerable range of usage, because steel is used in one form or another in structures made from all the common materials: wood, concrete, masonry, and, of course, steel. Erection of frames of wood requires the use of nails, screws, bolts, anchorage devices, and various metal connectors—all ordinarily made of steel. Modern concrete and masonry construction is typically achieved with major use of steel reinforcement as well as various anchorage and attachment devices of steel. However, the topic of steel structures as developed here is limited essentially to situations in which steel is used as the major material for the primary structure. Exceptions to this are the use of steel spanning systems in combination with vertical bearing structures of concrete or masonry and the use of composite spanning elements of steel and concrete or steel and wood.

A major use of steel is where a primary frame is erected, consisting of linear elements of structural steel. The term "structural steel" is usually applied to structures in which the major elements are those produced by the process of *hot rolling,* resulting in a linear element of constant cross section. Rolled sections—as produced in the United States—are ordinarily formed into shapes that conform with the standards established by the American Institute of Steel Construction and documented in its publication *Manual of Steel Construction* (Ref. 3), referred to commonly, and hereinafter in this book, as the AISC Manual.

Another category of steel structures is one that utilizes elements of a size and weight generally one notch below the rolled products, although use may also be made of the lightest of the rolled shapes. Other than rolled shapes, the elements commonly used are those produced by *cold forming* (stamping, folding, cold rolling, etc.) of relatively thin sheets of steel. Roof and floor decking and some wall paneling are produced in this manner, and the products are often described as being of *formed sheet steel.* Light framed elements, such as open-web joists, may also use cold-formed elements for some parts.

A third category of usage for steel is that generally described by the term *miscellaneous metals.* This includes elements of the construction that are not parts of the primary structural system but serve some secondary structural function, such as framing for curtain walls, suspended ceilings, door frames, and so on. As with the light steel structure, this construction may use the smaller rolled shapes or elements of cold-formed sheet steel.

Much of the material presented in this book deals with structural steel elements and systems—consisting of hot-rolled shapes. However, as mentioned in the Preface, the title of this book reflects a broader treatment of the subject, including major usage of cold-formed sheet steel products.

METHODS OF INVESTIGATION AND DESIGN

For the determination of design load conditions and the establishment of structural resistance of steel structures, there are currently three different methods in use:

1. *Load and resistance factor design (LRFD).* This method is based on the use of ultimate strength resistance of structural members and determination of failure load levels. The design loads and member resistances are found by multiplying the true (service) loads and the member ultimate resistances by adjustment factors; usually resulting in an increase in the load and a reduction of the member resistance.
2. *Allowable stress design (ASD—old).* Also known as working stress design or simply stress design, this method uses the true (service) loads and sets limits on stresses in members under these load conditions. This is the classic method developed and used in the nineteenth and early twentieth centuries for steel structures. The Empire State Building and Golden Gate Bridge were designed with this method, as were thousands of steel structures still in use.
3. *Allowable strength design (ASD—new).* This is a compromise method based on the use of the old stress design loadings and a factored resistance of structural members. Basically, it uses the service loads from the LRFD method without the increase factors and the strength resistance of members with larger reduction factors. Although the acronym ASD is still used, it is really the LRFD method with minor adjustments.

The LRFD method is now used by most structural engineering firms, and it is the principal focus of the current design aids provided by the AISC. Structural design software is based on this method, and most building codes either require its exclusive use or grant an exception to it in some cases. Factored loads are based on statistical likelihood, and member resistances are based on evaluations backed by laboratory tests to failure. It is unquestionably the most accurate method of design.

The old ADS method is still used by many engineers and by architects who do minor structural design work. It is also mostly used for teaching structures to architecture students and in general for teaching where short time is available for the subject (only part of a whole semester course).

Long-standing and easy-to-use design aids support the design work, providing shortcuts that are easily learned and used. It is not, however, as closely related to the research and development of design produced over the past 50 years and is sure to steadily fade from use.

Although choice of methods used by practicing engineers and architects can only be modified over time, the LRFD method will certainly eventually prevail. Anyone expecting to participate in design of steel structures in the future should learn the LRFD method. It is not that hard to learn, although it involves a few more steps in the design process. Analyses for resistance for some modes of failure are quite complex, but then they are also complex with the stress method. For most ordinary cases, design is simple and direct, and design aids can be used to shortcut the process.

Of course, teaching old dogs new tricks is certain to be met with resistance, especially with persons long experienced with the stress method. On the other hand, teaching students or other inexperienced persons is not really more difficult, because they have no point of comparison and no attachment to other methods. It can easily be accomplished in the same amount of classroom time.

We have chosen to use LRFD for steel design in this book, because it is clearly the important method for students to understand. Students don't have attachments to the "good old days," because they don't have any old days. Furthermore, in the whole process of design work, there is only a small fraction of the effort involved with computation of structural behaviors. It is an essential part of design work, but many other issues and circumstances must also be dealt with. Planning of structural systems, development of construction details, and integration of the building structure with the rest of the building construction and services involves much greater time and attention.

Inspection of this book will show that more space is devoted to issues other than computation of structural behaviors. Furthermore, when actual design work is performed for building cases, use is made wherever possible of shortcuts and available design aids. This is how professional designers work—not by spending the majority of their time cranking out data from complex equations. It helps in the development of understanding to know about the complex equations, but the shortcuts are much more practical for design work.

The procedures of the LRFD method are described more fully in Chapter 2, and its application is illustrated for various situations of structural investigation and design throughout the book.

REFERENCE SOURCES FOR DESIGN

Information about steel structures is forthcoming from a number of sources. These range from relatively unbiased textbooks and research reports to clearly biased promotional materials from the producers and suppliers of construction materials, equipment, and services. Bias is not to be considered as evil; it is merely to be acknowledged in evaluating the potential for completeness and neutrality in the materials presented. One cannot really expect people in the business of selling steel and fabricated steel products to provide unbiased information about the use of steel, especially information about any of its drawbacks or its true competitiveness with other materials.

The total published information about steel and all of the issues relating to it would fill a large building. Some is somewhat more essential and general in nature; most is highly detailed and narrowly directed. The material presented in this book is general in nature, specific to the interests of building designers, and represents a small, distilled essence of a number of general publications. Principal references used for this book are listed in the References located at the back of the book.

There are several notable industry-wide organizations in the United States that have some relation to steel structures. Although these organizations are largely industry supported, they do represent major sources of design codes and standards, as well as information about steel products. They also sponsor much of the research on which the design procedures are based. Materials from several of these sources are presented in this book, with the publications from which they are taken noted.

The most widely used sources for information for design of steel structures are the manuals of the American Institute of Steel Construction (AISC), reference 3 in the References. The AISC manuals have been in publication for many years, providing a copy of the latest design specification as well as much indispensable data regarding currently available steel products for structures. Also included in the manuals are many shortcut design aids, permitting rapid design of commonly used structural elements.

Throughout this book, as well as in the appendices, are many samples of materials from the AISC manuals. These are provided to help explain the use of the manual for investigation and design of structures. Readers can follow the text illustrations and perform the exercise problems in this book using the materials provided here, but they are strongly advised to obtain access to the current manual to become familiar with the scope of materials provided there.

Although the AISC is the principal industry-sponsored organization in the area of steel structures, many other industry and professional organizations also provide materials for designers. Some of these are the following:

The American Society for Testing and Materials (ASTM). This organization provides widely used standards for all sorts of materials, including many structural products. Just about every steel structural product is produced with some ASTM specification.

The American Society of Civil Engineers (ASCE). This organization includes a division that is a major affiliation of structural engineers, and it is the sponsor of many publications on structural design.

The Steel Deck Institute (SDI). This group provides information for design of formed sheet steel products, widely used for roof and floor decks and wall panel units. (See Ref. 5.)

The Steel Joist Institute (SJI). This group provides information regarding light fabricated trusses (called open-web joists) and other forms of steel spanning members. (See Ref. 4.)

There is, in fact, an industry or trade organization for just about every type of product used for construction. Any of them may be the source of useful information, and those referred to here are just a sampling of the mountain of available information.

Anyone intending to pursue the study of this subject beyond the scope of the material in this book should obtain access to a basic text, such as those used for courses in civil engineering schools. Reference 7 in the References is one such publication.

UNITS OF MEASUREMENT

Previous editions of this book have used U.S. units (feet, inches, pounds, etc.) for the basic presentation. In this edition, the basic work is developed with U.S. units with equivalent metric unit values in brackets [thus]. Although the building industry in the United States is still in the process of changing to metric units, our decision for the presentation here is a pragmatic one. Most of the references used for this book are still developed primarily in U.S. units.

Table 1 lists the standard units of measurement in the U.S. system with the abbreviations used in this work and a description of common usage

TABLE 1 Units of Measurement: U.S. System

Name of Unit	Abbreviation	Use in Building Design
Length		
Foot	ft	Large dimensions, building plans, beam spans
Inch	in.	Small dimensions, size of member cross sections
Area		
Square feet	ft^2	Large areas
Square inches	$in.^2$	Small areas, properties of cross sections
Volume		
Cubic yards	yd^3	Large volumes, of soil or concrete (commonly called simply "yards")
Cubic feet	ft^3	Quantities of materials
Cubic inches	$in.^3$	Small volumes
Force, Mass		
Pound	lb	Specific weight, force, load
Kip	kip, k	1000 pounds
Ton	ton	2000 pounds
Pounds per foot	lb/ft, plf	Linear load (as on a beam)
Kips per foot	kips/ft, klf	Linear load (as on a beam)
Pounds per square foot	lb/ft^2, psf	Distributed load on a surface, pressure
Kips per square foot	k/ft^2, ksf	Distributed load on a surface, pressure
Pounds per cubic foot	lb/ft^3	Relative density, unit weight
Moment		
Foot-pounds	ft-lb	Rotational or bending moment
Inch-pounds	in.-lb	Rotational or bending moment
Kip-feet	kip-ft	Rotational or bending moment
Kip-inches	kip-in.	Rotational or bending moment
Stress		
Pounds per square foot	lb/ft^2, psf	Soil pressure
Pounds per square inch	$lb/in.^2$, psi	Stresses in structures
Kips per square foot	$kips/ft^2$, ksf	Soil pressure
Kips per square inch	$kips/in.^2$, ksi	Stresses in structures
Temperature		
Degree Fahrenheit	°F	Temperature

in structural design work. In similar form, Table 2 gives the corresponding units in the metric system. Conversion factors to be used for shifting from one unit system to the other are given in Table 3. Direct use of the conversion factors will produce what is called a *hard conversion* of a reasonably precise form.

In the work in this book, many of the unit conversions presented are *soft conversions,* meaning ones in which the converted value is rounded off to produce an approximate equivalent value of some slightly more rel-

TABLE 2 Units of Measurement: SI System

Name of Unit	Abbreviation	Use in Building Design
Length		
Meter	m	Large dimensions, building plans, beam spans
Millimeter	mm	Small dimensions, size of member cross sections
Area		
Square meters	m^2	Large areas
Square millimeters	mm^2	Small areas, properties of member cross sections
Volume		
Cubic meters	m^3	Large volumes
Cubic millimeters	mm^3	Small volumes
Mass		
Kilogram	kg	Mass of material (equivalent to weight in U.S. units)
Kilograms per cubic meter	kg/m^3	Density (unit weight)
Force, Load		
Newton	N	Force or load on structure
Kilonewton	kN	1000 Newtons
Stress		
Pascal	Pa	Stress or pressure (1 pascal = 1 N/m^2)
Kilopascal	kPa	1000 pascals
Megapascal	MPa	1,000,000 pascals
Gigapascal	GPa	1,000,000,000 pascals
Temperature		
Degree Celsius	°C	Temperature

TABLE 3 Factors for Conversion of Units

To Convert from U.S. Units to SI Units, Multiply by:	U.S. Unit	SI Unit	To Convert from SI Units to U.S. Units, Multiply by:
25.4	in.	mm	0.03937
0.3048	ft	m	3.281
645.2	in.^{2}	mm^2	1.550×10^{-3}
16.39×10^3	in.^{3}	mm^3	61.02×10^{-6}
416.2×10^3	in.^{4}	mm^4	2.403×10^{-6}
0.09290	ft^2	m^2	10.76
0.02832	ft^3	m^3	35.31
0.4536	lb (mass)	kg	2.205
4.448	lb (force)	N	0.2248
4.448	kip (force)	kN	0.2248
1.356	ft-lb (moment)	N-m	0.7376
1.356	kip-ft (moment)	kN-m	0.7376
16.0185	lb/ft^3 (density)	kg/m^3	0.06243
14.59	lb/ft (load)	N/m	0.06853
14.59	kip/ft (load)	kN/m	0.06853
6.895	psi (stress)	kPa	0.1450
6.895	ksi (stress)	MPa	0.1450
0.04788	psf (load or pressure)	kPa	20.93
47.88	ksf (load or pressure)	kPa	0.02093
$0.566 \times (°F - 32)$	°F	°C	$(1.8 \times °C) + 32$

evant numerical significance to the unit system. Thus, a wood 2 × 4 (actually 1.5 × 3.5 inches in the U.S. system) is precisely 38.1 mm by 88.9 mm in the metric system. However, the metric equivalent "2 × 4" is more likely to be made 40 by 90 mm—close enough for most purposes in construction work.

ACCURACY OF COMPUTATIONS

Structures for buildings are seldom produced with a high degree of dimensional precision. Exact dimensions of some parts of the construction—such as window frames and elevator rails—must be reasonably precise; however, the basic structural framework is ordinarily achieved with only a very limited dimensional precision. Add this to considerations for the lack of precision in predicting loads for any structure, and the significance of highly precise structural computations becomes moot. This is not to be used for an argument to justify sloppy mathematical work, sloppy construction, or the use of vague theories of investigation of behaviors. Nevertheless, it makes a case for not being highly concerned with any numbers beyond about the second digit.

Although most professional design work these days is likely to be done with computer support, most of the work illustrated here is quite simple and was actually performed with a hand calculator (the eight-digit, scientific type is adequate). Rounding off of even these primitive computations is sometimes done, with no apologies.

SYMBOLS

The following shorthand symbols are frequently used:

Symbol	*Reading*
$>$	is greater than
$<$	is less than
$\geq$	equal to or greater than
$\leq$	equal to or less than
6′	6 feet
6″	6 inches
Σ	the sum of
ΔL	change in L

NOMENCLATURE

Nomenclature, or notation, used in this book complies generally with that used in the steel industry and the latest editions of standard specifications. The following list includes the notation used in this book and is compiled and adapted from more extensive lists in the references:

A = area, general
A_g = gross area of a section, defined by the outer dimensions
A_n = net area
C = compressive force
E = modulus of elasticity, general (of steel in this book)
F_a = allowable compressive stress due to axial load only
F_b = allowable compressive stress due to bending
F_y = specified yield strength of steel
I = moment of inertia

K = factor for modification of unbraced length of column, based on support conditions
L = length (usually of a span), or unbraced height of a column
M = bending moment
M_R = resisting moment capacity of a section
P = concentrated load
S = section modulus
T = tension force
V = (1) shear force; (2) vertical component of a force
W = (1) total gravity load; (2) weight, or dead load of an object; (3) total wind load force; (4) total of a uniformly distributed load or pressure due to gravity
b = width (general)
b_f = width of flange of W shape
c = in bending: distance from extreme fiber stress to the neutral axis
d = overall beam depth, out-to-out of flanges for a W shape
e = eccentricity of a nonaxial load, from the point of application of the load to the centroid of the section
f_a = computed stress due to axial load
f_b = computed bending stress
f'_c = specified compressive strength of concrete
f'_m = specified compressive strength of masonry
f_p = computed bearing stress
f_t = computed stress in tension
f_v = computed shear stress
s = spacing of objects, center to center
t = thickness, general
t_f = thickness of flange of W shape
t_w = thickness of web of W shape
w = unit of a distributed load on a beam (lb/ft, etc.)
Δ = deflection, usually maximum vertical deflection of a beam (also used to indicate "change of" in mathematical expression)
Σ = sum of
ϕ = strength reduction factor in strength design

1

CONSIDERATIONS FOR USE OF STEEL

The work of designing steel structures requires considerable understanding of the basic nature of steel as a material and of the products and processes from which steel structures are formed. This chapter treats some of the major concerns of this nature.

1.1 PROPERTIES OF STEEL

The strength, hardness, corrosion resistance, and some other properties of steel can be varied through a considerable range by changes in the production processes. Literally hundreds of different steels are produced, although only a few standard products are used for the majority of the elements of building structures. Working and forming processes such as rolling, drawing, machining, and forging may also alter some properties. However, certain properties such as density (unit weight), stiffness (modulus of elasticity), thermal expansion, and fire resistance tend to remain constant for all steels.

For various applications, other properties will be significant. Hardness affects the ease with which cutting, drilling, planing, and other working can be done. For welded connections the property of weldability of the base material must be considered. Resistance to rusting is normally low, but it can be enhanced by various materials added to the steel, producing various types of special steels, such as stainless steel and the so-called "rusting steels" that rust at a very slow rate.

These various properties of steel must be considered when working with the material and when designing for its use. However, it is the unique structural nature of steel with which we are most concerned in this book.

Structural Properties of Steel

Basic structural properties, such as strength, stiffness, ductility, and brittleness, can be interpreted from laboratory load tests on specimens of the material. Figure 1.1 displays characteristic forms of curves that are obtained by plotting stress and strain values from such tests. An important property of many structural steels is the plastic deformation (yield) phenomenon. This is demonstrated by curve 1 in Figure 1.1. For steels with this character, there are two different stress values of significance: the yield limit and the ultimate failure limit.

Generally, the higher the yield limit, the less the degree of ductility. The extent of ductility is measured as the ratio of the plastic deformation between first yield and strain hardening (see Figure 1.1) to the elastic deformation at the point of yield. Curve 1 in Figure 1.1 is representative of ordinary structural steel (ASTM A36), and curve 2 indicates the typical effect as the yield strength is raised a significant amount. Eventually, the significance of the yield phenomenon becomes virtually negligible when the yield strength approaches as much as three times the yield of ordinary steel (36 ksi for ASTM A36 steel).

Some of the highest-strength steels are produced only in thin sheet or drawn wire forms. Bridge strand is made from wire with strength as high as 300,000 psi. At this level, yield is almost nonexistent and the wires approach the brittle nature of glass rods.

For economical use of the expensive material, steel structures are generally composed of elements with relatively thin parts. This results in many situations in which the ultimate limiting strength of elements in bending, compression, and shear is determined by buckling, rather than by the stress limits of the material. Because buckling is a function of stiffness (modulus of elasticity) of the material, and because this property

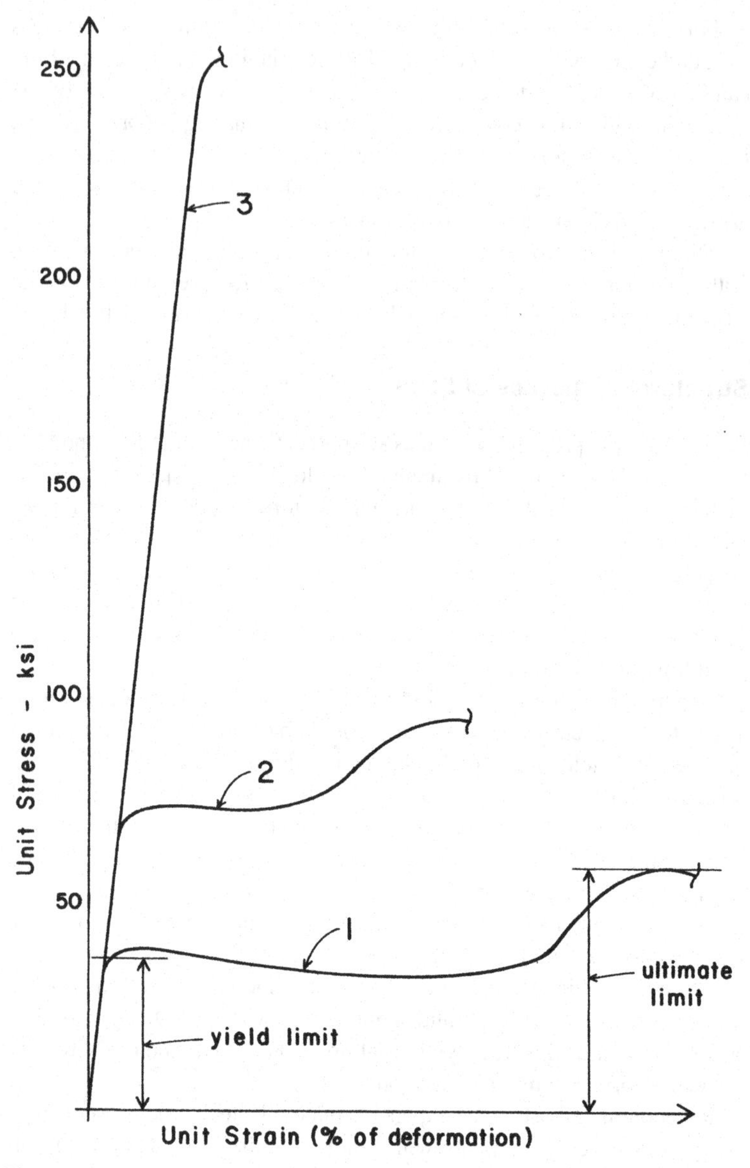

Figure 1.1 Stress/strain response: (1) ordinary structural steel; (2) high-strength steel for rolled shapes; (3) super-strength steel (usually in wire form).

remains the same for all steels, there is limited opportunity to make effective use of higher-strength steels in many situations. The grades of steel most commonly used are to some extent ones that have the optimal effective strength for most typical tasks.

Because many structural elements are produced as some manufacturer's product line, choices of basic materials are often mostly out of the hands of individual building designers. The proper steel for the task—on the basis of many properties—is determined as part of the product design, although a range of grades may be obtained for some products.

Steel that meets the requirements of the American Society for Testing and Materials (ASTM) Specification A36 is the grade of structural steel most commonly used to produce rolled steel elements for building construction. It must have an ultimate tensile strength of 58 to 80 ksi and a minimum yield point of 36 ksi. It may be used for bolted, riveted, or welded fabrication. This is the steel used for much of the work in this book, and it will be referred to simply as A36 steel.

Prior to 1963 a steel designated ASTM A7 was the basic product for structural purposes. It had a yield point of 33 ksi and was used primarily for riveted fabrication. With the increasing demand for bolted and welded construction, A7 steel became less useful, and in a short time, A36 steel was the material of choice for the majority of structural products.

For structural steel the AISC Specification expresses the allowable unit stresses used for ASD work in terms of some percent of the yield stress F_y or the ultimate stress F_u. Of course, LRFD work does not use allowable stresses, but rather the basic limiting stresses (yield and ultimate) of the material. Some modification for specific usage situations (tension, bending, shear, etc.) is made by use of different reduction factors (called resistance factors, as explained in Chapter 5).

Steel used for other purposes than the production of rolled products generally conforms to standards developed for the specific product. This is especially true for steel connectors, wire, cast and forged elements, and very high-strength steels produced in sheet, bar, and rod form for fabricated products. The properties and design stresses for some of these product applications are discussed in other sections of this book. Standards used typically conform to those established by industry-wide organizations, such as the Steel Joist Institute (SJI) and the Steel Deck Institute (SDI). In some cases, larger fabricated products make use of ordinary rolled products, produced from A36 steel or other grades of steel from which hot-rolled products can be obtained.

1.2 TYPES OF STEEL PRODUCTS

Steel itself is shapeless, coming basically in the form of a molten material or a softened lump. The structural products produced derive their basic forms from the general potentialities and limitations of the industrial processes of forming and fabricating. Standard raw stock elements—deriving from the various production processes—are the following:

1. *Rolled shapes.* These are formed by squeezing the heat-softened steel repeatedly through a set of rollers that shape it into a linear element with a constant cross section. Simple forms of round rods and flat bars, strips, plates, and sheets are formed, as well as more complex shapes of I, H, T, L, U, C, and Z. Special shapes, such as rails or sheet piling, can also be formed in this manner.
2. *Wire.* This is formed by pulling (called drawing) the steel through a small opening.
3. *Extrusion.* This is similar to drawing, although the sections produced are other than simple round shapes. This process is not much used for steel products of the sizes used for buildings.
4. *Casting.* This is done by pouring the molten steel into a form (mold). Casting is limited to objects of a three-dimensional form and is also not common for building construction elements.
5. *Forging.* This consists of pounding the softened steel into a mold until it takes the shape of the mold. This is preferred to casting because of the effects of the working on the properties of the finished material.

The raw stock steel elements produced by the basic forming processes may be reworked by various means, such as the following:

1. *Cutting.* Shearing, sawing, punching, or flame cutting can be used to trim and shape specific forms.
2. *Machining.* This may consist of drilling, planing, grinding, routing, or turning on a lathe.
3. *Bending.* Rods, sheets, plates, or other linear elements may be bent if made from steel with a ductile character.
4. *Stamping.* This is similar to forging; in this case, sheet steel is punched into a mold that forms it into some three-dimensional shape, such as a hemisphere.

5. *Rerolling.* This consists of reworking a linear element into a curved form (arched) or of forming a sheet or flat strip into a formed cross section.

Finally, raw stock or reformed elements can be assembled by various means into objects of multiple parts, such as a manufactured truss or a prefabricated wall panel. Basic means of assemblage include the following:

1. *Fitting.* Threaded parts may be screwed together or various interlocking techniques may be used, such as the tongue-and-groove joint or the bayonet twist lock.
2. *Friction.* Clamping, wedging, or squeezing with high-tensile bolts may be used to resist the sliding of parts in surface contact.
3. *Pinning.* Overlapping flat elements may have matching holes through which a pin-type device (bolt, rivet, or actual pin) is placed to prevent slipping of the parts at the contact face.
4. *Nailing, screwing.* Thin elements—mostly with some preformed holes—may be attached by nails or screws.
5. *Welding.* Gas or electric arc welding may be used to produce a bonded connection, achieved partly by melting the contacting elements together at the contact point.
6. *Adhesive bonding.* This usually consists of some form of chemical bonding that results in some fusion of the materials of the connected parts.

We are dealing here with industrial processes, which at any given time relate to the state of development of the technology, the availability of facilities, the existence of the necessary craft, and competition with other materials and products.

Rolled Structural Shapes

The products of the steel rolling mills used as beams, columns, and other structural members are called *sections* or *shapes,* and their designations are related to the profiles of their cross sections. American standard I-beams (Figure 1.2*a*) were the first beam sections rolled in the United States and are currently produced in sizes of 3 to 24 in. in depth. The W shapes (Figure 1.2*b*—originally called wide-flange shapes) are a modification of the I cross section and are characterized by parallel flange surfaces as contrasted with the tapered inside flange surfaces of standard

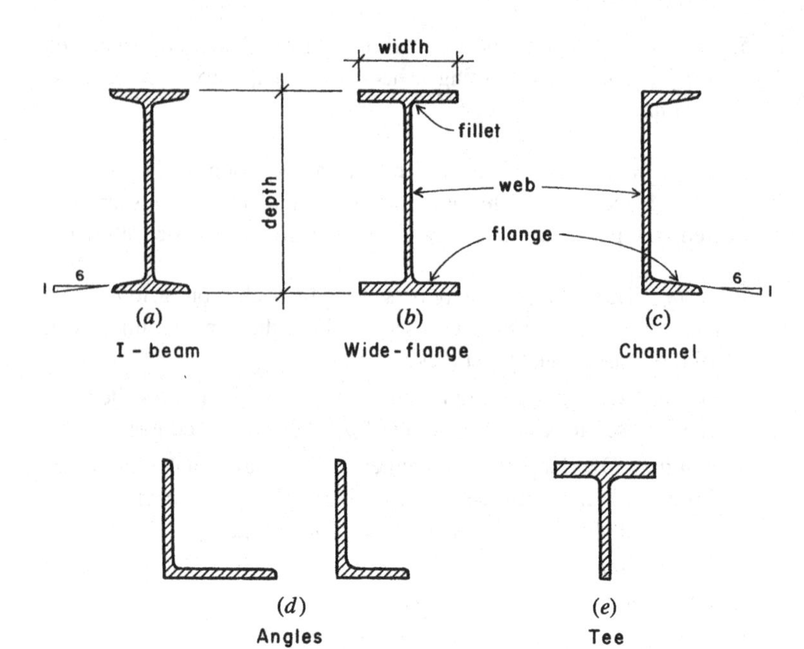

Figure 1.2 Shapes of typical hot-rolled products.

I-beams; they are available in depths of 4 to 44 in. In addition to the standard I and W sections, the structural steel shapes most commonly used in building construction are channels, angles, tees, plates, and bars. The tables in Appendix A list the dimensions, weights, and various properties of some of these shapes. Complete tables of structural shapes are given in the AISC Manual (Ref. 3).

W Shapes. In general, W shapes have greater flange widths and relatively thinner webs than standard I-beams; and, as noted above, the inner faces of the flanges are parallel to the outer faces. These sections are identified by the alphabetical symbol W, followed by the *nominal* depth in inches and the weight in pounds per linear foot. Thus, the designation W 12 × 26 indicates a wide-flange shape of nominal-12-in. depth, weighing 26 lb per linear foot.

The actual depths of W shapes vary within the nominal depth groupings. From Table A.3 in Appendix A, note that a W 12 × 26 has an actual depth of 12.22 in., whereas the depth of a W 12 × 30 is 12.34 in. This is a result of the rolling process during manufacture in which the cross-sectional areas of W shapes are increased by spreading the rollers both vertically and horizontally. The additional material is thereby added to the cross section by increasing flange and web thickness as well as flange width (Figure 1.2*b*). The resulting higher percentage of material in the flanges makes wide-flange shapes more efficient structurally than standard I-beams. A wide variety of weights is available within each nominal depth group.

In addition to shapes with profiles similar to the W 12 × 26, which has a flange width of 6.490 in., many W shapes are rolled with flange widths approximately equal to their depths. The resulting H configurations of these cross sections are much more suitable for use as columns than the I profiles. From Table A.3, note that the following shapes, among others, fall into this category: W 14 × 120, W 12 × 65, and W 10 × 49. It is recommended that the reader compare these shapes with others listed in their respective nominal depth groups in order to become familiar with the variety of geometrical relationships.

Standard I-Beams (S Shapes). American standard I-beams are identified by the alphabetical symbol S, the designations S 12 × 35 indicating a standard shape 12 in. deep weighing 35 lb per linear foot. Unlike W sections, standard I-beams in a given depth group have uniform depths, and shapes of greater cross-sectional area are made by spreading the rolls in one direction only. Thus, the depth remains constant, whereas the width of flange and thickness of web are increased.

All standard I-beams have a slope on the inside faces of the flanges of 16 percent or 1 in 6. In general, standard S shapes are not so efficient structurally as W shapes and consequently are not so widely used. Also the variety available is not so large as that for W shapes. Characteristics that may favor the use of S shapes in any particular situation are constant depth, narrow flanges, and thicker webs.

Standard Channels. The profile of an American standard channel is shown in Figure 1.2*c*. These shapes are identified by the alphabetical symbol C. The designation C 10 × 20 indicates a standard channel 10 in. deep and weighing 20 lb per linear foot. Table A.4 shows that this section has an area of 5.88 in.^{2}, a flange width of 2.739 in., and a web thickness of 0.379 in. Like the standard I-beams, the depth of a particular group remains constant and the cross-sectional area is increased by spreading the rolls to increase flange width and web thickness. Because of their tendency to buckle when used independently as beams or columns, channels require lateral support or bracing. They are generally used as elements of built-up sections such as columns and lintels. However, the absence of a flange on one side makes channels particularly suitable for framing around floor openings.

Angles. Structural angles are rolled sections in the shape of the letter L. Table A.5 gives dimensions, weights, and other properties of equal and unequal leg angles. Both legs of an angle have the same thickness.

Angles are designated by the alphabetical symbol L, followed by the dimensions of the legs and their thickness. Thus, the designation L 4 × 4 × ½ indicates an equal leg angle with 4-in. legs, ½ in. thick. From Table A.5, note that this section weighs 12.8 lb per linear foot and has a cross-sectional area of 3.75 in.^{2}. Similarly, the designation L 5 × 3½ × ½ indicates an unequal leg angle with one 5-in. and one 3½-in. leg, both ½ in. thick. Table A.5 shows that this angle weighs 13.6 lb per linear foot and has an area of 4 in.^{2}. To change the weight and area of an angle of a given leg length, the thickness of each leg is increased the same amount. Thus, if the leg thickness of the L 5 × 3½ × ½ is decreased to ⅜ in., Table A.5 shows that the resulting L 5 × 3½ × ⅜ has a weight of 10.4 lb per linear foot and an area of 3.05 in.^{2}. It should be noted that this method of spreading the rolls changes the leg lengths slightly.

Single angles are often used as lintels, and pairs of angles are often used as members of light steel trusses. Angles were formerly used as elements of built-up sections such as plate girders and heavy columns, but the advent of the heavier W shapes has largely eliminated their usefulness for this purpose. Short lengths of angles are commonly used as connecting members for beams and columns.

Structural Tees. A structural tee is made by splitting the web of a W shape (Figure 1.2*e*) or a standard I-beam (S shape). The cut, normally made along the center of the web, produces tees with a stem depth equal to half the depth of the original section. Structural tees cut from W shapes are identified by the symbol WT; those cut from standard S shapes by ST. The designation WT 6 × 53 indicates a structural tee with a 6-in. depth and a weight of 53 lb per linear foot. This shape is produced by splitting a W 12 × 106 shape. Similarly, ST 9 × 35 designates a structural tee 9 in. deep weighing 35 lb per linear foot and cut from an S 18 × 70.

Plates and Bars. Plates and bars are made in many different sizes and are available in many different structural steel specifications. Flat steel for structural uses is generally classified as follows:

Bars.	6 in. or less in width, and 0.203 in. or more in thickness.
Plates.	More than 8 in. in width, and 0.230 in. or more in thickness.
	More than 48 in. in width, and 0.180 in. or more in thickness.
Sheet.	Generally any flat material less than 0.180 in. in thickness.

Bars are available in varying widths and in virtually any required thickness and length. The usual practice is to specify bars in increments of $\frac{1}{4}$ in. for widths and $\frac{1}{8}$ in. in thickness.

For plates the preferred increments for width and thickness are the following:

Widths.	Vary by even inches, although smaller increments are obtainable.
Thickness.	$\frac{1}{32}$-in. increments up to $\frac{1}{2}$ in.
	$\frac{1}{16}$-in. increments of more than $\frac{1}{2}$ to 2 in.
	$\frac{1}{8}$-in. increments of more than 2 to 6 in.
	3-in. increments of more than 6 in.

The standard dimensional sequence when describing steel plate is

$$\text{Thickness} \times \text{Width} \times \text{Length}$$

$$\text{Example: PL } 1\tfrac{1}{2} \times 10 \times 16$$

All dimensions are given in inches, fractions of an inch, or decimals of an inch.

Column base plates and beam bearing plates may be obtained in the widths and thicknesses noted. For the design of beam bearing plates and column base plates see Sections 3.14 and 4.7, respectively.

Designations for Structural Steel Elements. As noted earlier, wide-flange shapes are identified by the symbol W and American standard beam shapes by S. It was also pointed out the W shapes have essentially parallel flange surfaces, whereas S shapes have a slope of approximately 16 percent on the inner flange faces. A third designation, M shapes, covers miscellaneous shapes that cannot be classified as W or S;

TABLE 1.1 Standard Designations for Structural Steel Elements

Elements	Designation
American standard I-beams, S-shapes	S 12 × 35
Wide flanges, W shapes	W 12 × 27
Miscellaneous shapes, M shapes	M 8 × 18.5
American standard channels, C shapes	C 10 × 20
Miscellaneous channels, MC shapes	MC 12 × 40
Bearing piles, HP-shapes	HP 14 × 117
Angles, L shapes	L 5 × 3 × ½
Structural tees, WT, ST, MT	WT 9 × 38
Plates	PL 1½ × 10 × 16
Structural tubing	HSS 10 × 6 × ½
Pipe, standard weight	Pipe 4 Std
Pipe, extra strong	Pipe 4 X-strong
Pipe, double extra strong	Pipe 4 XX-strong

these shapes have various slopes on their inner flange surfaces and many of them are of only limited availability. Similarly, some rolled channels cannot be classified as C shapes. These are designated by the symbol MC.

Table 1.1 lists the standard designations used for rolled shapes, formed rectangular tubing, and round pipe.

Cold-Formed Steel Products

Many structural elements are formed from sheet steel. Elements formed by the rolling process must be heat-softened, whereas those produced from sheet steel are ordinarily made without heating the steel; thus, the common description for these elements is *cold-formed.* Because they are typically formed from very thin sheets, they are also referred to as *light-gage* steel products.

Figure 1.3 illustrates the cross sections of some of these products. Large corrugated or fluted sheets are in wide use for all paneling and for structural decks for roofs and floors. Use of these elements for floor decking is discussed in Chapter 3. These products are made by a number of manufacturers, and information regarding their structural properties may be obtained directly from the manufacturer. General information on structural decks may also be obtained from the Steel Deck Institute (see Ref. 5).

Cold-formed shapes range from the simple L, C, U, and so on to the special forms produced for various construction systems. Structures for some buildings may be almost entirely composed of cold-formed prod-

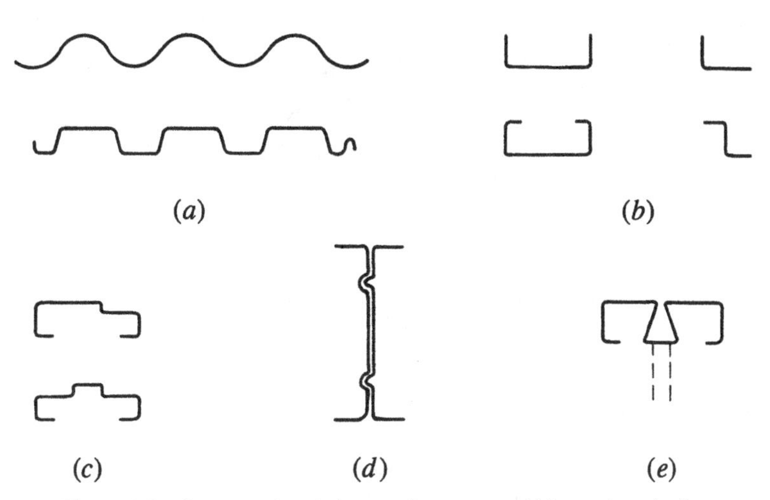

Figure 1.3 Cross-sectional shapes of common cold-formed products.

ucts. Several manufacturers produce patented systems of these components for the formation of predesigned, packaged building structures.

Fabricated Structural Components

A number of special products are formed of both hot-rolled and cold-formed elements for use as structural members in buildings. Open-web steel joists consist of prefabricated, light steel trusses. For short spans and light loads, a common design is that shown in Figure 1.4*a* in which the web consists of a single, continuous bent steel rod and the chords of steel rods or cold-formed elements. For larger spans or heavier loads, the forms more closely resemble those for ordinary light steel trusses; single angles, double angles, and structural tees constitute the truss members. Open-web joists for floor framing are discussed in Section 3.9.

Another type of fabricated joist is shown in Figure 1.4*b*. This member is formed from standard rolled shapes by cutting the web in a zigzag fashion. The resulting product has a greatly reduced weight-to-depth ratio when compared with the lightest of the rolled shapes.

Other fabricated steel products range from those used to produce whole building systems to individual elements for construction of windows, doors, curtain wall systems, and the framing for interior partition walls. Many components and systems are produced as proprietary items by a single manufacturer, although some are developed under controls of

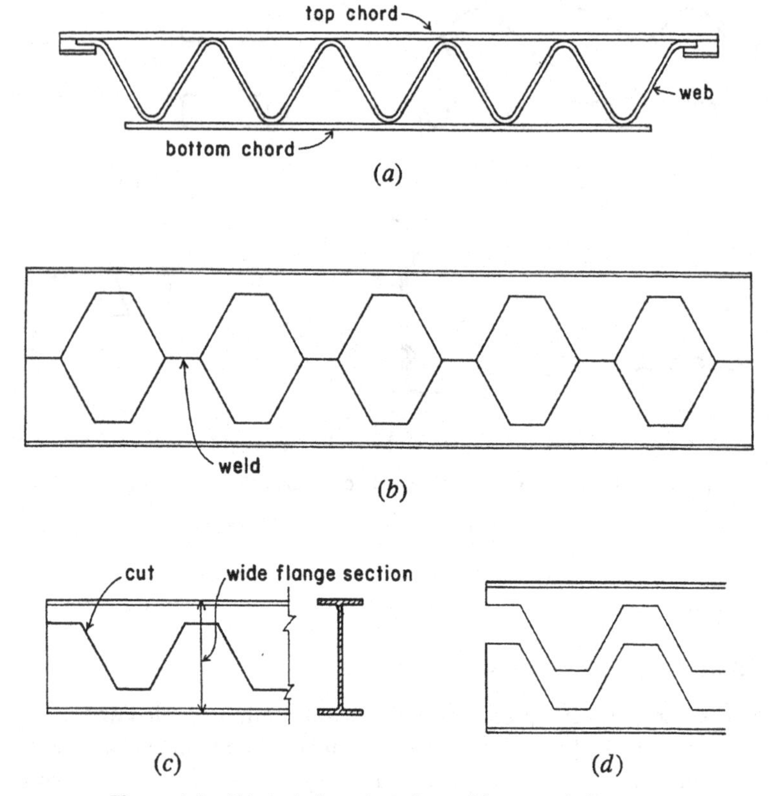

Figure 1.4 Fabricated products formed from steel elements.

industry-wide standards, such as those published by the Steel Joist Institute and Steel Deck Institute.

1.3 DEVELOPMENT OF STRUCTURAL SYSTEMS

Structural systems that compose entire roof, floor, or wall constructions—or even entire buildings—are typically assembled from many individual elements. These individual elements may be of some variety, as in the case of the typical floor using rolled steel shapes for beams and a formed sheet steel deck. Selection of individual elements is often done with some data from structural investigations but is also often largely a matter of practical development of the form of the construction.

It is common for a building to incorporate more than a single material for its entire structural system. Various combinations are possible, such as a wood deck on steel beams, or masonry bearing walls for a steel spanning floor or roof structure. This book deals primarily with structures of steel, but some of these mixed-material situations are very common and are discussed in various parts of the book.

1.4 CONNECTION METHODS

Connection of structural steel members that consist of rolled elements is typically achieved by direct welding or by steel rivets or bolts. Riveting for building structures has become generally obsolete in favor of high-strength bolts. The design of bolted connections and simple welded connections are discussed in Chapter 8. In general, welding is preferred for shop fabrication and bolting for field connections.

Thin elements of cold-formed steel may be attached by welding, by bolting, or by sheet metal screws. Thin deck and wall paneling elements are sometimes attached to one another by simple interlocking at their abutting edges; the interlocked parts are sometimes folded or crimped to give further security to the connection.

Adhesives or sealants may be used to seal joints or to bond thin sheet materials in laminated fabrications. Some elements used in connections may be attached to connected parts by adhesion to facilitate the work of fabrication and erection, but adhesive connection is not used for major structural joints.

A major structural design problem is that of the connections of columns and beams in heavy frames for multistory buildings. For rigid frame action to resist lateral loads, these connections are achieved with very large welds, generally developed to transfer the full strength of the connected members. Design of these connections is beyond the scope of this book, although lighter framing connections of various form are discussed in Chapter 8.

1.5 DATA FOR STEEL PRODUCTS

Information in general regarding steel products used for building structures must be obtained from steel industry publications. The AISC is the primary source of design information regarding structural rolled products, which are the principal elements used for major structural components: columns, beams, large trusses, and so on. Several other industry-wide organizations also publish documents that provide information about particular products, such as manufactured trusses (open-web joists), cold-formed sections, and formed sheet steel decks. Many of these organizations and their publications are described in the appropriate chapters and sections of this book.

Individual manufacturers of steel products usually conform to some industry-wide standards in the design and fabrication of their particular products. Still, there is often some room for variation of products; it is therefore advised that the manufacturers' own publications be used for specific data and details of the actual products. As in other similar situations in building design, the designer should strive to design and specify components of the building construction so that only those controls that are critical are predetermined, leaving flexibility in the choice of a particular manufactured product.

Some of the tabulated data presented in this book has been reproduced or abstracted from industry publications. In many cases the data presented here are abbreviated and limited to uses pertinent to the work displayed in the text example computations and necessary for the exercise problems. The reference sources cited should be consulted for more complete information, particularly because change occurs frequently due to growth of the technology, advances in research, and modification of codes and industry standards.

1.6 USAGE CONSIDERATIONS FOR STEEL STRUCTURES

Steel is a relatively expensive, industrialized product. Use of steel for structures must generally be made with very careful consideration of the limitations of the material, attention to high efficiency in the volume of material used, and design with clear understanding of the practical aspects of production and erection of steel products. In this chapter, we present discussions of a number of aspects regarding intelligent use of steel for structural applications.

Stress and Strain

Steel is one of the strongest materials used for building structures, but it has limitations for various forms of stress development. Unlike wood, stress response tends to be nondirectional, and unlike concrete or masonry, stress resistance is high for all the basic stresses: tension, compression, and shear. Some specific stress limits for steel are the following:

1. Stress beyond the yield point will produce permanent deformations, which may be tolerable in the small dimensions of a joint but may create major problems within the general form of structural elements. Even though ultimate strength may be high, the much lower yield stress must be used for a limit of acceptable behavior in most situations.
2. Ordinary steel is formed by molten casting, resulting in a crystalline structure of the material. Certain forms of stress failure may be precipitated by fracture along crystalline fault lines, especially those related to dynamic, repetitive force actions. This is more of a problem in machinery, but dynamically loaded structures may need consideration for this effect.
3. Various actions, such as cold-forming, machining, or welding, may change the character of the material, resulting in hardening, loss of ductility, or locked-in stresses within the material. The processes of fabrication and erection must be carefully studied to be sure these do not produce undesirable conditions to complicate stress behavior under service load situations.

In some cases the anticipated stress-strain responses may cause certain actions to occur that affect the overall structural resistance of a steel structure. An example of this is the formation of plastic hinges in rigid frames or in frames with eccentric bracing. The adjusted behavior of the structure in load response that occurs when a plastic hinge yields is a major element in visualization and computation of the structure's response. Some considerations for this action are discussed in Chapters 3 and 4.

Because of its strength, steel tends to carry a disproportionately high share of the load when sharing loads with other materials, such as wood or concrete. This is a major factor in design of composite structural elements, such as flitched beams of steel and wood and composite deck systems of steel and concrete (see Section 6.2). It is also a major consideration in the design of reinforced concrete and masonry.

Although stress resistance is subject to variation, strain resistance—as measured by the direct stress modulus of elasticity—is not. This makes for a shifting relationship when stress capability is raised to produce higher grades of steel. Although load resistance—as measured by stress capacity—may be increased, resistance to deformations is not. Thus, deflection or buckling—both affected by the stiffness of the material—may become relatively more critical for structures made with higher grades of steel.

Stability

Unlike the solid forms common with timber and concrete, elements of steel are often composed of relatively thin parts. In addition, framed assemblages often consist of fairly slender linear components. All of this thin and slender character results in a condition in which buckling collapse, rather than crushing or tension cracking, is a common limiting failure behavior. This situation requires that designers pay special attention to the potential for various types of buckling failure. These forms of failure are treated in the various chapters in this book that treat individual types of structural elements and systems.

Another type of stability problem results from the usual type of assemblage connections used with steel structures. These connections have generally the character of very little moment-resistance, often qualifying essentially as pinned connections, rather than fixed connections. In truth the typical connection is one that is partially fixed, with some limited moment-transmitting capacity. In any event, the problem is that most assemblages do not derive much stability from the connections. Attention must therefore be paid to what *does* stabilize the assembled structure.

The problem of stability just described often relates to the resistance of lateral loads—that is, horizontal forces by comparison to the vertical forces of gravity. However, the problem may be a general one of giving the structure some degree of three-dimensional stability. The general means of achieving this are as follows:

1. Modify the usual connections to more fully resist moments, producing what is described as *rigid frame* action.
2. Arrange the frame so that the overall assemblage works for stability without rigid joints, as with triangulation that produces truss action.
3. Add extra bracing elements (guys, struts, X-bracing, flying buttresses, etc.) for the specific added purpose of achieving stability.
4. Borrow stability from other parts of the building construction; for example, masonry walls.

The point here is that this problem requires some extra consideration beyond that given to ordinary resistance of gravity loads. Specific situations of this type are discussed in the various chapters of the book.

Deformation Limits and Controls

The most critical limit for a structure for purposes of establishing safety is usually the magnitude of load it can resist, as measured in terms of strength. However, as discussed in the preceding section, failure may be precipitated by a loss of stability, before stress levels achieve limiting magnitudes. True safety must therefore be established by consideration for both strength and stability.

Although it does not often relate to safety, a practical limitation that must also be considered is that of the amount of *deformation*—literally, shape change—that can be tolerated. Stress resistance cannot be developed without some accompanying strain, so deformation is inevitable for any structure.

For structural members, the most critical deformations are usually those caused by bending, due simply to their larger dimension. A column may be heavily loaded but shorten by a virtually imperceptible amount, whereas even a short-span beam will deflect noticeably when loaded. The most common deformation problem is thus the vertical deflection of beams.

Practical deflection limitations derive mostly from consideration of effects on the general building construction. Cracking of tiled floors or plastered ceilings may provide such limitations. However, many other situations also limit the tolerable movement of the structure. Although the real need for control of deformation may easily be understood, the practical means for establishing design criteria is elusive, and much professional judgement is involved. These problems are treated in some depth in the chapters that deal with spanning structures.

Another type of deformation problem occurs within structural connections. Here also development of stress is unavoidably accompanied by strain and deformation. This may add to stability problems or simply increase overall movement of an assembled structure. Some of these issues are discussed in Chapter 8.

Rust

Exposed to air and moisture, most steels will rust at the surface of the steel mass. Rusting will generally continue at some rate until the entire steel mass is eventually rusted away. Response to this problem may involve one or more of the following actions:

1. Do nothing, if there is essentially no exposure, as when the steel element is encased in cast concrete or other encasing construction.
2. Paint the steel surface with rust-inhibiting material.
3. Coat the surface with nonrusting metal, such as zinc or aluminum.
4. Use a steel that contains ingredients in the basic material that prevent or retard the rusting action (see discussion in Section 1.1 on corrosion-resistant steels).

Rusting is generally of greater concern when exposure conditions are more severe. It is also of greatest concern for the thinner elements, especially those formed of thin sheet steel, such as formed roof decks.

In some cases it may be necessary to leave the steel in an essentially bare condition, such as when field (on-site) welding is to be done or when steel items are to be encased in concrete. These are standard practices in building construction, but they can be difficult to deal with when appearance is important due to the final exposed condition of the structure.

When structures are exposed to conditions likely to cause serious rusting, designers tend to avoid use of excessively thin parts. This reduces vulnerability of the structure to failure by loss of material in cross sections, in the event that rust prevention methods are less than totally successful.

Deterioration of steel can also be caused by exposure to various corrosive chemicals, such as acid rain, seacoast salt air, or air heavily polluted with various industrial wastes. Special protection or simply avoiding exposure as much as possible may be required for such conditions. Where not just appearance but actual structural safety is at risk, these matters require serious attention by the structural designer.

Fire

As with all materials, the stress and strain response of steel varies with its temperature. The rapid loss of strength (and stiffness, which may be more important when buckling is critical) at high temperatures, coupled with rapid heat gain due to the high conductivity of the material and the common use of thin parts, makes steel structures highly susceptible to fire. On the other hand, the material is noncombustible and less critical for some considerations compared with constructions with thin elements of wood.

The chief strategy for improving fire safety with steel structures is to prevent the fire (and the rapid heat buildup) from getting to the steel by providing some coating or encasement with fire-resistant, insulative materials. Ordinary means for this include use of concrete, masonry, plaster, mineral fiber, or gypsum plasterboard elements. The general problem and some specific design situations are presented in Chapter 10.

Concrete is often used with steel framing as a fill on top of formed steel deck or as a structural concrete slab bearing directly on steel beams. In some situations the concrete may also be used to encase steel columns or beams, although building code acceptance of other means for achieving necessary fire ratings have largely eliminated this practice. One easy form of this construction occurs with the steel beam-plus-concrete slab construction shown in Figure 1.5. A problem for concern in design of such a system, however, is the considerable added weight of the construction due to the concrete encasement.

System Assemblage Considerations

As whole systems, steel structures consist of many individual parts. Assemblage of the complete structure—that is, simply bringing together all of the parts and connecting them—is a major design concern. Designers must deal with the work of design for each of the individual parts in many cases, but they must also consider the assemblage of the complete structure. Following are two major aspects of the development of the assemblage:

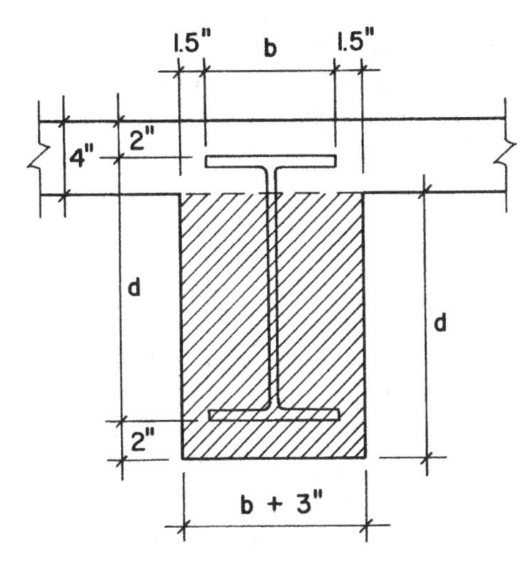

Figure 1.5 Steel beam encased in concrete for fire protection. Shaded portion indicates concrete in excess of that required for the supported concrete slab.

1. Planning of the structural arrangement. This consists of decisions for the overall form of the structure and for dimensions of spans, story heights, sizes of openings, and so on. A frequent decision involves choice of repetitive modules, such as the spacing for sets of beams and columns.
2. Mating of the parts. Decks must be attached to beams, beams attached to columns, columns attached to footings, and so on. This involves geometric considerations for the shape of individual parts and the development of individual connections for necessary transfer of loads.

Assemblage is mostly achieved with standard methods of connection. These relate to the form of the connected elements and to the type of loads being transferred among them. The amount of connecting work required for a typical steel structure is considerable, so methods used should be practical and economical—and above all, familiar to the construction assemblage crew.

A critical factor has to do with the assemblage that is performed as fabrication in the factory (called the *shop*) and that is performed at the job site (called the *field*). To some degree different methods are employed at these locations, relating to working conditions. Part of the designer's task is to visualize where the assemblage occurs, because it may well affect choices for individual members and for connecting methods.

Assemblage problems are discussed for individual types of structural elements (beams, decks, columns, etc.) in the various chapters of this book. Overall problems of structural assemblages are treated in Section 1.7 and in the discussions for the building case examples in Chapter 10.

Cost of Construction

Steel is relatively expensive, on a volume basis. The real dollar cost of concern, however, is the final *installed cost,* that is, the total cost for the erected structure. Economy concerns begin with attempts to use the least volume of the material, but this is applicable only within the design of a single type of item. Rolled structural shapes do not cost the same per pound as fabricated open web joists. Furthermore, each item must be transported to the site and erected, using various auxiliary devices to complete the structure, such as connecting elements for structural components and bridging for joists.

Cost concerns for structures as a whole, and for the total building construction, are discussed in Chapter 9. In other parts of the book, when design of single structural components are discussed, the usual approach is to generally seek to use the lightest-weight (least volume of material) elements that will satisfy the design criteria.

1.7 CHOICE AND PLANNING OF STEEL STRUCTURAL SYSTEMS

Elements of steel may be used to provide a variety of horizontal spanning floor or roof structures. The two primary spanning systems treated in this book are the rolled steel beam and the light, prefabricated truss. In this chapter we deal with some of the general issues involved in development of spanning systems. Design of beams and decks is treated in Chapter 3, and design of trusses is treated in Chapter 7. The special cases of rigid frames and bents are discussed in Chapter 5.

Deck–Beam–Girder Systems

A framing system extensively used for buildings with large roof or floor areas is that in which columns are arranged in orderly rows for the support of a rectangular grid of steel beams or trusses. The actual roof or floor surface is then generated by a solid deck of wood, steel, or concrete, which spans in multiple, continuous spans over a parallel set of supports. Planning for such a system must begin with consideration of the general architectural design of the building, but should also respond to logical considerations for the development of the structure.

Consider the system shown in the partial framing plan in Figure 1.6*a*. In developing the layout for the system and choosing its components, considerations such as the following must be made:

1. *Deck span.* The type of deck as well as its specific variation (thickness of plywood, gage of steel sheet, etc.) will relate to the deck span.
2. *Joist spacing.* This determines the deck span and the magnitude of load on the joist. The type of joist selected may limit the spacing, based on the joist capacity. The type and spacing of joists must be coordinated with the selection of the deck.
3. *Beam span.* For systems with some plan regularity, the joist spacing should be some full-number division of the beam span.

4. *Column spacing.* The spacing of the columns determines the spans for the beams and joists, and is thus related to the planning modules for all the other components.

For a system such as that shown in Figure 1.6*a,* the basic planning begins with the location of the system supports, usually columns or bearing walls. The character of the spanning system is closely related to the magnitude of the spans it must achieve. Decks are mostly quite short in span, requiring relatively close spacing of the elements that provide their direct support. Joists and beams may be small or large, depending mostly on their spans. The larger they are, the less likely they will be very closely spaced. Thus, very long-span systems may have several levels of components, ranging in size down from the elements that achieve the longest span to the elements that directly support the deck.

Concerns for the design of individual components of the deck–beam–girder system are discussed in Chapter 3. General discussion of design in the context of whole building system development is presented in the building system design examples in Chapter 10.

Figure 1.6*b* shows a plan and elevation of a system that uses trusses for the major span. If the trusses are very large and the purlin spans quite long, the purlins may have to be quite widely spaced. A constraint on the purlin locations is usually that they coincide with the joints in the top of the truss, so as to avoid high shear and bending in the truss top chord. If purlins are widely spaced, it may be advisable to use joists between the purlins to provide support for the deck. On the other hand, if the truss spacing is a modest distance, it may be possible to use a long-span deck with no purlins. The basic nature of the system can thus be seen to change with different magnitudes of spans deriving from the locations of supports. In any event, the truss span and panel module, the column spacing, the purlin span and spacing, the joist span and spacing, and the deck span are interrelated and the selection of the components is a highly interactive exercise.

For systems with multiple elements, some consideration must be given to the various intersections and connections of the components. For the framing plan shown in Figure 1.6*a,* there is a five-member intersection at the column, involving the column, the two beams, and the two joists (plus an upper column, if the building is multistory). Depending on the materials and forms of the members, the forms of connections, and the types of force transfer at the joint, this may be a routine matter of construction or a real mess. Some relief of the traffic congestion may be

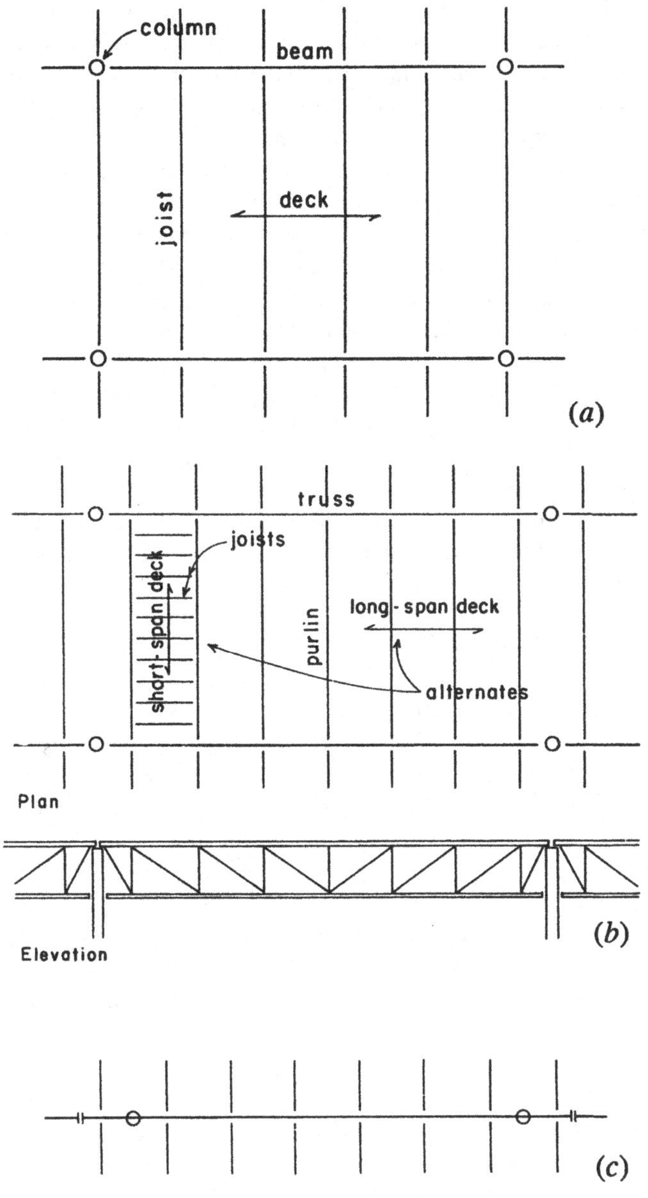

Figure 1.6 Planning considerations for beam framing.

achieved by the plan layout shown in Figure 1.6*c*, in which the module of the joist spacing is offset at the columns, leaving only the column and beam connections at the column location. A further reduction possible is that shown in Figure 1.6*c*, where the beam is made continuous through the column, with the beam splice occurring off the column. In the plan in Figure 1.6*c*, the connections are all only two-member relationships: column to beam, beam to beam, and beam to joist.

Bridging, blocking, and cross-bracing for trusses must also be planned with care. These members may interfere with continuous piping or ducting or may create complex connection problems similar to those just discussed. Use of required bracing elements for multiple purposes should be considered. Blocking required for plywood nailing may also function as edge nailing for ceiling panels and as lateral bracing for slender joists or rafters. The cross-bracing required to brace tall trusses may be used to support ceilings, ducts, building equipment, catwalks, and so on.

In the end, structural planning must be carefully coordinated with the general planning of the building and with its various subsystems. True optimization of the structure may need to yield to other, more pragmatic concerns.

Cantilevered Edges

A problem that occurs frequently is allowing for the extension of the horizontal structure beyond the plane of the building's exterior walls, providing a cantilevered edge. This most often occurs with an overhanging roof, but it can also be required for balconies or exterior walkways for a floor. Figure 1.7*a* shows one possibility for achieving this, by simply extending the ends of joists or rafters that are perpendicular to the wall. With steel framing, this type of cantilever is most easily achieved if the extended roof framing members perpendicular to the wall simply rest on top of the supporting beam or bearing wall at the wall plane. With an exterior column system, an alternate is shown in Figure 1.7*b*, in which the column line members are extended to support a member at the cantilevered edge, which in turn supports simple span members between the column lines. Loading, member size and type, and the magnitude of the cantilever would all affect the choice of one of these schemes over the other.

A special problem with the cantilevered edge occurs at a building outside corner, when both sides of the building have the cantilever condition, as shown in Figure 1.8*a*. With the framing system shown in Figure 1.7*a*, a possibility for the corner is that shown in Figure 1.8*b*, where the sup-

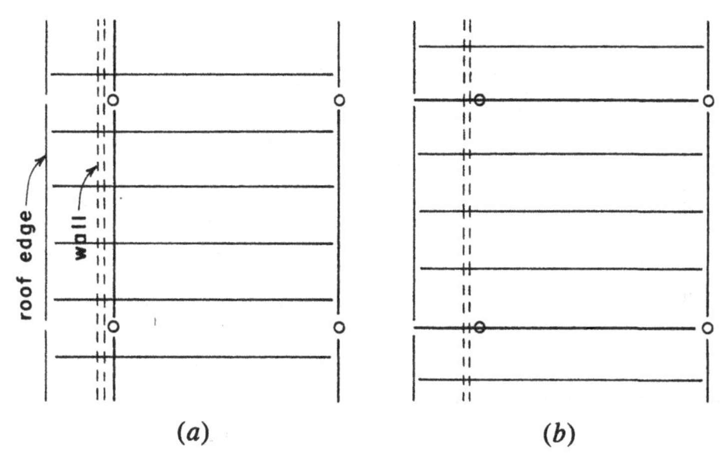

Figure 1.7 Framing at cantilevered edge.

ported beam is cantilevered to support edge member 1 and the joists are cantilevered to support edge member 2.

For the system shown in Figure 1.7*b*, a way of achieving the corner is depicted in Figure 1.8*c*. In this case the column-line member is cantilevered as usual to support edge member 1, which in turn cantilevers to the corner to support edge member 2.

A third possibility for the corner is the use of a diagonal member, as shown in Figure 1.8*d*. A feature of this solution is the reorientation of the framing system as the corner is turned. This layout is more often utilized in wood than in steel and is commonly used for sloping roofs when the diagonal member defines a ridge as the roof slopes to both edges. Note that there is a rather busy intersection at the interior column in the plan in Figure 1.8*d*.

Special Concerns

Various general planning concerns for structures are discussed in the examples in Chapter 10. The following are some particular issues that relate to design of steel framing systems.

Ceilings. Where ceilings exist, they are generally provided for in one of three ways: by direct attachment to the overhead structure (underside of the roof or floor above), by some independent structure that

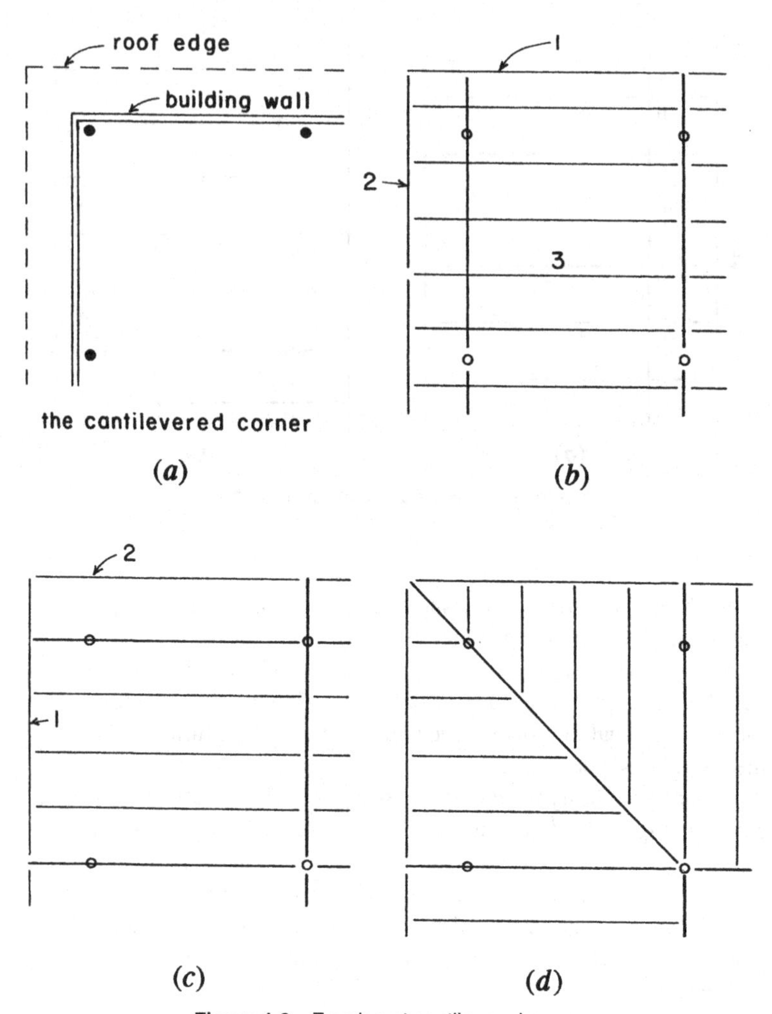

Figure 1.8 Framing at cantilevered corner.

achieves its own span, or by suspension from the overhead structure. Suspended ceilings are quite common, because the space created between the ceiling and the structure is often used for concealment of ducting and registers of the HVAC system, wiring and recessed fixtures of the lighting system, and various other items of building equipment. If a joist or rafter system is used with closely spaced members (4 ft center to center or less), the structure for the ceiling is usually suspended from these members. The other means for suspension is to use hangers attached to the deck, an advantage being the freeing of the modules of the spanning structure and the ceiling framing from each other. The suspended ceiling is also used when the form of the ceiling does not correspond to that of the overhead structure (for example, sloped rafters with a horizontal ceiling).

Roof Drainage. Providing for minimum slopes required for drainage of flat roofs is always a problem for a roof framing system. The most direct means is simply to tilt the framing to provide the slope patterns required. For a complex roof, this gets to be quite complicated with regard to the specification of the levels of the various framing members. The desired patterns of slopes and the locations of drains may not relate well to the layout of the roof framing members. Another possibility is to keep the framing flat but vary the thickness of the deck (applicable only to cast-in-place concrete decks) or to use tapered insulation fill on top of the deck. The latter technique simplifies the framing details but is usually capable of developing only a few inches of slope differential. If a flat ceiling is required and is to be attached directly to the roof structure, this must be considered in facilitating the drainage.

For some types of structural members—most notably the manufactured trusses—it is possible to slope the top of the member while keeping the bottom flat. This makes it possible to have a sloping roof surface and a flat ceiling, with both the roof deck and the ceiling surfacing directly attached to the truss chords.

Dynamic Behavior. Roof structures may usually be optimized for light weight without major restriction, resulting in a benefit of dead-load reduction for both the spanning structure and its supports. Lightweight floor structures, on the other hand, tend to be bouncy, which is generally not a desirable characteristic. Bounciness can also be a result of an excessive span-to-depth ratio for the spanning elements. Experience is the primary guide in this matter, but following are some general rules:

1. Restrict live-load deflections to a conservative ratio of the span (usually not greater than 1/360 of the span for any floor).
2. Limit span-to-depth ratios well below those of the maximum permitted. Suggestions: maximum of 20 for solid members, 15 for trussed joists.

3. Use a very stiff deck for its load-distributing function (to achieve the repetitive member effect, as described for wood joists).
4. Even if load distribution is not significant, do not use decks for the longest spans listed in design data references.

A major factor in reducing bounciness is the presence of the concrete fill on top of steel decks. This fill is now commonly also used on top of wood decks.

Holes. Both floor and roof surfaces are commonly pierced by a number of passages for various items. Large openings are required for stairs and elevators, medium-sized ones for ducts and chimneys, and small ones for piping and wiring. The structure must be planned and detailed to accommodate these openings, which entails some of the following considerations:

1. *Location of openings.* Openings may often occur at locations not convenient for the framing. This may indicate some poor planning of the framing or may be essentially unavoidable. For structures that utilize column line rigid frame bents for lateral bracing, the integrity of the bents generally requires that openings be kept off of the column lines. For regularly spaced systems in general, the layout of the framing and locations of required openings should be coordinated to maintain a maximum regularity of the system. Openings should not interrupt the major elements of the system (large trusses or girders).
2. *Size of openings.* Large openings must have some framing around their perimeters, which are also likely to be locations of supported wall construction. Small openings (for single pipes, for example) may simply pierce the deck with no special provision. For sizes of openings between these extremes, the accommodation requirements depend on the form and size of the elements of the structure. For closely spaced joists, provision for openings of a size that fits between the joists is usually quite simple; when the size requires the interruption of one or more joists, it entails some more difficult measures, such as doubling the joists on each side of the opening.
3. *Openings near columns.* For efficiency in architectural planning, it is sometimes convenient to locate duct shafts or chases for piping or wiring next to a structural column. If this can be done without

interrupting a major spanning member, it may not present a problem. If the opening must be on the column line, it may require straddling of the opening with a double framing member of two spaced elements.

4. *Loss of effectiveness of diaphragms.* Presence of large openings must be considered with regard to effects on the functioning of the floor or roof system as a horizontal diaphragm for lateral bracing. It may be necessary to provide special framing or connections to develop collector functions, drag struts, or the subdivision of the diaphragm, as described in Section 10.4.

The Three-Dimensional Frame

Steel elements are frequently used to obtain what is essentially a two-dimensional structure, often constituting a single plane of a floor, a roof, or a wall. However, steel elements can also be arranged in three-dimensional systems, producing a skeleton structure for a tower or a multistory building. One of the early major uses of rolled shapes was for the early skyscrapers of the late nineteenth century.

Product development and usage applications often grow interactively. Such was the case for development of the W shapes (as mentioned, originally called wide-flange shapes). These shapes have relatively wide flanges, and most of the flange surface is flat rather than tapered. This geometry particularly facilitates the assemblage of frameworks that use a common joint configuration as shown in Figure 1.9. Here a multistory steel column is shown as continuous between two stories, with steel beams framing into it from three horizontal directions. The W-shape column (actually I- or H-shaped in cross section) is ideally formed to accommodate this framing. To achieve the joint shown in Figure 1.9, the column needs the following attributes:

1. The flanges must be wide enough to accept the framing connection of the beam on the flange side of the column.
2. The distance from flange to flange (or the *depth* of the W shape) must be large enough to accommodate the framing connection of the beams that frame into the column web.
3. For the connection detail as shown, any splice joints in the column must be located above or below the beam level.

This results in some common practices, such as the following.

Minimum Column Depth. Steel columns for multistory construction are usually a minimum of 10 in. in nominal depth. For heavier loads and larger beams depths may be 12 or 14 in.

Minimum Flange Width. Shapes used for columns are mostly those with flanges at least 6-in. wide. Wider flanges are also available, with an approximately square column shape being commonly produced for column use in the 10-, 12-, and 14-in. nominal shapes series.

Common Column Splice Location. This is usually about 3 ft above the tops of the beams, also a handy height for the steel erection crew.

Planning of three-dimensional frames involves many considerations, including those mentioned in the preceding section for two-dimensional systems. Obviously, it is generally desirable for columns in one story to be located over columns in the story below whenever possible. Another possible consideration involves the use of so-called framed bents. In the three-dimensional system, these may be constituted by the columns and beams in a single vertical plane. The nature and problems of such bents are discussed in Chapter 5.

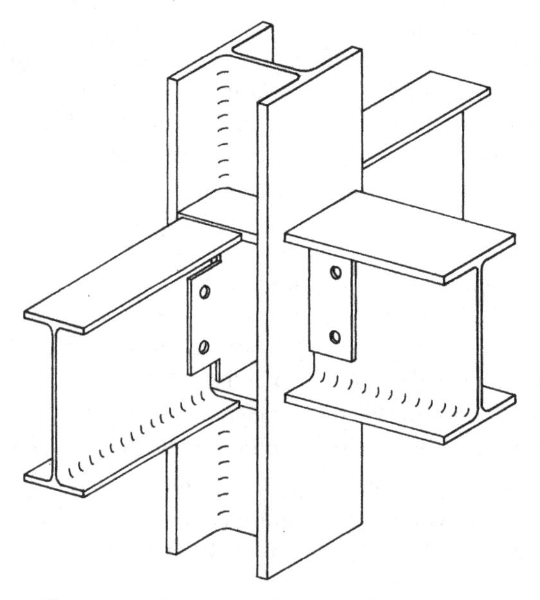

Figure 1.9 Connections for column/beam framing.

Truss Systems

Trussing can be used to produce some of the lightest of steel-framed structures. Two basic principles are involved. The first is the use of the basic planar stability of the three-sided triangle held rigid simply by resistance of change in the length of its sides. The second is the isolation of highly efficient concentrations of material with great distances of separation—a feat only possible with a relatively strong material. The specific usage and problems of steel trusses are discussed at length in Chapter 7.

In steel frameworks, trusses may be substituted for beams in some situations. This may be done to achieve spans more efficiently, especially when spans are great. However, the trussed joist of more modest dimensions is also frequently used in place of beams with solid webs. A benefit deriving from the absence of a solid web is the ease of passing of ducting, piping, or wiring through the system—a value that may well favor the selection of trusses over solid-web beams.

Another major use of trussing consists of adding diagonal members to a vertical planar arrangement of steel beams and columns to produce a *trussed bent.* This is a common means of producing three-dimensional stability for steel frameworks and is one of the major options for development of bracing for resistance to the horizontal force effects of wind or earthquakes.

Trussing is also highly adaptable to the generation of many forms other than rectilinear frameworks. Linear rolled shapes can be bent or curved, but arrangements of truss members can more easily be shaped to achieve just about any form.

Rigid Frames

Rigid frames are frameworks in which rigid (moment-resistive) connections are made between members. This is not the "normal" way of connecting linear steel elements for building structural frames or trusses, so the joints must be specially developed. In Figure 1.9, for example, the "normal" connection between the beams and the supporting column involves the use of some connecting device that is attached to the beam web and then to the column. This is used essentially to transfer only vertical load. To make this joint capable of transmitting bending moment to the column, the beam flanges must be connected to the column. If this is done, a rigid, or moment-resistive, joint will be produced.

One way of achieving the rigid joint is to weld the beam flanges directly to the column. For the beams on the flange side of the column, this is a direct connection, achieved with a butt weld at the end of the beam flange. Because bending moment in the column is most effectively developed by the column flanges, this is a reasonably direct transfer between the beam and the column.

However, the beam thus grabs only one of the column flanges, so the joint is typically enhanced by welding of filler plates (as shown in position in Figure 1.9) on the inside of the column at the level of the beam flanges. This helps transfer the bending across the whole column section.

The most effective column-beam rigid frame bent is obviously achieved by turning the W-shape columns in plan so that the column flanges are perpendicular to the plane of the planar framed bent. However, in some situations, it may be necessary to achieve bent action in both directions and thus to attach the beams that intersect the open side of the columns for moment transfer. This is not so easily achieved, although possibilities do exist.

Various problems of achieving framing connections are discussed in Chapter 8. General problems of rigid frames are discussed in Chapter 5. Some general planning problems for rigid frame bents in multistory construction are discussed in Section 10.3.

The rigid bent can also be used for simple planar structures. Examples of such structures are presented in Sections 10.1 and 10.5.

Mixed Systems

The all-steel structure is sometimes possible—and is surely ideal in the eyes of people in the steel business. However, the typical building utilizes a variety of materials, so that the mixed-material structure occurs frequently. Spanning roof or floor systems of steel may be supported by steel columns but are also frequently supported by structural walls of concrete or masonry. Steel frameworks are sometimes surfaced with formed sheet steel but are also surfaced with plywood panels, cast-in-place concrete decks, and precast concrete panels.

The planning of elements in a mixed-material system must relate to problems of all the materials and types of structural elements involved. Because this book is devoted essentially to steel structures, it is not possible to develop all possible combinations here. The case study examples in Chapter 10 involve mixed materials in various situations and discussions there deal with some planning concerns.

2

STRUCTURAL INVESTIGATION AND DESIGN

Investigation of structures, whether simply to acquire general knowledge or useful data or for a more intelligent solution of some specific problem, is itself a major field of study. The material in this chapter consists of discussions of the nature, purposes, and various techniques of the investigation of structures. As in all of the work in this book, the primary focus is on material relevant to the tasks of structural design.

2.1 SITUATIONS FOR INVESTIGATION AND DESIGN

Most structures exist to perform a task. Their evaluation must therefore begin with consideration of the effectiveness with which they satisfy the usage requirements determined by that task. Three factors of this effectiveness may be considered: the structure's functionality, feasibility, and safety.

Functionality refers to the various attributes of the structure, such as its shape, detail, durability, fire resistance, weather resistance, and so on, because these relate to its use. Feasibility includes considerations of dollar cost, availability of materials, and overall practicality of achieving its construction. Safety in terms of structural actions is generally obtained in the form of some margin between the structure's capacity for resistance and the demands placed on it. Investigation of a structure, as related to the effectiveness of design work, must deal with all of these concerns.

Investigation for safety, consisting of an analysis of a structure's actions under some loading condition, is intended to establish an understanding of its behavior. In this way, its working responses to service demand conditions can be understood and evaluated. If the loading condition is extended to the point of failure of the structure, the ultimate limit of the structure's capability can be quantified. It is useful for various design purposes to understand both of these: the actions under service load conditions and the ultimate resistance of the structure.

Analysis for investigation may progress with the following considerations:

1. Determination of the structure's physical nature with regard to material, form, scale, detail, orientation, location, support conditions, and internal character.
2. Determination of the demands placed on the structure—that is, the loads and the manner of their application and any usage limits on the structure's deformation.
3. Determination of the nature of the structure's responses to the demands (loads) in terms of support forces, deformations, and internal stresses.
4. Determination of the limits of the structure's capabilities, including its ultimate (maximum) load resistance.
5. General evaluation of the structure's effectiveness.

Analysis can be performed in several ways. The nature of the structure's deformation under load can be visualized, with mental images or sketches. Using available theories and techniques, mathematical models of the structure can be created and manipulated. Finally, the structure itself can be built and tested under a simulated load, or a scaled model can be built and tested in a laboratory.

When reasonably precise quantitative evaluations are required, the most useful tools are direct measurements of physical responses or careful mathematical modeling with theories and procedures that have been demonstrated to be reliable. Ordinarily, mathematical modeling of some kind must precede actual construction, even that of a test model. Direct measurement is usually limited to experimental studies or to efforts to verify questionable theories or techniques.

Physical testing in the design process for building structures is rare. Most design work is based on experience, with some support from computations using commonly accepted procedures and data. However, to support the procedures for mathematical analysis and provide data for computations, a great amount of testing has been done on laboratory specimens. Thus, although individual design projects seldom utilize physical testing, the design procedures and data used have been developed with extensive verification from physical tests.

Techniques and Aids for Investigation

In this book, major use is made of graphical visualization, because the use of sketches as learning and problem-solving aids cannot be overemphasized. Two types of graphical devices are most useful: the free-body diagram and the exaggerated profile of the load-deformed structure.

A free-body diagram consists of a picture of an isolated physical element together with representations of all of the forces that act externally on the element. The isolated element may be a whole structure or any fractional part of it. Consider the structure shown in Figure 2.1. Figure 2.1*a* shows the entire structure, consisting of a beam and column rigid bent. The free-body diagram in Figure 2.1*a* shows the entire structure with forces acting externally on it represented by arrows; these include the forces of gravity, wind, and the reactive resistance of the supports (called the *reactions*). The structure is held in static equilibrium (as a free-body) by this force system.

Shown in Figure 2.1*b* is a free-body diagram of a single beam from the bent. Operating on the beam are the forces of its own weight plus the interaction between the beam ends and the columns to which it is attached. These interactions are not visible in the free-body diagram of the whole bent, so one purpose of the diagram for the beam is simply to visualize the nature of these interactions. It can now be seen that the columns transmit to the ends of the beams horizontal and vertical forces as well as rotational bending actions.

Figure 2.1*c* shows an isolated portion of the beam length, which is produced by slicing vertical planes a short distance apart and removing the beam portion between them. Operating on this free body are its own

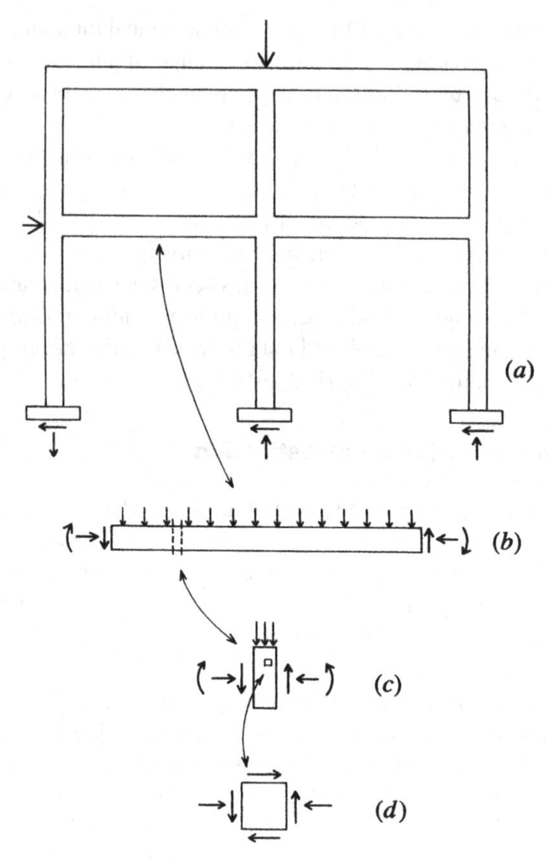

Figure 2.1 Free-body diagrams.

weight and the actions of the beam segments on the opposite sides of the slicing planes, because it is these actions that hold the removed portion in place in the whole beam. This device, called the *cut section,* is used to visualize the internal force actions in beams, because they are not visible in the diagram of the whole beam (Figure 2.1*b*).

Finally, in Figure 2.1*d,* a tiny segment, or particle, of the material of the beam is isolated and the external effects consisting of interactions between this particle and those adjacent to it are visualized. This is a basic device for the visualization of stress; in this case, because of its location in the beam, the particle is subjected to a combination of shear and linear compression stresses.

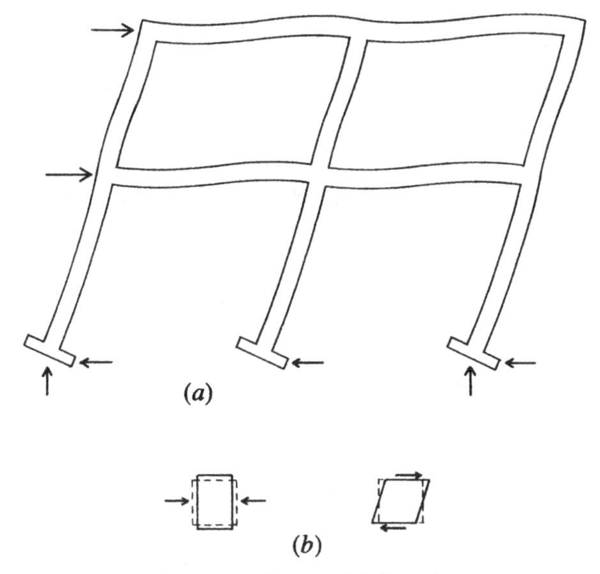

Figure 2.2 Structural deformation.

Figure 2.2*a* shows the exaggerated deformation of the bent in Figure 2.1 under wind loading. The nature of bending action in each member of the frame can be visualized from this figure. The nature of deformation of individual particles under various types of stress can also be visualized, as shown in Figure 2.2*b*. These diagrams are very helpful in establishing the qualitative nature of the relationships between force actions and shape changes. Indeed, one can be inferred from the other: deformation due to force or stress; type of force or stress implied by a particular form of deformation.

Another useful graphical device is the scaled plot of some mathematical relationship or of the data from some observed physical phenomenon. Considerable use is made of this technique in presenting ideas in this book. The graph in Figure 2.3 represents the form of a damped vibration of an elastic spring. It consists of the plot of the displacement s against elapsed time t and represents the graph of the expression

$$s = \frac{1}{e^t} P \sin (Q^t + R)$$

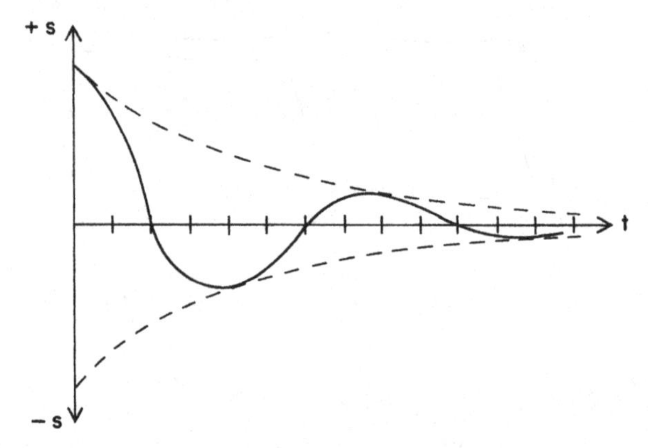

Figure 2.3 Graphical plot of a damped cyclic motion.

Although the equation is technically sufficient for the description of the phenomenon, the graph helps in the visualization of many aspects of the relationship, such as the rate of decay of the displacement, the interval (period) of the vibration, the specific position at some specific elapsed time, and so on.

2.2 METHODS OF INVESTIGATION AND DESIGN

Traditional structural design was developed primarily with a method now referred to as *stress design.* This method utilizes basic relationships derived from classic theories of elastic behavior of materials, and the adequacy or safety of designs are measured by comparison with two primary limits: an acceptable level for maximum stress and a tolerable limit for the extent of deformation (deflection, stretch, etc.). These limits are calculated as they occur in response to the service loads—that is, the loads caused by the normal usage conditions visualized for the structure. This method is also called the *working stress method,* the stress limits are called *allowable working stresses,* and the tolerable movements are called *allowable deflection, allowable elongation,* and so forth.

To convincingly establish both stress and strain limits, it was necessary to perform tests on actual structures. This was done extensively, in both the field (on real structures) and in testing laboratories (on specimen prototypes or models). When nature provides its own tests in the form of structural failures, forensic studies are typically made extensively by various people—for research or for establishment of liability.

Testing has helped to prove, or disprove, the design theories and to provide data for the shaping of the processes into an intelligent operation. The limits for stress levels and for magnitudes of deformation—essential to the working stress method—have been established in this manner. Thus, although a difference is clearly seen between the stress and strength methods, they are actually both based on evaluations of the total capacity of structures tested to their failure limits. The difference is not minor, but it is really mostly one of procedure.

The Stress Method

The stress method generally consists of the following:

1. The service (working) load conditions are visualized and quantified as intelligently as possible. Adjustments may be made here by the determination of various statistically likely load combinations (dead load plus live load plus wind load, etc.), by consideration of load duration, and so on.
2. Stress, stability, and deformation limits are set by standards for the various responses of the structure to the loads: in tension, bending, shear, buckling, deflection, and so on.
3. The structure is then evaluated (investigated) for its adequacy or is proposed (designed) for an adequate response.

An advantage obtained in working with the stress method is that the real usage condition (or at least an intelligent guess about it) is kept continuously in mind. The principal disadvantage comes from its detached nature regarding real failure conditions, because most structures develop much different forms of stress and strain as they approach their failure limits.

The Strength Method

In essence, the working stress method consists of designing a structure to work at some established appropriate percentage of its total capacity. The strength method consists of designing a structure to fail, but at a load condition well beyond what it should have to experience in use. A major reason for favoring of strength methods is that the failure of a structure is relatively easily demonstrated by physical testing. What is truly appropriate as a working condition, however, is pretty much a theoretical speculation. In any event, the strength method is now largely preferred in professional design work. It was first mostly developed for design of reinforced concrete structures but is now generally taking over all areas of structural design work.

Nevertheless, it is considered necessary to study the classic theories of elastic behavior as a basis for visualization of the general ways that structures work. Ultimate responses are usually some form of variant from the classic responses (because of inelastic materials, secondary effects, multimode responses, etc.). In other words, the usual study procedure is to first consider a classic, elastic response and then to observe (or speculate about) what happens as failure limits are approached.

For the strength method, the process is as follows:

1. The service loads are quantified as in Step 1 for the stress method and then are multiplied by an adjustment factor (essentially a safety factor) to produce the factored ultimate load.
2. The form of response of the structure is visualized, and its ultimate (maximum, failure) resistance is quantified in appropriate terms (resistance to compression, to buckling, to bending, etc.). This quantified resistance is also subject to an adjustment factor called the *resistance factor.*
3. The usable (factored) resistance of the structure is then compared to the ultimate resistance required (an investigation procedure), or a structure with an appropriate resistance is proposed (a design procedure).

When the design process using the strength method employs both load and resistance factors, it is now called *load and resistance factor design* (abbreviated LRFD).

2.3 INVESTIGATION OF COLUMNS AND BEAMS

Structural investigation begins with an overall analysis of the entire structure to determine responses at supports (reactions) and the type and magnitudes of interior force actions (tension, compression, shear, bending, and torsion). For systems composed of simple beams and individual, pin-ended columns (as in most wood frames, for example), this analysis is quite easily performed. Most concrete frame structures, on the other hand, have members that are continuous through many spans and beam and column groups that constitute rigid frames. Concrete frames are thus commonly quite statically indeterminate and their investigation is complex. Analysis of complex indeterminate structures is beyond the scope of this book, but the following materials in this chapter explain some aspects of their behavior. For approximate design, various simplified approximation methods may be used, as discussed in Section 2.5.

Investigation of Columns

Elementary considerations for columns begin with basic concern for direct, axial compression, with possible concern for any potential for buckling (failure due partly to slenderness of the column). In rigid frame structures, however, columns are also subjected to bending and shear. In the case of the three-dimensional rigid frame—the common situation for building frame structures—there is typically bending in two directions and sometimes additional torsional twisting.

For all columns, there is also concern for the feasibility of precise construction, so that even for simple, axially loaded columns, some bending is often assumed. The general concerns for analysis and design of columns are discussed in Chapter 4. Some aspects of column-beam frame behavior are discussed in this chapter and also in Chapter 5.

Investigation of Beams

The simple, single-span beam occurs commonly in steel structures. As shown in Figure 2.4, the simple beam may exist when a single span is supported on bearing-type supports that offer little restraint (Figure 2.4*a*), or when beams are connected to columns with connections that offer little moment resistance (Figure 2.4*b*). Although these situations are common in structures of steel and wood, they seldom occur in concrete structures, except when precast elements are used.

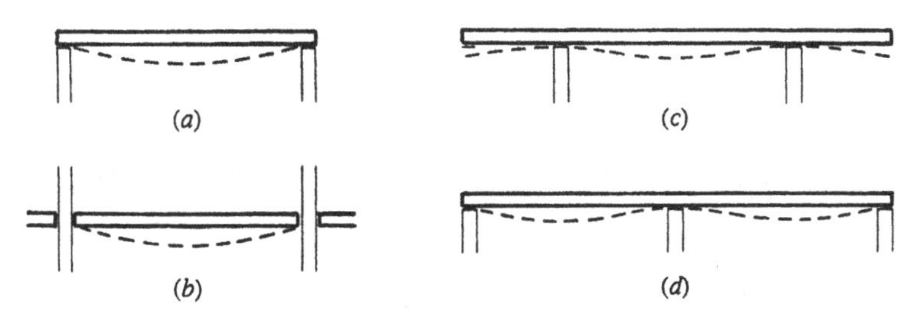

Figure 2.4 Flexural deformation of various beam forms.

For single-story structures, supported on bearing-type supports, continuity resulting in complex bending can occur when the spanning members are extended over the supports. This may occur in the form of cantilevered ends (Figure 2.4*c*) or of multiple spans (Figure 2.4*d*). These conditions are common in wood and steel structures, and can also occur in reinforced concrete structures. Members of steel and wood are usually constant in cross section throughout their length; thus, it is necessary only to find the single maximum value for shear and the single maximum value for moment. For the concrete member, however, the variations of shear and moment along the beam length must be considered, and several different cross sections must be investigated.

Figure 2.5 shows conditions that occur in welded steel frames and in concrete structures when beams and columns are cast monolithically. For the single-story structure (Figure 2.5*a*), the rigid joint between the beam and its supporting columns will result in behavior shown in Figure 2.5*b*, with the columns offering some degree of restraint to the rotation of the beam ends. Thus, some moment will be added to the tops of the columns and the beam will behave as for the center portion of the span in Figure 2.4*c*, with both positive and negative moments.

For the multiple-story, multiple-span frame of concrete or welded steel, the typical behavior will be as shown in Figures 2.5*c* and 2.5*d*. The columns above and below, plus the beams in adjacent spans, will contribute to the development of restraint for the ends of an individual beam span. This condition occurs in steel structures only when welded or heavily bolted moment-resisting connections are used. In concrete structures, it is the normal condition.

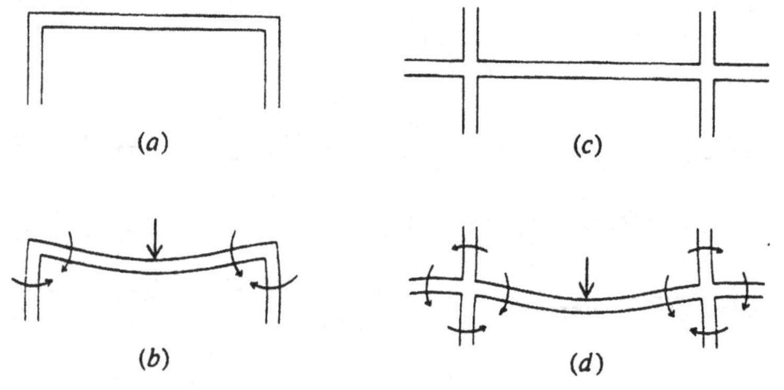

Figure 2.5 Beam actions in rigid beam/column frames.

The structures shown in Figures 2.4*d*, 2.5*a*, and 2.5*c* are statically indeterminate. This means that their investigation cannot be performed using only the conditions of static equilibrium. Although a complete consideration of statically indeterminate behaviors is well beyond the scope of this book, some treatment must be given for a realistic development of the topic of design of reinforced concrete structures. The discussions that follow will serve to illustrate the various factors in the behavior of continuous frames and will provide material for approximate analysis of common situations.

Effects of Beam End Restraint

Figures 2.6*a* to *d* show the effects of various end support conditions on a single-span beam with a uniformly distributed load. Similarly, Figures 2.6*e* to *h* show the conditions for a beam with a single concentrated load. Values are indicated for the maximum shears, moments, and deflection for each case. (Values for end reaction forces are not indicated, because they are the same as the end shears.)

Note the following for the four cases of end support conditions:

1. Figures 2.6*a* and *e* show the cantilever beam, supported at only one end with a *fixed-end* condition. Both shear and moment are critical at the fixed end, and maximum deflection occurs at the unsupported end.
2. Figures 2.6*b* and *f* show the classic "simple" beam, with supports offering only vertical force resistance. We will refer to this type of support as a *free end* (meaning that it is free of rotational restraint). Shear is critical at the supports, and both moment and deflection are maximum at the center of the span.
3. Figures 2.6*c* and *g* show a beam with one free end and one fixed end. This support condition produces an unsymmetrical situation for the vertical reactions and the shear. The critical shear occurs at the fixed end, but both ends must be investigated separately for the concrete beam. Both positive and negative moments occur, with the maximum moment being the negative one at the fixed end. Maximum deflection will occur at some point slightly closer to the free end.
4. Figures 2.6*d* and *h* show the beam with both ends fixed. This symmetrical support condition results in a symmetrical situation for the reactions, shear, and moments, with the maximum deflection occurring at midspan. It may be noted that the shear diagram is the same as for the simple beams in Figures 2.6*b* and *f*.

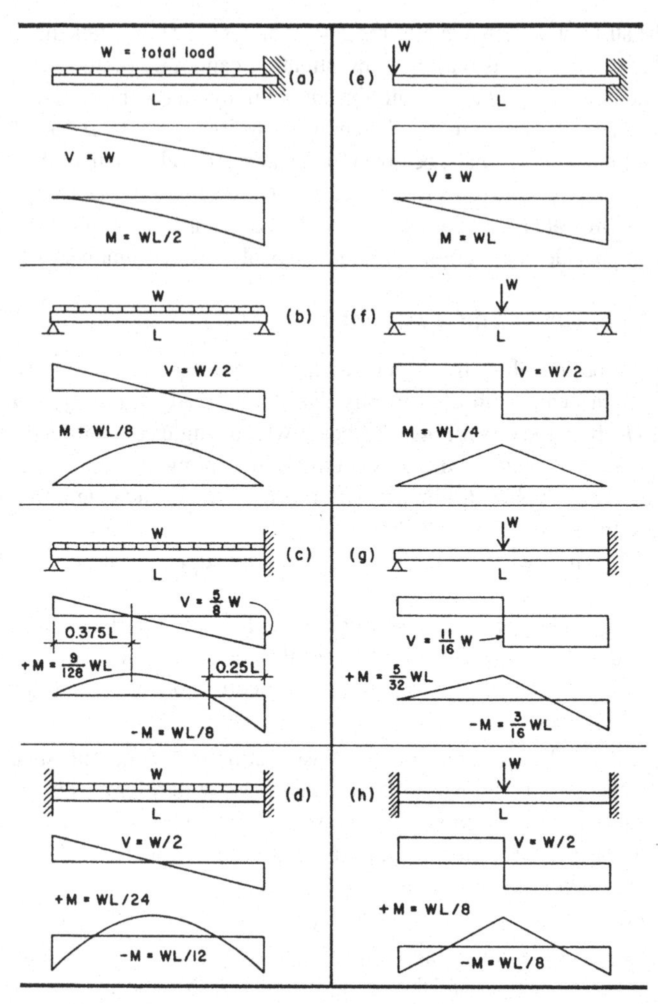

Figure 2.6 Response values for beams with uniformly distributed loading and single concentrated loading.

Continuity and end restraint have both positive and negative effects with regard to various considerations. The most positive gain is in the form of reduction of deflections, which is generally more significant for steel and wood structures, because deflections are less often critical for concrete members. For the beam with one fixed end (Figure 2.6*c*), it may be noted that the value for maximum shear is increased and the maximum moment is the same as for the simple span (no gain in those regards). For full end fixity (Figure 2.6*d*), the shear is unchanged, whereas both moment and deflection are quite substantially reduced in magnitude.

For the rigid frames shown in Figure 2.5, the restraints will reduce moment and deflection for the beam, but the cost is at the expense of the columns, which must take some moment in addition to axial force. Rigid frames are often utilized to resist lateral loads due to wind and earthquakes, presenting complex combinations of lateral and gravity loading that must be investigated.

Effects of Concentrated Loads

Framing systems for roofs and floors often consist of series of evenly spaced beams that are supported by other beams placed at right angles to them. The beams supporting beams are thus subjected to a series of spaced, concentrated loads—the end reactions of the supported beams. The effects

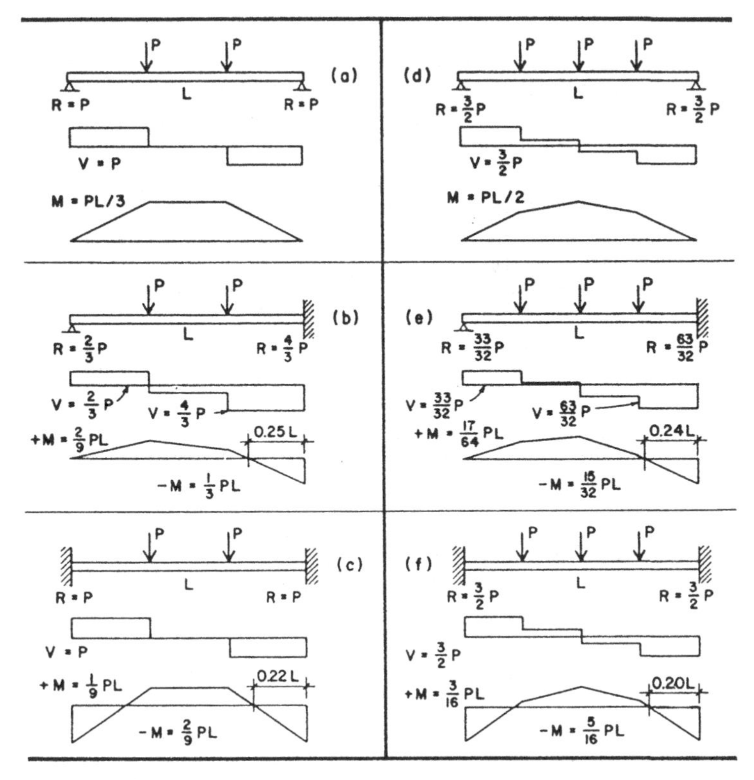

Figure 2.7 Response values for beams with multiple concentrated loads.

of a single such load at the center of a beam span are shown in Figure 2.6*f* to *h*. Two additional situations of evenly spaced concentrated loading are shown in Figure 2.7. When more than three such loads occur, it is usually adequate to consider the sum of the concentrated loads as a uniformly distributed load and to use the values given for Figures 2.6*b* to *d*.

Multiple Beam Spans

Figure 2.8 shows various loading conditions for a beam that is continuous through two equal spans. When continuous spans occur, it is usually necessary to give some consideration to the possibilities of partial beam loading, as shown in Figures 2.8*b* and *d*. It may be noted for Figure 2.8*b* that although there is less total load on the beam, the values for maximum positive moment, for deflection, and for shear at the free end are all higher than for the fully loaded beam in Figure 2.8*a*. This condition of

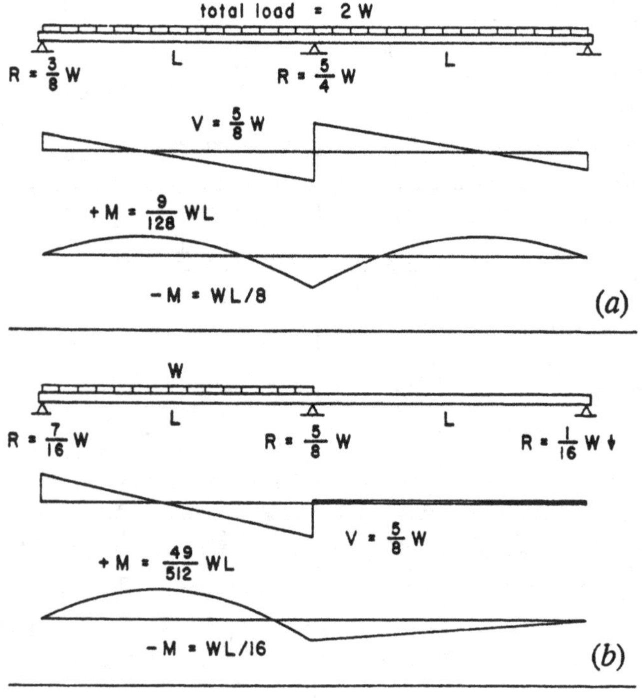

Figure 2.8 Response values for two-span beams.

partial loading must be considered for *live loads* (people, furniture, snow, etc.). For design, the partial loading effects due to the live load must be combined with those produced by *dead load* (permanent weight of the construction) for the full action of the beam.

Figure 2.9 shows a beam that is continuous through three equal spans, with various situations of uniform load on the beam spans. Figure 2.9*a* gives the loading condition for dead load (*always* present in *all* spans). Figures 2.9*b* to *d* show the several possibilities for partial loading, each of which produces some specific critical values for the reactions, shears, moments, and deflections.

Complex Loading and Span Conditions

Although values have been given for many common situations in Figures 2.6 through 2.9, there are numerous other possibilities in terms of un-

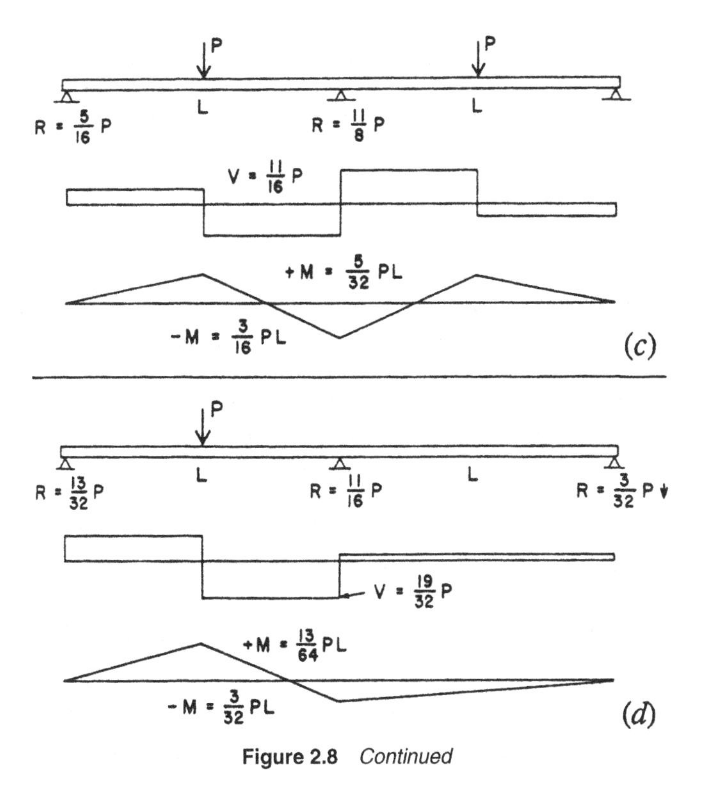

Figure 2.8 *Continued*

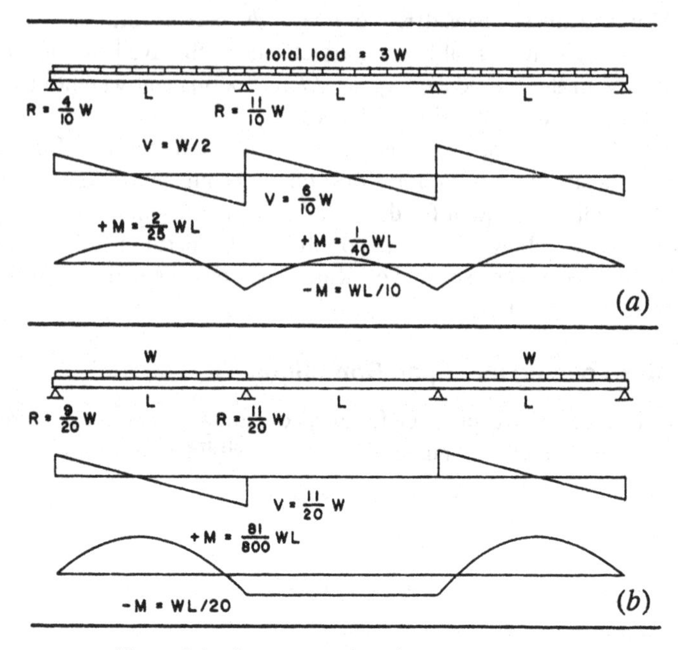

Figure 2.9 Response values for three-span beams.

symmetrical loadings, unequal spans, cantilevered free ends, and so on. Where these occur, an analysis of the indeterminate structure must be performed. For some additional conditions, various handbooks are available that contain tabulations similar to those presented here, such as the AISC Manual (Ref. 3). Appendix Figure B.1 contains a summary of values for common beam loading and support conditions.

Loads and their required combinations for design are discussed in Chapter 9. Resisting loads is basically what structures exist to do. It is very important therefore to understand the source and derivation of loads for design, as well as the reliability of their quantification.

2.4 INVESTIGATION OF COLUMN AND BEAM FRAMES

Frames in which two or more of the members are attached to each other with connections that are capable of transmitting bending between the ends of the members are called *rigid frames.* The connections used to

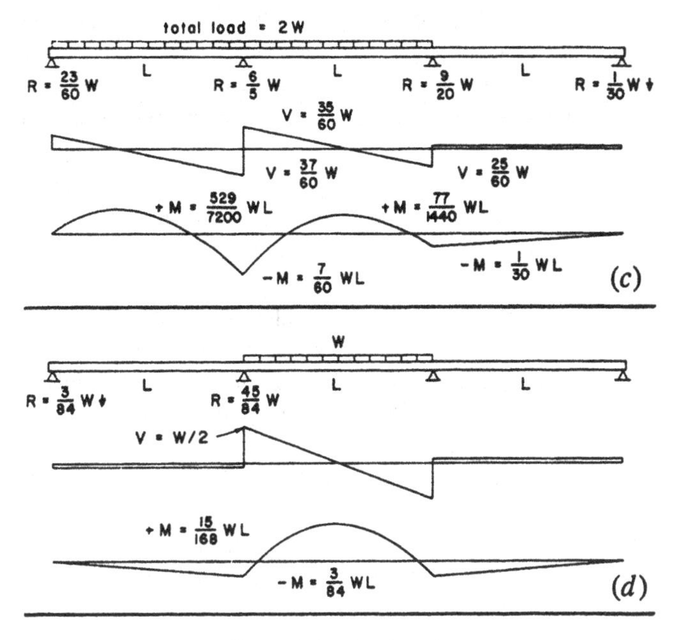

Figure 2.9 *Continued*

achieve such a frame are called *moment connections* or *moment-resisting connections.* Most rigid frame structures are statically indeterminate and do not yield to investigation by consideration of static equilibrium alone. The examples presented in this section are all rigid frames that have conditions that make them statically determinate and thus capable of being fully investigated by methods developed in this book.

Cantilever Frames

Consider the frame shown in Figure 2.10*a,* consisting of two members rigidly joined at their intersection. The vertical member is fixed at its base, providing the necessary support condition for stability of the frame. The horizontal member is loaded with a uniformly distributed loading and functions as a simple cantilever beam. The frame is described as a *cantilever frame* because of the single fixed support. The five sets of figures shown in Figures 2.10*b* through *f* are useful elements for the investigation of the behavior of the frame. They consist of the following:

1. The free-body diagram of the entire frame, showing the loads and the components of the reactions (Figure 2.10*b*). Studying this diagram will help in establishing the nature of the reactions and in the determination of the conditions necessary for stability of the frame as a whole.
2. The free-body diagrams of the individual elements (Figure 2.10*c*). These are of great value in visualizing the interaction of the parts of the frame. They are also useful in the computations for the internal forces in the frame.
3. The shear diagrams of the individual elements (Figure 2.10*d*). These are sometimes useful for visualizing, or for actually computing, the variations of moment in the individual elements. No particular sign convention is necessary unless in conformity with the sign used for moment.
4. The moment diagrams for the individual elements (Figure 2.10*e*). These are very useful, especially in determining the deformation of the frame. The sign convention used is that of plotting the moment on the compression side of the element.
5. The deformed shape of the loaded frame (Figure 2.10*f*). This is the exaggerated profile of the bent frame, usually superimposed on an outline of the unloaded frame for reference. This is very useful for the general visualization of the frame behavior. It is particularly useful for determining the character of the external reactions and the form of interaction between the parts of the frame. Correlation between the deformed shape and the form of the moment diagram is a useful check.

In investigations, these elements are not usually produced in the sequence just described. In fact, it is generally recommended that the deformed shape be sketched first so that its correlation with other factors in the investigation may be used as a check on the work. The following examples illustrate the process of investigation for simple cantilever frames.

Example 1. Find the components of the reactions and draw the free-body diagrams, shear and moment diagrams, and the deformed shape of the frame shown in Figure 2.11*a*.

Solution: The first step is to determine the reactions. Considering the free-body diagram of the whole frame (Figure 2.11*b*), we compute the reactions as follows:

$$\Sigma F = 0 = +8 - R_v,\ R_v = 8 \text{ kips (up)}$$

and with respect to the support:

$$\Sigma M = 0 = M_R - (8 \times 4),\ M_R = 32 \text{ kip-ft (counterclockwise)}$$

Note that the sense, or sign, of the reaction components is visualized from the logical development of the free-body diagram.

Consideration of the free-body diagrams of the individual members will yield the actions required to be transmitted by the moment connec-

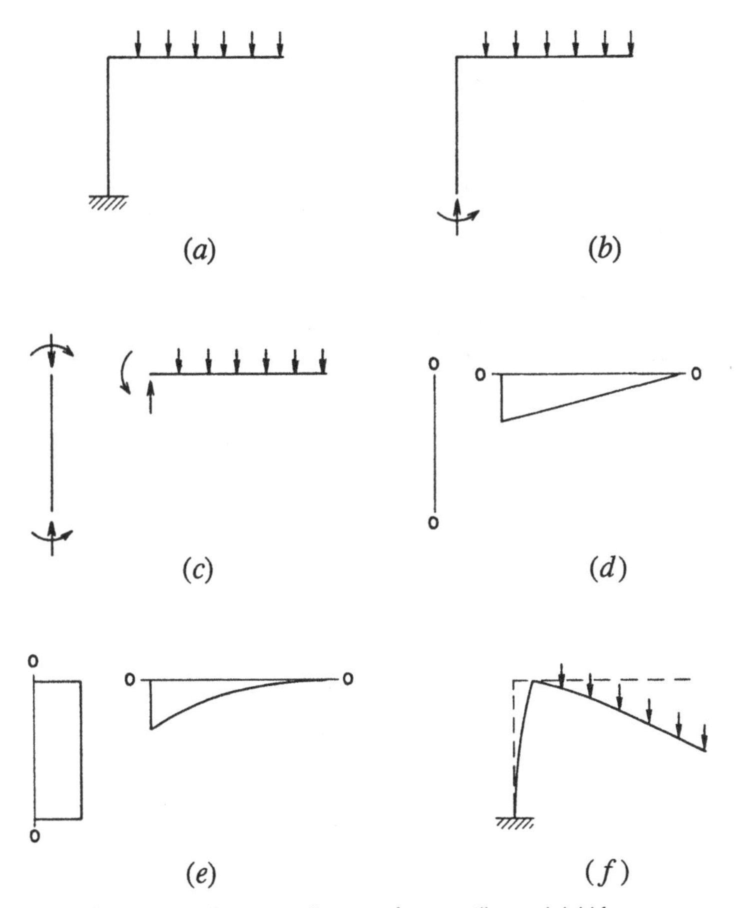

Figure 2.10 Response diagrams for a cantilevered rigid frame.

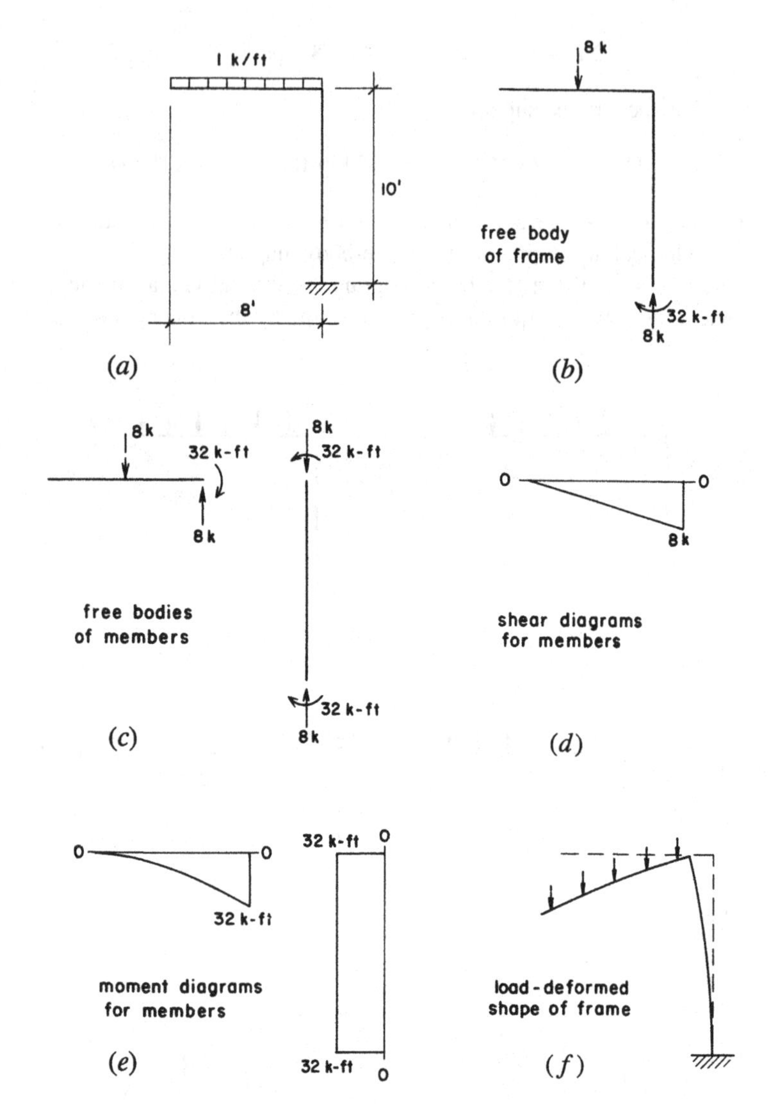

Figure 2.11 Behavior of the frame in Example 1.

tion. These may be computed by application of the conditions for equilibrium for either of the members of the frame. Note that the sense of the force and moment is opposite for the two members, simply indicating that what one does to the other is the opposite of what is done to it.

In this example there is no shear in the vertical member. As a result, there is no variation in the moment from the top to the bottom of the member. The free-body diagram of the member, the shear and moment diagrams, and the deformed shape should all corroborate this fact. The shear and moment diagrams for the horizontal member are simply those for a cantilever beam.

It is possible with this example, as with many simple frames, to visualize the nature of the deformed shape without recourse to any mathematical computations. This is advisable as a first step in investigation, as well as checking continually during the work that individual computations are logical with regard to the nature of the deformed structure.

Example 2. Find the components of the reactions and draw the shear and moment diagrams and the deformed shape of the frame in Figure 2.12*a*.

Solution: In this frame, there are three reaction components required for stability, because the loads and reactions constitute a general coplanar force system. Referring to the free-body diagram of the whole frame shown in Figure 2.12*b*, note that the three conditions for equilibrium for a coplanar system are used to find the horizontal and vertical reaction components and the moment component. If necessary, the reaction force components could be combined into a single-force vector, although this is seldom required for design purposes.

Note that the inflection occurs in the larger vertical member because the moment of the horizontal load about the support is greater than that of the vertical load. In this case, you must compute this value before you can draw the deformed shape accurately.

Example 3. Investigate the frame shown in Figure 2.13 for the reactions and internal conditions. Note that the right-hand support allows for an upward vertical reaction only, whereas the left-hand support allows for vertical and horizontal components. Neither support provides moment resistance.

Solution: Do the following:

1. Sketch the deflected shape (a little tricky, but a good exercise).
2. Consider the equilibrium of the free-body diagram of the whole frame to find the reactions.
3. Consider the equilibrium of the left-hand vertical member to find the internal actions at its top.
4. Consider the equilibrium of the horizontal member.
5. Consider the equilibrium of the right-hand vertical member.
6. Draw the shear and moment diagrams for the members. Then check to see that the required correlation exists.

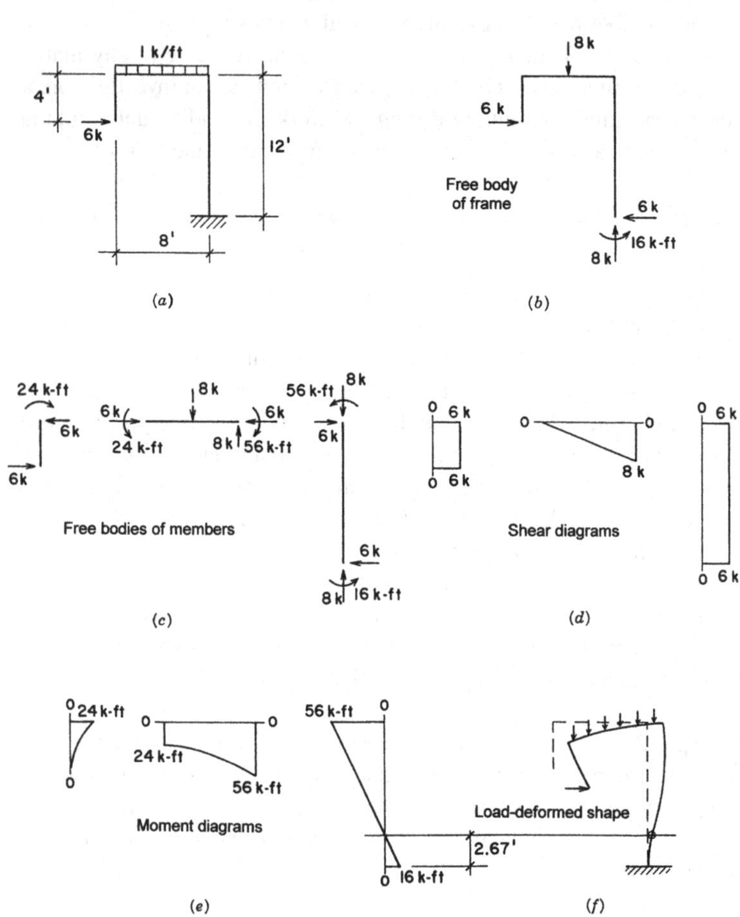

Figure 2.12 Behavior of the frame in Example 2.

Note: Before attempting the following exercise problems, you should produce the results shown in Figure 2.13 independently.

Problems 2.4.A–C. For the frames shown in Figure 2.14*a* through *c*, find the components of the reactions, draw the free-body diagrams of the

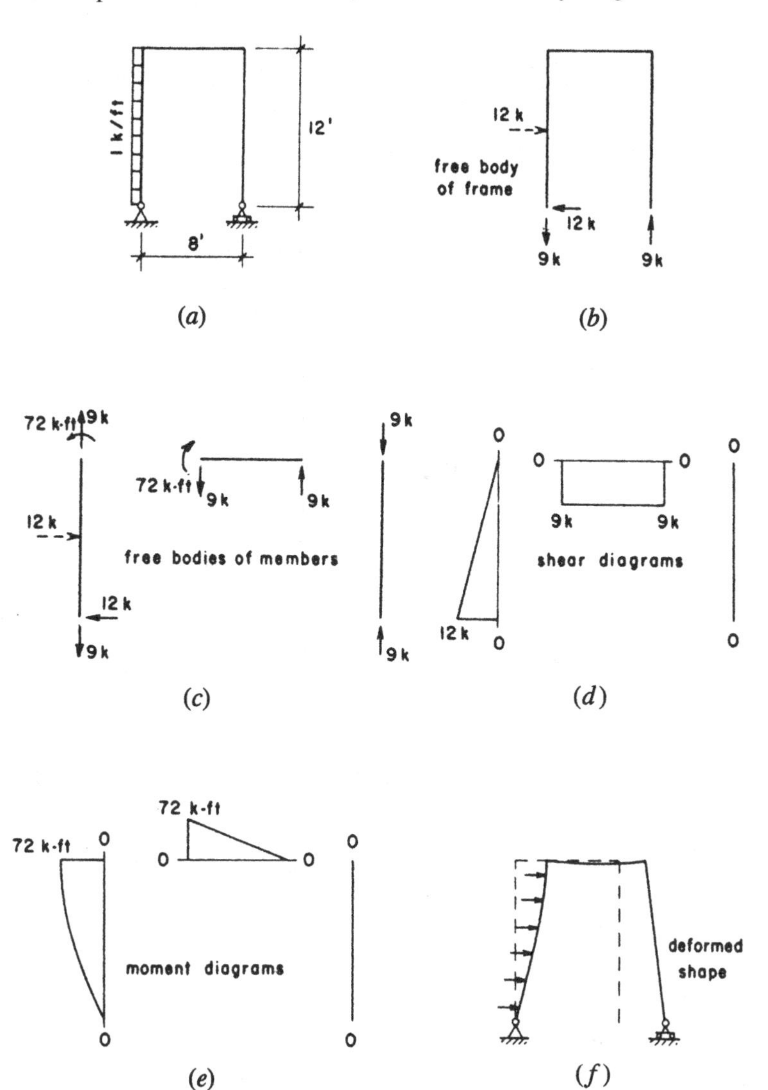

Figure 2.13 Behavior of the frame in Example 3.

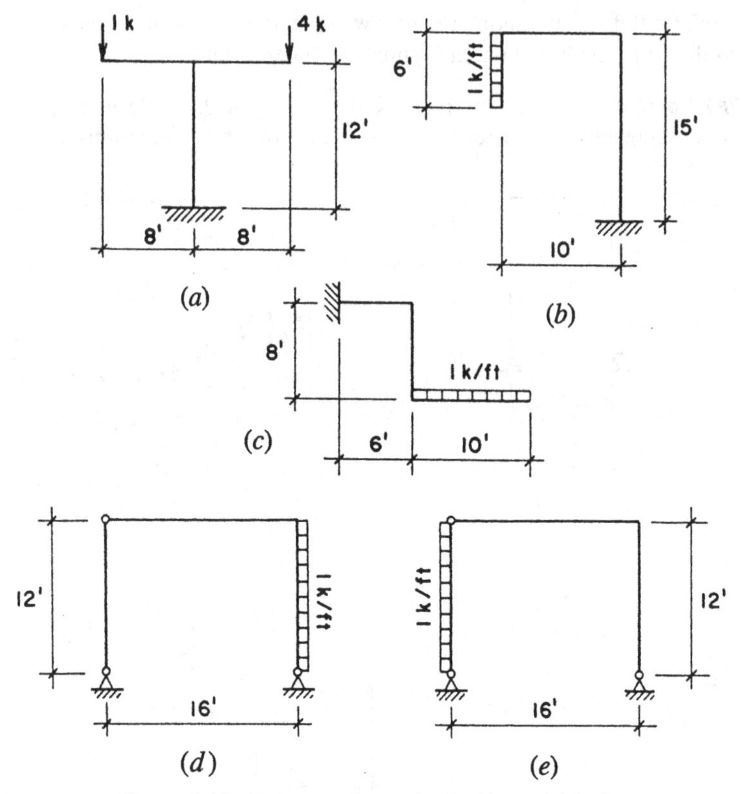

Figure 2.14 Reference figures for Problems 2.4.A–E.

whole frame and the individual members, draw the shear and moment diagrams for the individual members, and sketch the deformed shape of the loaded structure.

Problems 2.4.D, E. Investigate the frames shown in Figure 2.14*d* and *e* for reactions and internal conditions, using the procedure shown for the preceding examples.

2.5 APPROXIMATE INVESTIGATION OF INDETERMINATE STRUCTURES

There are many possibilities for the development of rigid frames for building structures. Two common types of frames are the single-span bent and the vertical, planar bent, consisting of the multistory columns and multispan beams in a single plane in a multistory building.

As with other structures of a complex nature, the highly indeterminate rigid frame presents a good case for use of computer-aided methods. Programs utilizing the finite-element method are available and are used frequently by professional designers. So-called shortcut hand-computation methods such as the moment distribution method were popular in the past. They are "shortcut" only in reference to more laborious hand-computation methods; applied to a complex frame, they constitute a considerable effort—and then produce answers for only one loading condition.

Rigid-frame behavior is much simplified when the joints of the frame are not displaced; that is, they move only by rotating. This is usually only true for the case of gravity loading on a symmetrical frame—and only with a symmetrical gravity load. If the frame is not symmetrical, or the load is not uniformly distributed, or lateral loads are applied, frame joints will move sideways (called *sidesway* of the frame) and additional forces will be generated by the joint displacements.

If joint displacement is considerable, there may be significant increases of force effects in vertical members due to the *P*-delta effect. (See Section 4.5.) In relatively stiff frames, with quite heavy members, this is usually not critical. In a highly flexible frame, however, the effects may be serious. In this case, the actual lateral movements of the joints must be computed to obtain the eccentricities used for determination of the *P*-delta effect. Reinforced concrete frames are typically quite stiff, so this effect is often less critical than for more flexible frames of wood or steel.

Lateral deflection of a rigid frame is related to the general stiffness of the frame. When several frames share a loading, as in the case of a multistory building with several bents, the relative stiffness of the frames must be determined. This is done by considering their relative deflection resistances.

The Single-Span Bent

Figure 2.15 shows two possibilities for a rigid frame for a single-span bent. In Figure 2.15*a* the frame has pinned bases for the columns, resulting in the load-deformed shape shown in Figure 2.15*c,* and the reaction components as shown in the free-body diagram for the whole frame in Figure 2.15*e*. The frame in Figure 2.15*b* has fixed bases for the columns, resulting in the slightly modified behavior indicated. These are common

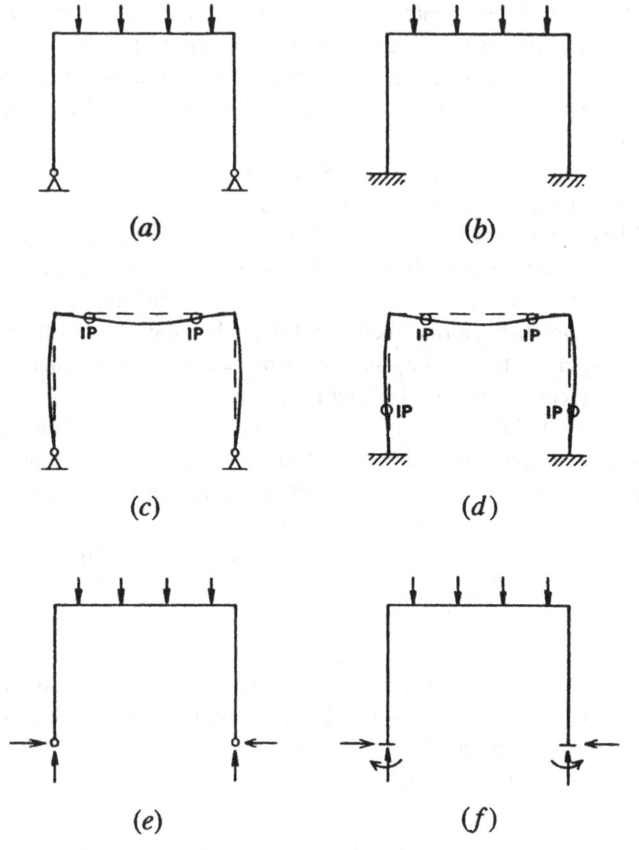

Figure 2.15 Responses of a rigid-framed bent under gravity load with (a) pinned column bases and (b) fixed column bases.

situations, the base condition depending on the supporting structure as well as the frame itself.

The frames in Figure 2.15 are both technically not statically determinate and require analysis by something more than statics. However, if the frame is symmetrical and the loading is uniform, the upper joints do not move sideways and the behavior is of a classic form. For this condition, analysis by moment area, three-moment equation, or moment distribution may be performed, although tabulated values for behaviors can also be obtained for this common form of structure.

Figure 2.16 shows the single-span bent under a lateral load applied at

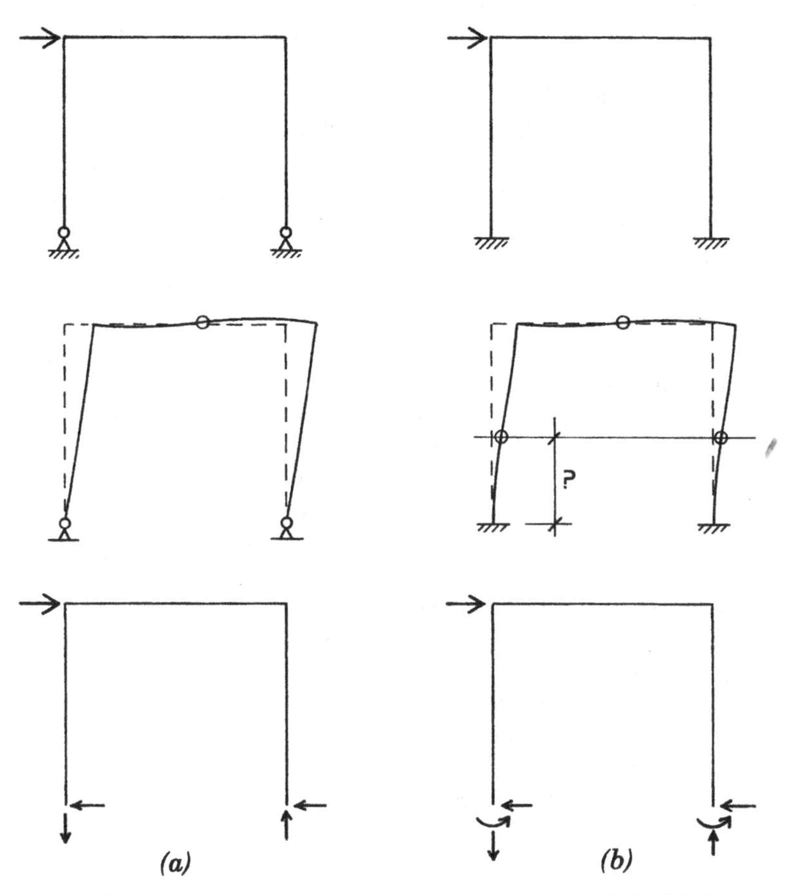

Figure 2.16 Responses of a rigid-framed bent under lateral load with (a) pinned column bases and (b) fixed column bases.

the upper joint. In this case, the upper joints move sideways, the frame taking the shape indicated with reaction components as shown. This also presents a statically indeterminate situation, although some aspects of the solution may be evident. For the pinned base frame in Figure 2.16, for example, a moment equation about one column base will cancel out the vertical reaction at that location, plus the two horizontal reactions, leaving a single equation for finding the value of the other vertical reaction. Then if the bases are considered to have equal resistance, the horizontal reactions will each simply be equal to one-half of the load. The behavior of the frame is thus completely determined, even though it is technically indeterminate.

For the frame with fixed column bases in Figure 2.16*b,* we may use a similar procedure to find the value of the direct force components of the reactions. However, the value of the moment at the fixed base is not subject to such simplified procedures. For this investigation of this frame—as well as that of the frames in Figure 2.15—it is necessary to consider the relative stiffness of the members, as is done in the moment distribution method or in any method for solution of the indeterminate structure.

The rigid-frame structure occurs quite frequently as a multiple-level, multiple-span bent, constituting part of the structure for a multistory building. In most cases, such a bent is used as a lateral bracing element; although once it is formed as a moment-resistive framework, it will respond as such for all types of loads.

The multistory rigid bent is quite indeterminate, and its investigation is complex, requiring considerations of several different loading combinations. When loaded or formed unsymmetrically, it will experience sideways movements that further complicate the analysis for internal forces. Except for very early design approximations, the analysis is now sure to be done with a computer-aided system. The software for such a system is quite readily available.

For preliminary design purposes, it is sometimes possible to use approximate analysis methods to obtain member sizes of reasonable accuracy. Actually, many of the older high-rise buildings still standing were completely designed with these techniques—a reasonable testimonial to their effectiveness.

2.6 SERVICE CONDITIONS

The effects that structures must resist are called the *service conditions.* In general, these conditions are established by the various loads on the structure. When identified and evaluated, these are called the *service loads.* Identification is related to the load source and to its character. A complete discussion of loads is given in Chapter 9.

2.7 LIMIT STATES VERSUS SERVICE CONDITIONS

The differences between the stress method and strength method were discussed in Section 2.2. Stress methods emphasize behavior at service load conditions, whereas strength methods relate primarily to the limits of resistance of structures. The service (actual anticipated usage) condition is not ignored in strength design. Some service load behavior must be considered, such as that pertaining to deflections. In addition, the service load is visualized as accurately as possible, because it serves as the basis for derivation of the factored load.

In fact, the LRFD method as currently implemented combines elements of the old stress method with newer elements of limit states analyses and risk analysis for a total design procedure. The objective is to kill two structural birds with one stone: Make it work for service conditions and make it safe by intelligent evaluation of its limiting capacity.

2.8 RESISTANCE FACTORS

Factoring (modifying) the loads is one form of adjustment for control of safety in strength design. The second basic adjustment is in modifying the quantified resistance of the structure. This amounts to first determining its strength in some terms (compression resistance, moment capacity, buckling limit, etc.) and then reducing it by some percentage. The reduction percentage (obtained by using the resistance factor) is based on various considerations, including concerns for the reliability of theories, quality control in production, ability to accurately predict behaviors, and so on.

Strength design usually consists of comparing the factored load (the load increased by some percentage) to the factored resistance (the resistance reduced by some percentage) of the loaded structure. The factored resistance, called the *design strength,* is obtained by using the following values for the resistance factor, ϕ:

0.90 for limit states involving yielding
0.75 for limit states involving rupture
0.85 for limit states involving compression buckling

When combined with the load factors, application of the resistance factors amounts to a magnification of the safety percentage level.

2.9 CHOICE OF DESIGN METHOD

Application of design procedures in the working stress method tends to be simpler and more direct-appearing than in the strength methods. For example, the design of a beam may amount to simply inverting a few basic stress or strain equations to derive some required properties (section modulus for bending, area for shear, moment of inertia for deflection, etc.). Strength method applications tend to be more obscure, mostly because the mathematical formulations and data are usually more complex and less direct-appearing.

Extensive experience with either method will eventually produce some degree of intuitive judgment, allowing designers to make quick approximations even before the derivation of any specific requirements. Thus, an experienced designer can look merely at the basic form, dimensions, and general usage of a structure and quickly determine an approximate solution—probably quite close in most regards to what a highly exact investigation will produce. Having designed many similar structures before is the essential basis for this quick solution. This can be useful when some error in computation causes the exact design process to produce a weird answer.

Extensive use of strength methods has required careful visualizations and analyses of modes of failures. After many such studies, one develops some ability to quickly ascertain the single, major response characteristic that is the primary design determinant for a particular structure. Concentration on those selected, critical design factors helps to quickly establish a reasonable design, which can then be tested by many basically routine investigative procedures for other responses.

Use of multiple load combinations, multimode structural failures, multiple stress and strain analyses, and generally complex investigative or design procedures is much assisted by computers in most professional design work. A treacherous condition, however, is to be a slave to the computer, accepting its answers with no ability to judge their true appropriateness or correctness. Grinding it out by hand—at least a few times—helps one appreciate the process.

3

HORIZONTAL-SPAN FRAMING SYSTEMS

There are many steel elements that can be used for the basic functions of spanning, including rolled shapes, cold-formed shapes, and fabricated beams and trusses. This chapter deals with fundamental considerations for these elements, with an emphasis on rolled shapes. For simplicity, it is assumed that all the rolled shapes used for the work in this chapter are of steel with F_y = 36 ksi [250 Mpa] or F_y = 50 ksi [350 Mpa].

3.1 FACTORS IN BEAM DESIGN

Various rolled shapes may serve beam functions, although the most widely used shape is the *wide-flange shape*—that is, the member with an I-shaped cross section that bears the standard designation of W shape. Except for those members of the W series that approach a square in cross section (flange width approximately equal to nominal depth), the proportions of members in this series are developed for optimal use in flexure about their major axis (designated as x-x). Design for beam use may involve any combination of the following considerations:

Flexural Stress. Flexural stresses generated by bending moments are the primary stress concern in beams. There are several failure modes for steel beams that define our approach to designing with them, but the general equation for the design of bending members is as follows:

$$\phi_b M_n \geq M_u$$

where ϕ_b = 0.9 for rolled sections
M_n = nominal moment capacity of the member
M_u = maximum moment due to the factored loading

Shear Stress. Whereas shear stress is quite critical in wood and concrete beams, it is less often a problem in steel beams, except for situations where buckling of the thin beam web may be part of a general buckling failure of the cross section. For beam shear alone, the actual critical stress is the diagonal compression, which is the direct cause of buckling in beam action. The general equation for the design for shear is as follows:

$$\phi_v V_n \geq V_u$$

where:

ϕ_v = 0.9 for rolled sections
V_n = nominal shear capacity of the member
V_u = maximum shear due to the factored loading

Buckling. In general, beams that are not adequately braced may be subject to various forms of buckling. Especially critical are beams with very thin webs or narrow flanges or with cross sections especially weak in the lateral direction (on the minor, or y-y axis). Buckling controls the failure mechanism in inadequately braced members and greatly reduces bending capacity. The most effective solution is to provide adequate bracing to eliminate this mode of failure.

Deflection. Although steel is the stiffest material used for ordinary structures, steel structures tend to be quite flexible; thus, vertical deflection of beams must be carefully investigated. A significant value to monitor is the span-to-depth ratio of beams; if this is kept within certain limits, deflection is much less likely to be critical.

Connections and Supports. Framed structures contain many joints between separate pieces, and the details of the connections and supports must be developed for proper construction as well as for the transfer of necessary structural forces through the joints.

Individual beams are often parts of a system in which they play an interactive role. Besides their basic beam functions, there are often design considerations that derive from the overall system actions and interactions. Discussions in this chapter focus mostly on individual beam actions, but discussions in many other chapters treat the usage and overall incorporation of beams in structural systems.

There are several hundred different W shapes for which properties are listed in the AISC manuals (Table A.3). In addition, there are several other shapes that frequently serve beam functions in special circumstances. Selection of the optimal shape for a given situation involves many considerations; frequently, an overriding consideration is the choice of the most economical shape for the task. In general, the least costly shape is usually the one that weighs the least—other things being equal—because steel is priced by unit weight. In most design cases, therefore, the *least weight* selection is typically considered the most economical. Just as a beam may be asked also to develop other actions, such as tension, compression, or torsion, other structural elements may also develop beam actions. Walls may span for bending against wind pressures; columns may receive bending moments as well as compression loads; truss chords may span as beams as well as function for basic truss actions. The basic beam functions described in this chapter may thus be part of the design work for various structural elements besides the real, singular-purpose beam.

3.2 INELASTIC VERSUS ELASTIC BEHAVIOR

There are two principal competing methods of design for steel: Allowable Stress Design (ASD) and Load and Resistance Factor Design (LRFD). This text has adopted LRFD for steel because it has become the standard for the construction industry. The basic difference between these two methods is rooted in the distinction between elastic or inelastic theory of member failure. ASD is rooted in elastic theory, and LRFD is rooted in inelastic theory. The purpose of this section is to compare these two theories so you can better understand inelastic theory and therefore the LRFD method of design.

The maximum resisting moment by elastic theory is predicted to occur when the stress at the extreme fiber reaches the elastic yield value, F_y, and it may be expressed as:

$$M_y = F_y \times S$$

Beyond this condition, the resisting moment can no longer be expressed by elastic theory equations because an inelastic, or *plastic,* stress condition will start to develop on the beam cross section.

Figure 3.1 represents an idealized form of a load-test response for a specimen of ductile steel. The graph shows that up to the yield point the deformations are proportional to the applied stress and that beyond the yield point there is a deformation without an increase in stress. For A36 steel, this additional deformation, called the *plastic range,* is approximately 15 times that produced just before yield occurs. This relative magnitude of the plastic range is the basis for qualification of the material as significantly ductile.

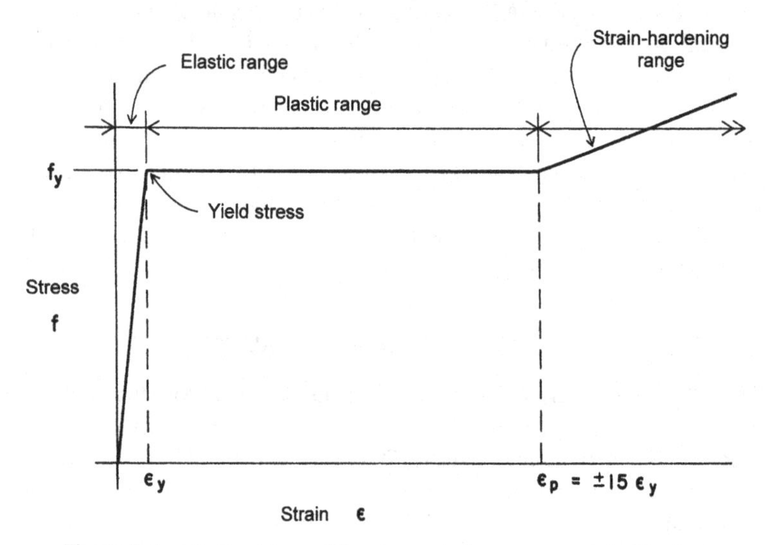

Figure 3.1 Idealized form of the stress-strain response of ductile steel.

Note that beyond the plastic range, the material once again stiffens, called the *strain hardening* effect, which indicates a loss of the ductility and the onset of a second range in which additional deformation is produced only by additional increase in stress. The end of this range establishes the *ultimate stress* limit for the material.

The following example illustrates the application of the elastic theory and will be used for comparison with an analysis of plastic behavior.

Example 1. A simple beam has a span of 16 ft [4.88 m] and supports a single concentrated load of 18 kips [80 kN] at its center. If the beam is a W 12 × 30, compute the maximum flexural stress.

Solution: See Figure 3.2. For the maximum value of the bending moment:

$$M = \frac{PL}{4} = \frac{18 \times 16}{4} = 72 \text{ kip-ft [98 kNm]}$$

In Table A.3, find the value of S for the shape as 38.6 in.^{3} [632 × 10^3 mm^3]. Thus, the maximum stress is

$$f = \frac{M}{S} = \frac{72 \times 12}{38.6} = 22.4 \text{ ksi [154 MPa]}$$

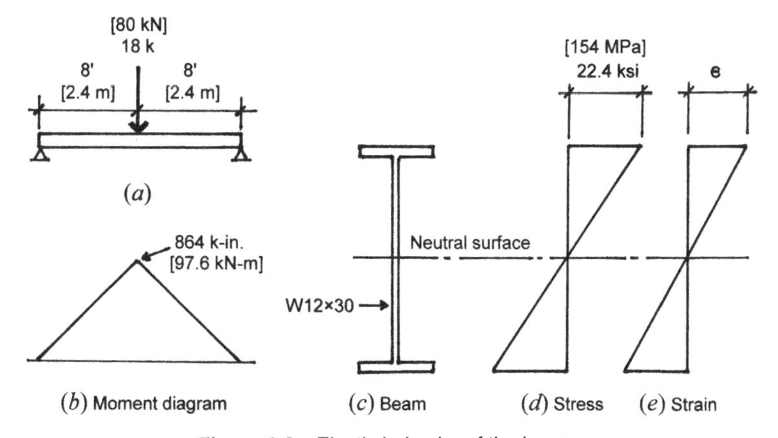

Figure 3.2 Elastic behavior of the beam.

and it occurs as shown in Figure 3.2*d*. Note that this stress condition occurs only at the beam section at midspan. Figure 3.2*e* shows the form of the deformations that accompany the stress condition. This stress level is well below the elastic stress limit (yield point).

The limiting moment for elastic stress occurs when the maximum flexural stress reaches the yield limit, as stated before in the expression for M_y. This condition is illustrated by the stress diagram in Figure 3.3*a*.

If the loading (and the bending moment) that causes the yield limit flexural stress is increased, a stress condition like that illustrated in Figure 3.3*b* begins to develop as the ductile material deforms plastically. This spread of the higher stress level over the beam cross section indicates the development of a resisting moment in excess of M_y. With a high level of ductility, a limit for this condition takes a form as shown in Figure 3.3*c*, and the limiting resisting moment is described as the *plastic moment*, designated M_p. Although a small percentage of the cross section near the beam's neutral axis remains in an elastic stress condition, its effect on the development of the resisting moment is quite negligible. Thus, it is assumed that the full plastic limit is developed by the condition shown in Figure 3.3*d*.

Attempts to increase the bending moment beyond the value of M_p will result in large rotational deformation, with the beam acting as though it were hinged (pinned) at this location. For practical purposes, therefore, the resisting moment capacity of the ductile beam is considered to be exhausted with the attaining of the plastic moment; additional loading will merely cause a free rotation at the location of the plastic moment. This

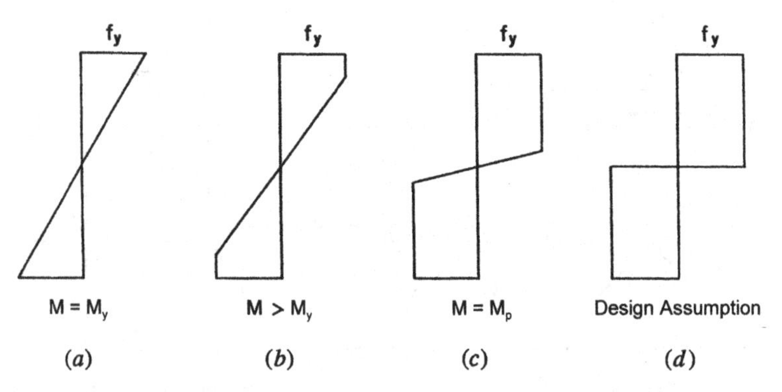

Figure 3.3 Progression of development of flexural stress, from the elastic to the plastic range.

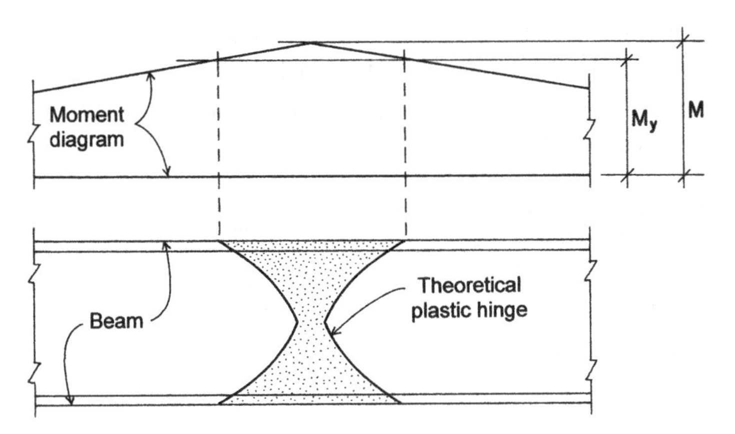

Figure 3.4 Development of the plastic hinge.

location is thus described as a *plastic hinge* (see Figure 3.4), and its effect on beams and frames will be discussed further.

In a manner similar to that for elastic stress conditions, the value of the resisting plastic moment is expressed as

$$M_p = F_y \times Z$$

The term Z is called the *plastic section modulus* and its value is determined as follows.

Referring to Figure 3.5, which shows a W shape subjected to a level of flexural stress corresponding to the fully plastic section (Figures 3.3*d* and 3.4), note the following:

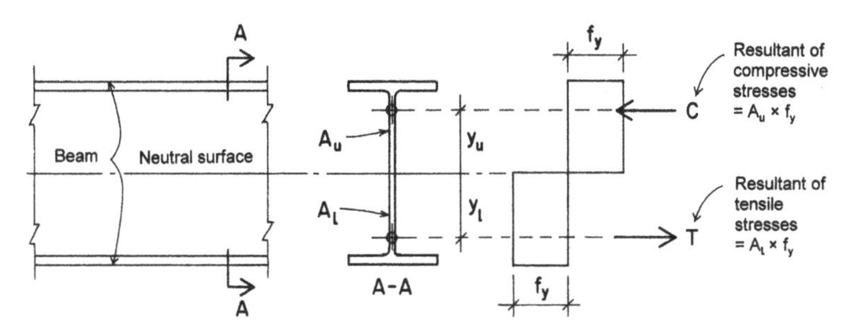

Figure 3.5 Development of the plastic resisting moment.

A_u = upper area of the cross section, above the neutral axis
y_u = distance of the centroid of A_u from the neutral axis
A_l = lower area of the cross section, below the neutral axis
y_l = distance of the centroid of A_l from the neutral axis

For equilibrium of the internal forces on the cross section (the resulting forces C and T developed by the flexural stresses), the condition can be expressed as

$$\Sigma F_h = 0$$

or

$$[A_u \times (+F_y)] + [A_l \times (-F_y)] = 0$$

and, thus:

$$A_u = A_l$$

This shows that the plastic stress neutral axis divides the cross section into equal areas, which is apparent for symmetrical sections, but it applies to unsymmetrical sections as well. The resisting moment equals the sum of the moments of the stresses; thus, the value for M_p may be expressed as

$$M_p = C \times y_u + T \times y_l$$

or

$$M_p = (A_u \times F_y \times y_u) + (A_l \times F_y \times y_l)$$

or

$$M_p = F_y [(A_u \times y_u) + (A_l \times y_l)]$$

or

$$M_p = F_y \times Z$$

and the quantity $[(A_u \times y_u) + (A_l \times y_l)]$ is the property of the cross section defined as the plastic section modulus, designated Z.

Using the expression for Z just derived, its value for any cross section can be computed. However, values of Z are tabulated in the AISC Manual (Ref. 3) for all rolled sections used as beams. See Table A.3 in Appendix. A.

Comparison of the values for S_x and Z_x for the same W shape will show that the values for Z are larger. This presents an opportunity to compare the fully plastic resisting moment to the yield stress limiting moment by elastic stress—that is, the advantage of using plastic analysis.

Example 2. A simple beam consisting of a W 21 × 57 is subjected to bending. Find the limiting moments (a) based on elastic stress conditions and a limiting stress of $F_y = 36$ ksi, and (b) based on full development of the plastic moment.

Solution: For (a) the limiting moment is expressed as:

$$M_y = F_y \times S_x$$

From Table A.3, for the W 21 × 57, S_x is 111 in.^{3}, so the limiting moment is

$$M_y = (36) \times (111) = 3996 \text{ kip-in.} \quad \text{or} \quad \frac{3996}{12} = 333 \text{ kip-ft [452 kN-m]}$$

For (b) the limiting plastic moment, using the value of $Z_x = 129$ in.^{3} from Table A.3, is

$$M_p = (36) \times (129) = 4644 \text{ kip-in.} \quad \text{or} \quad \frac{4644}{12} = 387 \text{ kip-ft [525 kN-m]}$$

The increase in moment resistance represented by the plastic moment indicates an increase of 387 − 333 = 54 kip-ft, or a percentage gain of (54/333)(100) = 16.2 percent.

Problem 3.2.A. A simple-span, uniformly loaded beam consists of a W 18 × 50 with $F_y = 36$ ksi. Find the percentage of gain in the limiting bending moment if a fully plastic condition is assumed, instead of a condition limited by elastic stress.

Problem 3.2.B. A simple-span, uniformly loaded beam consists of a W 16 × 45 with $F_y = 36$ ksi. Find the percentage of gain in the limiting bending moment if a fully plastic condition is assumed, instead of a condition limited by elastic stress.

3.3 NOMINAL MOMENT CAPACITY OF STEEL BEAMS

The first step in design for bending using LRFD is the determination of the bending capacity (M_n) of a steel section. The capacity of a steel section is based upon the cross-sectional properties, its yield stress, and bracing of the member from out of plane buckling. Each of these parameters affect how the beam will ultimately fail and thus how much capacity it will have for bending.

Ideally, the failure mode, which every beam will be controlled by, is the inelastic failure described in Section 3.2. If a member section is capable of failing in a plastic hinge, it is considered a "compact" cross section. A compact shape is one that meets the following criteria:

$$\frac{b_f}{2t_f} \leq \frac{65}{\sqrt{F_y}} \quad \text{and} \quad \frac{h_c}{t_w} \leq \frac{640}{\sqrt{F_y}}$$

where b_f = flange width in inches
t_f = flange thickness in inches
F_y = minimum yield stress in ksi
h_c = height of the web in inches
t_w = web thickness in inches

For A36 steel, this translates to

$$\frac{b_f}{2t_f} \leq 10.8 \quad \text{and} \quad \frac{h_c}{t_w} \leq 107$$

and for steel with a yield stress of 50 ksi it translates to

$$\frac{b_f}{2t_f} \leq 9.19 \quad \text{and} \quad \frac{h_c}{t_w} \leq 90.5$$

In the AISC manual (Ref. 35) and in Table 3.1, the ratios needed to determine compactness are computed for each structural shape, and in many of the AISC tables, the noncompact sections are clearly labeled. It should be noted that most shapes used for beams are compact by design.

Compact beams that are adequately laterally supported will fail with a plastic hinge, and therefore the moment capacity is the yield moment for the section:

$$M_n = M_p = F_y \times Z$$

Example 3. Determine the moment capacity of an A36 W 24 × 76 steel beam that is adequately laterally supported.

Solution: First check to make sure that it is a compact section. Compact section criteria is taken from Table 3.1.

$$\frac{b_f}{2t_f} \leq \frac{65}{\sqrt{F_y}} \quad \text{and} \quad \frac{h}{t_w} \leq \frac{640}{\sqrt{F_y}}$$

$$6.61 \leq 10.8 \qquad\qquad 49 \leq 107$$

The test for compactness was correct; therefore, the moment capacity will be equal to the plastic moment.

$$M_n = M_y = F_y \times Z = 36 \text{ ksi} \times 200 \text{ in.}^3 = 7200 \text{ kip-in.} = \frac{7200}{12}$$
$$= 600 \text{ kip-ft } [814 \text{ kN-m}]$$

To ensure that a compact section fails plastically, the maximum spacing between lateral supports of the beam (L_b) must be less than a limiting laterally unbraced length for full plastic flexural strength (L_p) which is defined as:

$$L_p = \frac{300 \times r_y}{\sqrt{F_y}}$$

where r_y = radius of gyration about the y-axis in inches
F_y = minimum yield stress in ksi

If $L_b > L_p$, the beam will fail due to the buckling of the compression side of the beam similar to buckling in columns. The buckling added to the load on the beam causes a torsional failure of the beam. Plastic deformation continues on the cross section prior to buckling as long as the unbraced lateral length (L_b) does not exceed a limiting laterally unbraced length for inelastic lateral-torsional buckling (L_r). When $L_b > L_r$, the cross section will fail in buckling with the section in the elastic range. The limiting laterally unbraced length for inelastic lateral-torsional buckling (L_r) is defined as:

$$L_r = \left(\frac{r_y \times X_1}{(F_y - F_r)}\right) \times \sqrt{1 + \sqrt{1 + X_2 \times (F_y - F_r)^2}}$$

where r_y = radius of gyration about the y-axis in inches
X_1 = beam buckling factor
F_y = minimum yield stress in ksi
F_r = compressive residual stress in section = 10 ksi (for rolled sections)
X_2 = beam buckling factor

Example 4. Determine the limiting lateral lengths L_p and L_r for an A36 W 24 × 76 steel beam.

Solution: From Table 3.1 for a W 24 × 76:

$r_y = 1.92$ in.
$X_1 = 1760$ ksi
$X_2 = 0.0186\ (1/\text{ksi})^2$

$$L_p = \frac{300 \times r_y}{\sqrt{F_y}} = \frac{300 \times 1.92 \text{ in.}}{\sqrt{36 \text{ ksi}}} = 96 \text{ in.} = 8 \text{ ft } [2.44 \text{ m}]$$

$$L_r = \left(\frac{r_x \times X_1}{F_y - F_r}\right) \times \sqrt{1 + \sqrt{1 + X_2 \times (F_y - F_r)^2}}$$

$$= \left(\frac{1.92 \text{ in.} \times 1760 \text{ ksi}}{(36 \text{ ksi} - 10 \text{ ksi})}\right)$$

$$\times \sqrt{1 + \sqrt{1 + 0.0186\ (1/\text{ksi})^2 \times (36 \text{ ksi} - 10 \text{ ksi})^2}}$$

$$= 281 \text{ in.} = 23.4 \text{ ft } [7.14 \text{ m}]$$

Knowing the relationship between the unbraced length (L_b) and the limiting lateral unbraced lengths (L_p and L_r), you can determine the nominal moment capacity (M_n) for any beam. Figure 3.6 shows the form of the relation between M_n and L_b using an example of a W 18 × 50 with F_y = 50 ksi. The AISC Manual (Ref. 3) contains a series of such graphs for shapes commonly used as beams.

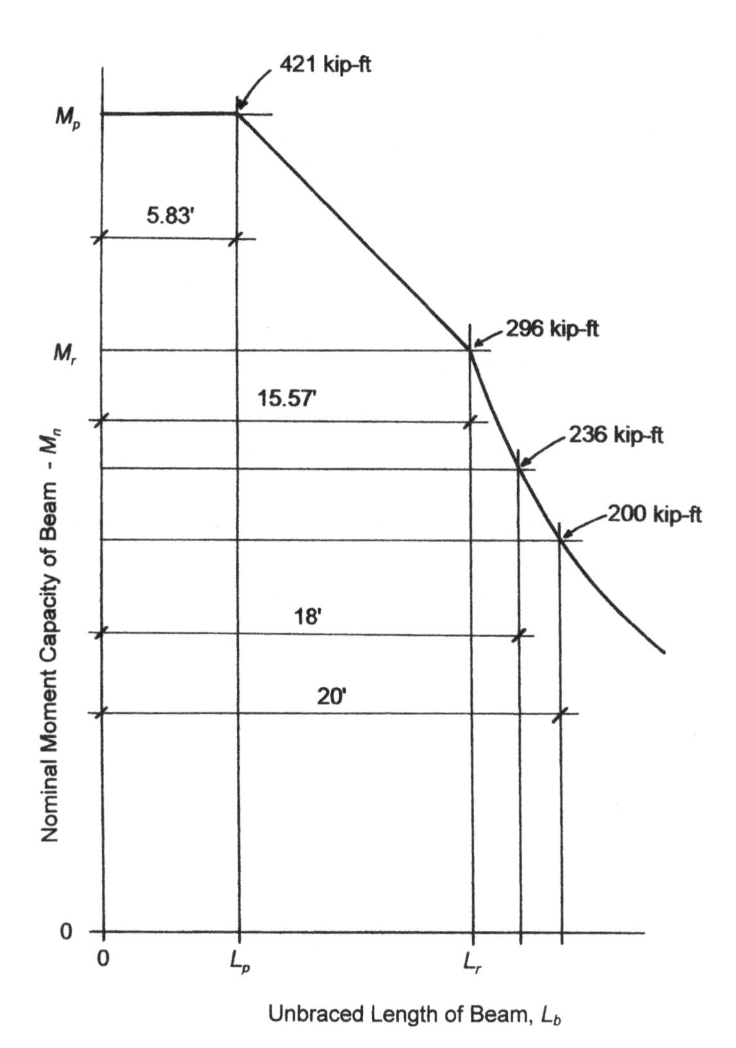

Figure 3.6 Relation between nominal moment capacity M_n and lateral unbraced length L_b for a W 18 × 50 steel beam with F_y = 50 ksi.

If $L_b \leq L_p$

$$M_n = M_p = F_y \times Z$$

If $L_p < L_b \leq L_r$

$$M_n = M_p - (M_p - M_r) \times \left(\frac{L_b - L_p}{L_r - L_p}\right)$$

where $M_r = (F_y - F_r)\, S_x$
F_r = compressive residual stress = 10 ksi for rolled shapes
S_x = section modulus about x-axis in in.^{3}

If $L_b > L_r$

$$M_n = \left(\frac{S_x \times X_1 \times \sqrt{2}}{(L_b/r_y)}\right) \times \sqrt{1 + \frac{(X_1)^2 \times X_2}{2 \times (L_b/r_y)^2}}$$

where X_1 and X_2 = beam buckling factors
r_y = radius of gyration about the y axis

Example 5. Determine the moment capacity of an A36 W 24 × 76 steel beam that is laterally supported every 10 feet.

Solution: (L_p = 8 ft) < (L_b = 10 ft) < (L_r = 23.4 ft) (from Example 4).

Determine M_p:

$$M_p = F_y \times Z = 36 \times 200 = 7200 \text{ kip-in.} = 600 \text{ kip-ft } [8754 \text{ kN-m}]$$

Determine M_r:

$$M_r = (F_y - F_r) \times S_x = (36 \text{ ksi} - 10 \text{ ksi}) \times 176 \text{ in.}^3$$
$$= 4576 \text{ kip-in.} = 381 \text{ kip-ft } [571 \text{ kN-m}]$$

Determine M_n:

$$M_n = M_p - (M_p - M_r) \times \left(\frac{L_b - L_p}{L_r - L_p}\right)$$

$$= 600 \text{ kip-ft} - (600 \text{ kip-ft} - 381 \text{ kip-ft}) \times \left(\frac{10 \text{ ft} - 8 \text{ ft}}{23.4 \text{ ft} - 10 \text{ ft}}\right)$$

$$= 567 \text{ kip-ft } [769 \text{ kN-m}]$$

Example 6. Determine the moment capacity of a 25 ft [7.63 m] A36 W 24 × 76 steel beam that is laterally unsupported.

Solution: $L_b = 25$ ft $> L_r = 23.4$ ft (from Example 4). Next, determine M_n:

$$M_n = \left(\frac{S_x \times X_1 \times \sqrt{2}}{L_b/r_y}\right) \times \sqrt{1 + \frac{(X_1)^2 \times X_2}{2 \times (L_b/r_y)^2}}$$

$$= \left(\frac{176 \text{ in.}^3 \times 1760 \text{ ksi} \times \sqrt{2}}{\left\{\left[25 \text{ ft} \times \left(\frac{12 \text{ in.}}{1 \text{ ft}}\right)\right] \Big/ 1.92 \text{ in.}\right\}}\right)$$

$$\times \sqrt{1 + \frac{(1760 \text{ ksi})^2 \times (0.0186 \text{ ksi})}{2 \times \left\{\left[25 \text{ ft} \times \left(\frac{12 \text{ in.}}{1 \text{ ft}}\right)\right] \Big/ 1.92 \text{ in.}\right\}^2}}$$

$$= \left(\frac{438{,}000 \text{ kip-in.}}{156}\right) \times \sqrt{1 + \frac{57{,}600}{48{,}800}}$$

$$= 2810 \text{ kip-in.} \times 1.48 = 4160 \text{ kip-in.} = 347 \text{ kip-ft } [471 \text{ kN-m}]$$

Problem 3.3.A. Determine the nominal moment capacity (M_n) for a W 30 × 90 made of A36 steel and the following unbraced lengths: (1) 5 ft, (2) 15 ft, (3) 30 ft.

Problem 3.3.B. Determine the nominal moment capacity (M_n) for a W 16 × 36 made of A36 steel and the following unbraced lengths: (1) 5 ft, (2) 10 ft, and (3) 20 ft.

3.4 DESIGN FOR BENDING

Design for bending usually involves the determination of the ultimate bending moment (M_u) that the beam must resist and the use of formulas derived in Section 3.3 for definition of bending resistance of the member (M_n). The basic formulation of this equation is

$$\phi_b M_n = M_u$$

where $\phi_b = 0.9$

Design for Plastic Failure Mode

The formula $F_y = M_n/Z_x$ or $Z_x = M_n/F_y$ is used to determine the minimum plastic section modulus required. Because weight is determined by area, not plastic section modulus, the beam chosen may have more plastic section modulus than required and still be the most economical choice. The following example illustrates the basic procedure.

Example 7. Design a simply supported floor beam to carry a superimposed load of 2 kips per ft [29.2 kN/m] over a span of 24 ft [7.3 m]. (The term *superimposed load* is used to denote any load other than the weight of a structural member itself.) The superimposed load is 25 percent dead load and 75 percent live load. The yield stress is 36 ksi [250 MPa]. The floor beam is continuously supported along its length against lateral buckling.

Solution: The load must first be factored for the load combination to determine the maximum factored load. The combined load factors are given in Section 9.2. The applicable factors are as follows:

$$\begin{aligned} w_u &= 1.4 \times (\text{dead load}) \\ &= 1.4 \times (0.5 \text{ kip/ft}) = 0.7 \text{ kip/ft } [10.2 \text{ kN/m}] \end{aligned}$$

or

$$\begin{aligned} w_u &= 1.2 \times (\text{dead load}) + 1.6 \times (\text{live load}) \\ &= 1.2 \times (0.5 \text{ kip/ft}) + 1.6 \times (1.5 \text{ kip/ft}) = 3.0 \text{ kip/ft } [43.8 \text{ kN/m}] \end{aligned}$$

The bending moment due to the maximum factored superimposed load is

$$M_u = \frac{wL^2}{8} = \frac{3 \times (24)^2}{8} = 216 \text{ kip-ft } [293 \text{ kN-m}]$$

The required bending resistance of the member is

$$M_u = \phi_b \times M_n$$

$$M_n = \frac{M_u}{\phi_b} = \frac{216 \text{ kip-ft}}{0.9} = 240 \text{ kip-ft [325 kN-m]}$$

The required plastic section modulus for this moment is

$$Z_x = \frac{M_n}{F_y} = \frac{240 \text{ kip-ft} \times (12 \text{ in./1 ft})}{36 \text{ ksi}} = 80.0 \text{ in.}^3 \, [1.31 \times 10^6 \text{ mm}^3]$$

Table 3.1 lists a number of shapes commonly used as beams in descending order of the value of their plastic modulus. Also listed in Table 3.1 are values for L_p, L_r, M_p, and M_r. From Table 3.1, a W 16 × 45 is found with a plastic section modulus of 82.3 in.^{3} [1192 ×10^3 mm^3]. Further scanning of the table reveals a W 14 × 53 with a Z_x of 87.1 in.^{3} [1328 × 10^3 mm^3], a W 18 × 46 with a Z_x of 90.7 in.^{3} [1291 × 10^3 mm^3], a W 10 × 58 with a Z_x of 85.3 in.^{3} [1398 × 10^3 mm^3], and a W 21 × 44 with a Z_x of 95.4 in.^{3} [1564 × 10^3 mm^3]. In the absence of any known restriction on the beam depth, try the lightest section (W 21 × 44). The weight of the beam, which was not taken into consideration in the earlier calculations, must now be considered. First the weight of the member must be factored:

$$w_u = 1.2 \times (\text{dead load}) + 1.6 \times (\text{live load})$$

$$= 1.2 \times (44 \text{ lb/ft}) + 1.6 \times (0 \text{ lb/ft}) = 53.0 \text{ lb/ft [773 N/m]}$$

The bending moment at the center of the span, due to the beam weight is

$$M_u = \frac{wL^2}{8} = \frac{53 \text{ lb/ft} \times (24 \text{ ft})^2}{8}$$

$$= 3816 \text{ ft-lb or } 3.82 \text{ kip-ft [5.18 kN-m]}$$

TABLE 3.1 Load Factor Resistance Design Selection for Shapes Used as Beams

		$F_y = 36$ ksi				$F_y = 50$ ksi								
Designation	Z_x in.^{3}	L_p ft	L_r ft	M_p kip-ft	M_r kip-ft	L_p ft	L_r ft	M_p kip-ft	M_r kip-ft	r_y in.	$b_f/2t_f$	h/t_w	X_1 ksi	$X_2 \times 10^6$ $(1/\text{ksi})^2$
W 33 × 141	**514**	**10.1**	**30.1**	**1,542**	**971**	**8.59**	**23.1**	**2,142**	**1,493**	**2.43**	**6.01**	**49.6**	**1,800**	**17,800**
W 24 × 162	468	12.7	45.2	1,404	897	10.8	32.4	1,950	1,380	3.05	5.31	30.6	2,870	2,260
W 24 × 146	418	12.5	42.0	1,254	804	10.6	30.6	1,742	1,237	3.01	5.92	33.2	2,590	3,420
W 33 × 118	**415**	**9.67**	**27.8**	**1,245**	**778**	**8.20**	**21.7**	**1,729**	**1,197**	**2.32**	**7.76**	**54.5**	**1,510**	**37,700**
W 30 × 124	408	9.29	28.2	1,224	769	7.88	21.5	1,700	1,183	2.23	5.65	46.2	1,930	13,500
W 21 × 147	373	12.3	46.4	1,119	713	10.4	32.8	1,554	1,097	2.95	5.44	26.1	3,140	1,590
W 24 × 131	370	12.4	39.3	1,110	713	10.5	29.1	1,542	1,097	2.97	6.70	35.6	2,330	5,290
W 30 × 108	**346**	**8.96**	**26.3**	**1,038**	**648**	**7.60**	**20.3**	**1,442**	**997**	**2.15**	**6.89**	**49.6**	**1,680**	**24,200**
W 27 × 114	343	9.08	28.2	1,029	648	7.71	21.3	1,429	997	2.18	5.41	42.5	2,100	9,220
W 24 × 117	327	12.3	37.1	981	631	10.4	27.9	1,363	970	2.94	7.53	39.2	2,090	8,190
W 21 × 122	307	12.2	41.0	921	592	10.3	29.8	1,279	910	2.92	6.45	31.3	2,630	3,160
W 30 × 90	**283**	**8.71**	**24.8**	**849**	**531**	**7.39**	**19.4**	**1,179**	**817**	**2.09**	**8.52**	**57.5**	**1,410**	**49,600**
W 27 × 94	278	8.83	25.9	834	527	7.50	19.9	1,158	810	2.12	6.70	49.5	1,740	19,900
W 14 × 145	260	16.6	81.6	780	503	14.1	54.7	1,083	773	3.98	7.11	16.8	4,400	348
W 24 × 94	254	8.25	25.9	762	481	7.00	19.4	1,058	740	1.98	5.18	41.9	2,180	7,800

W 21 × 101	**253**	**12.0**	**37.1**	**759**	**492**	**10.2**	**27.6**	**1,054**	**757**	**2.89**	**7.68**	**37.5**	**2,200**	**6,400**
W 12 × 152	243	13.3	94.8	729	453	11.3	62.1	1,013	697	3.19	4.46	11.2	6,510	79
W 18 × 106	230	11.1	40.4	690	442	9.40	28.7	958	680	2.66	5.96	27.2	2,990	1,880
W 14 × 120	212	15.6	67.9	636	412	13.2	46.2	883	633	3.74	7.80	19.3	3,830	601
W 24 × 76	**200**	**8.00**	**23.4**	**600**	**381**	**6.79**	**18.0**	**833**	**587**	**1.92**	**6.61**	**49.0**	**1,760**	**18,600**
W 16 × 100	200	10.4	42.7	600	384	8.84	29.6	833	590	2.5	5.29	23.2	3,530	947
W 21 × 83	196	7.63	24.9	588	371	6.47	18.5	817	570	1.83	5.00	36.4	2,400	5,250
W 18 × 86	186	11.0	35.5	558	360	9.30	26.1	775	553	2.63	7.20	33.4	2,460	4,060
W 12 × 120	186	13.0	75.5	558	353	11.1	50.0	775	543	3.13	5.57	13.7	5,240	184
W 21 × 68	**160**	**7.50**	**22.8**	**480**	**303**	**6.36**	**17.3**	**667**	**467**	**1.8**	**6.04**	**43.6**	**2,000**	**10,900**
W 24 × 62	**154**	**5.71**	**17.2**	**462**	**286**	**4.84**	**13.3**	**642**	**440**	**1.37**	**5.97**	**49.7**	**1,730**	**23,800**
W 16 × 77	152	10.3	35.4	456	295	8.70	25.5	633	453	2.46	6.77	29.9	2,770	2,460
W 12 × 96	147	12.9	61.4	441	284	10.9	41.4	613	437	3.09	6.76	17.7	4,250	407
W 10 × 112	147	11.2	86.4	441	273	9.48	56.5	613	420	2.68	4.17	10.4	7,080	57
W 18 × 71	146	7.08	24.5	438	275	6.01	17.8	608	423	1.7	4.71	32.4	2,690	3,290
W 14 × 82	139	10.3	42.8	417	267	8.77	29.5	579	410	2.48	5.92	22.4	3,590	849
W 24 × 55	**135**	**5.58**	**16.6**	**405**	**249**	**4.74**	**12.9**	**563**	**383**	**1.34**	**6.94**	**54.1**	**1,570**	**36,500**
W 21 × 57	129	5.63	17.3	387	241	4.77	13.1	538	370	1.35	5.04	46.3	1,960	13,100
W 18 × 60	123	7.00	22.3	369	234	5.94	16.6	513	360	1.68	5.44	38.7	2,290	6,080
W 12 × 79	119	12.7	51.8	357	232	10.8	35.7	496	357	3.05	8.22	20.7	3,530	839
W 14 × 68	115	10.3	37.3	345	223	8.70	26.4	479	343	2.46	6.97	27.5	3,020	1,660
W 10 × 88	113	11.0	68.4	339	213	9.30	45.1	471	328	2.63	5.18	13.0	5,680	132

(continued)

TABLE 3.1 ***(Continued)***

		F_y = 36 ksi				F_y = 50 ksi								
Designation	Z_x in.^{3}	L_p ft	L_r ft	M_p kip-ft	M_r kip-ft	L_p ft	L_r ft	M_p kip-ft	M_r kip-ft	r_y in.	$b_f/2t_f$	h/t_w	X_1 ksi	$X_2 \times 10^6$ $(1/\text{ksi})^2$
W 21 × 50	**110**	**5.42**	**16.2**	**330**	**205**	**4.60**	**12.5**	**458**	**315**	**1.3**	**6.10**	**49.4**	**1,730**	**22,600**
W 16 × 57	105	6.67	22.8	315	200	5.66	16.6	438	307	1.6	4.98	33.0	2,650	3,400
W 18 × 50	**101**	**6.88**	**20.5**	**303**	**193**	**5.83**	**15.6**	**421**	**296**	**1.65**	**6.57**	**45.2**	**1,920**	**12,400**
W 21 × 44	**95.4**	**5.25**	**15.4**	**286**	**177**	**4.45**	**12.0**	**398**	**272**	**1.26**	**7.22**	**53.6**	**1,550**	**36,600**
W 18 × 46	90.7	5.38	16.6	272	171	4.56	12.6	378	263	1.29	5.01	44.6	2,060	10,100
W 14 × 53	87.1	8.00	28.0	261	169	6.79	20.1	363	259	1.92	6.11	30.9	2,830	2,250
W 10 × 68	85.3	10.8	53.7	256	164	9.16	36.0	355	252	2.59	6.58	16.7	4,460	334
W 16 × 45	82.3	6.54	20.2	247	158	5.55	15.2	343	242	1.57	6.23	41.1	2,120	8,280
W 18 × 40	**78.4**	**5.29**	**15.7**	**235**	**148**	**4.49**	**12.1**	**327**	**228**	**1.27**	**5.73**	**50.9**	**1,810**	**17,200**
W 12 × 53	77.9	10.3	35.8	234	153	8.77	25.6	325	235	2.48	8.69	28.1	2,820	2,100
W 14 × 43	69.6	7.88	24.7	209	136	6.68	18.3	290	209	1.89	7.54	37.4	2,330	4,880
W 10 × 54	66.6	10.7	43.9	200	130	9.05	30.2	278	200	2.56	8.15	21.2	3,580	778
W 12 × 45	64.2	8.13	28.3	193	125	6.89	20.3	268	192	1.95	7.00	29.6	2,820	2,210

W 16 × 36	**64.0**	**6.33**	**18.2**	**192**	**122**	**5.37**	**14.0**	**267**	**188**	**1.52**	**8.12**	**48.1**	**1,700**	**20,400**
W 10 × 45	54.9	8.38	35.1	165	106	7.11	24.1	229	164	2.01	6.47	22.5	3,650	758
W 14 × 34	**54.6**	**6.38**	**19.0**	**164**	**105**	**5.41**	**14.4**	**228**	**162**	**1.53**	**7.41**	**43.1**	**1,970**	**10,600**
W 12 × 35	51.2	6.42	20.7	154	99	5.44	15.2	213	152	1.54	6.31	36.2	2,430	4,330
W 16 × 26	**44.2**	**4.67**	**13.4**	**133**	**83**	**3.96**	**10.4**	**184**	**128**	**1.12**	**7.97**	**56.8**	**1,480**	**40,300**
W 14 × 26	**40.2**	**4.50**	**13.4**	**121**	**76**	**3.82**	**10.2**	**168**	**118**	**1.08**	**5.98**	**48.1**	**1,880**	**14,100**
W 10 × 33	38.8	8.08	27.5	116	76	6.86	19.8	162	117	1.94	9.15	27.1	2,720	2,480
W 12 × 26	**37.2**	**6.29**	**18.1**	**112**	**72**	**5.34**	**13.8**	**155**	**111**	**1.51**	**8.54**	**47.2**	**1,820**	**13,900**
W 10 × 26	**31.3**	**5.67**	**18.6**	**94**	**60**	**4.81**	**13.6**	**130**	**93**	**1.36**	**6.56**	**34.0**	**2,510**	**3,760**
W 12 × 22	**29.3**	**3.53**	**11.2**	**88**	**55**	**3.00**	**8.41**	**122**	**85**	**0.848**	**4.74**	**41.8**	**2,170**	**8,460**
W 10 × 19	**21.6**	**3.64**	**12.0**	**65**	**41**	**3.09**	**8.89**	**90**	**63**	**0.874**	**5.09**	**35.4**	**2,440**	**5,030**

Source: Compiled from data in the *Manual of Steel Construction* with permission of the publishers, American Institute of Steel Construction.

The required bending capacity due to the beam weight is

$$M_n = \frac{M_u}{\phi_b} = \frac{3.82 \text{ kip-ft}}{0.9} = 4.24 \text{ kip-ft [5.75 kN-m]}$$

Thus, the total bending moment at midspan is

$$M = 240 + 4.32 = 244 \text{ kip-ft [331 kN-m]}$$

The plastic section modulus required for this moment is

$$Z_x = \frac{M_n}{F_y} = \frac{244 \text{ kip-ft} \times (12 \text{ in./1 ft})}{36 \text{ ksi}} = 81.3 \text{ in.}^3 \text{ [}1.33 \times 10^6 \text{ mm}^3\text{]}$$

Because this required value is less than that of the W 21 × 44, this section is still acceptable.

Use of Plastic Section Modulus Tables

Selection of rolled shapes on the basis of required plastic section modulus may be achieved by the use of tables in the AISC Manual (Ref. 3), in which beam shapes are listed in descending order of their section modulus values. Table 3.1 is similar to these tables and presents a small sample of the reference table data. Note that certain shapes have their designations listed in boldface type. These are sections that have an especially efficient bending moment resistance, indicated by the fact that there are other sections of greater weight but the same or smaller section modulus. Thus, for a savings of material cost, these *least-weight* sections offer an advantage. Other beam design factors, however, may sometimes make this a less important concern.

Data are also supplied in Table 3.1 for lateral support for beams. For lateral support, the values are given for the two limiting lengths L_p and L_r. If a calculation has been made by assuming the minimum yield stress of 36 ksi [250 MPa], the required plastic section modulus obtained will be proper only for beams in which the lateral unsupported length is equal to or less than L_p under the column labeled 36 ksi. Similarly, if the required plastic section modulus was obtained using a minimum yield stress of 50 ksi [350 MPa], the lateral unsupported length needs to be equal to or less than L_p under the 50 ksi column.

A second method of using Table 3.1 for beams omits the calculation of a required plastic section modulus and refers directly to the listed values for the plastic bending moment of the sections, given as M_p in the tables. If $M_p \leq M_u/\phi$, then it is an appropriate section if $L_b < L_p$.

Example 8. Rework the problem in the preceding example in this section by using Table 3.1.

Solution: As before, the bending moment due to the superimposed loading is found to be 240 kip-ft [325 kN-m]. Noting that some additional bending capacity will be required because of the beam's own weight, scan the tables for shapes with an M_p of more than 240 kip-ft [325 kN-m]. Thus:

Shape	M_p *(kip-ft)*	M_p *(kN-m)*
W 21 × 44	286	388
W 18 × 46	272	369
W 14 × 53	261	354
W 10 × 68	256	347
W 16 × 45	247	335

Although the W 21 × 44 is the least-weight section, other design considerations, such as restricted depth or lateral support, may make any of the other shapes the appropriate choice. In the original problem, the lateral support was said to be continuous along the beam. The reality is that continuous support is rare and that it is usually given at various points along the beam. The W 21 × 44 would remain the most appropriate section if the maximum distance between lateral support (L_b) was less than or equal to 5.25 ft (L_p). If the distance was greater other members should be chosen. An example of this is as follows:

Lateral Supports at:	*Maximum Unbraced Length*	*Most Economical Section*
Quarter points	6 ft [1.83 m]	W 16 × 45
Third points	8 ft [2.44 m]	W 14 × 53
Midpoint	12 ft [3.66 m]	W 12 × 79
Ends of beam only	24 ft [7.32 m]	No section qualifies

Note that not all of the available W shapes listed in Table A.3 are included in Table 3.1. Specifically excluded are the shapes that are approximately square (depth equal to flange width) and are ordinarily used for columns rather than beams.

The following problems involve design for bending under plastic failure mode only. Use A36 steel and assume that least-weight members are desired for each case.

Problem 3.4.A. Design for flexure a simple beam 14 ft [4.3 m] in length and having a total uniformly distributed dead load of 13.2 kips [59 kN] and a total uniformly distributed live load of 26.4 kips [108 kN].

Problem 3.4.B. Design for flexure a beam having a span of 16 ft [4.9 m] with a concentrated live load of 40 kips [178 kN] at the center of the span.

Problem 3.4.C. A beam of 15 ft [4.6 m] in length has three concentrated live loads of 6 kips, 7.5 kips, and 9 kips at 4 ft, 10 ft, and 12 ft [26.7 kN, 33.4 kN, and 40.0 kN at 1.2 m, 3 m, and 3.6 m], respectively, from the left-hand support. Design the beam for flexure.

Problem 3.4.D. A beam 30 ft [9 m] long has concentrated live loads of 9 kips [40 kN] each at the third points and also a total uniformly distributed dead load of 20 kips [89 kN] and a total uniformly distributed live load of 10 kips [44 kN]. Design the beam for flexure.

Problem 3.4.E. Design for flexure a beam 12 ft [3.6 m] in length, having a uniformly distributed dead load of 1 kip/ft [14.6 kN/m], a total uniformly distributed live load of 1 kip/ft [14.6 kN/m], and a concentrated dead load of 8.4 kips [37.4 kN] a distance of 5 ft [1.5 m] from one support.

Problem 3.4.F. A beam of 19 ft [5.8 m] in length has concentrated live loads of 6 kips [26.7 kN] and 9 kips [40 kN] at 5 ft [1.5 m] and 13 ft [4 m], respectively, from the left-hand support. In addition, there is a uniformly distributed dead load of 1.2 kip/ft [17.5 kN/m] beginning 5 ft [1.5 m] from the left support and continuing to the right support. Design the beam for flexure.

Problem 3.4.G. A steel beam 16 ft [4.9 m] long has a uniformly distributed dead load of 100 lb/ft [1.46 kN/m] extending over the entire span and a uniformly distributed live load of 100 lb/ft [1.46 kN/m] extending 10 ft [3 m] from the left support. In addition, there is a concentrated live load of 8 kips [35.6 kN] at 10 ft [3 m] from the left support. Design the beam for flexure.

Problem 3.4.H. Design for flexure a simple beam 21 ft [6.4 m] in length, having two concentrated loads of 20 kips [89 kN] each, one 7 ft [2.13 m] from the left end and the other 7 ft [2.13 m] from the right end. The concentrated loads are each made up of equal parts dead load and live load.

Problem 3.4.I. A cantilever beam 8 ft [2.4 m] long has a uniformly distributed dead load of 600 lb/ft [8.8 kN/m] and a uniformly distributed dead load of 1000 lb/ft [14.6 kN/m]. Design the beam for flexure.

Problem 3.4.J. A cantilever beam 6 ft [1.8 m] long has a concentrated live load of 12.3 kips [54.7 kN] at its unsupported end. Design the beam for flexure.

3.5 DESIGN OF BEAMS FOR BUCKLING FAILURE

Although it is preferable to design beams to fail under the plastic hinge mode discussed in Section 3.4, it is not always possible to do so. This is commonly caused by excessive unbraced lengths for lateral support of the beam (L_b). The simplest solution is to decrease the maximum unbraced length to make it less than the plastic limit length (L_p). If this is not possible, the solution is to accept that the beam failure mode is buckling and use the appropriate equations to determine the nominal moment capacity (M_n) of the beam as described in Section 3.3.

If $L_p < L_b \leq L_r$

$$M_n = M_p - (M_p - M_r) \times \left(\frac{L_b - L_p}{L_r - L_p}\right)$$

If $L_b > L_r$

$$M_n = \left(\frac{S_x \times X_1 \times \sqrt{2}}{(L_b/r_y)}\right) \times \sqrt{1 + \frac{(X_1)^2 \times X_2}{2 \times (L_b/r_y)^2}}$$

Example 9. A 14-ft- [4.7-m-] long simply supported beam has a uniform live load of 3 kip/ft [43.3 kN/m] and a dead load of 2 kip/ft [29.2 kN/m]. It is laterally supported only at its ends. Determine the most economical W shape available.

Solution: First determine the appropriate load combination and maximum factored moment:

$$w_u = 1.4 \times (\text{dead load})$$
$$= 1.4 \times (2 \text{ kip/ft}) = 2.8 \text{ kip/ft } [40.9 \text{ kN/m}]$$

or

$$w_u = 1.2 \times (\text{dead load}) + 1.6 \times (\text{live load})$$
$$= 1.2 \times (2 \text{ kip/ft}) + 1.6 \times (3 \text{ kip/ft}) = 7.2 \text{ kip/ft } [105 \text{ N/m}]$$

The bending moment due to the maximum factored superimposed load is

$$M_u = \frac{wL^2}{8} = \frac{7.2 \times (14)^2}{8} = 176 \text{ kip-ft } [239 \text{ kNm}]$$

The required bending resistance of the member is

$$M_u = \phi_b \times M_n$$

$$M_n = \frac{M_u}{\phi_b} = \frac{176 \text{ kip-ft}}{0.9} = 196 \text{ kip-ft } [266 \text{ kNm}]$$

The required plastic section modulus for this moment is

$$Z_x = \frac{M_n}{F_y} = \frac{196 \text{ kip-ft} \times (12 \text{ in./1 ft})}{36 \text{ ksi}} = 65.3 \text{ in.}^3 \ [4213 \text{ mm}^3]$$

From Table 3.1 note that the most economical cross section not taking into account unbraced length is a W 18 × 40, but this section has a plastic limit on length (L_p) of 5.29 ft and therefore does not work for this problem. It should be investigated to determine if it is possible to support this beam at its third points. If it can be supported, a W 18 × 40 would be the most economical shape available. If it cannot be supported, look in Table 3.1 to see if there are any beams that have $Z_x > 65.3$ in.^{3} and $L_p \geq 14$ ft. There are two sections that match this criteria: a W 14 × 120 and a W 14 × 145. The only problem with these sections is that they require three times the amount of steel to do the work than the amount required by the plastic section modulus Z_x. Next, we will look for a section that has $Z_x > 65.3$ in.^{3} and $L_p < 14$ ft $< L_r$ and whose moment capacity $M_n >$ 196 kip-ft.

Begin with the W 18 × 40 that was chosen before, because it would be the most economical if it works. The values for M_p, M_r, L_p, and L_r will be taken from Table 3.1.

$$M_n = M_p - (M_p - M_r) \times \left(\frac{L_b - L_p}{L_r - L_p}\right)$$

$$= 235 \text{ kip-ft} - (235 \text{ kip-ft} - 148 \text{ kip-ft}) \times \left(\frac{14 \text{ ft} - 5.29 \text{ ft}}{15.7 \text{ ft} - 5.29 \text{ ft}}\right)$$

$$= 162 \text{ kip-ft} < 196 \text{ kip-ft}$$

Next, choose the W 14 × 43 (the next most economical section) and check to see if it works:

$$M_n = 209 \text{ kip-ft} - (209 \text{ kip-ft} - 136 \text{ kip-ft}) \times \left(\frac{14 \text{ ft} - 7.88 \text{ ft}}{24.7 \text{ ft} - 7.88 \text{ ft}}\right)$$

$$= 182 \text{ kip-ft} < 196 \text{ kip-ft}$$

Trying a W 21 × 44:

$$M_n = 286 \text{ kip-ft} - (286 \text{ kip-ft} - 177 \text{ kip-ft}) \times \left(\frac{14 \text{ ft} - 5.25 \text{ ft}}{15.4 \text{ ft} - 5.25 \text{ ft}}\right)$$

$$= 192 \text{ kip-ft} < 196 \text{ kip-ft}$$

Trying a W 16 × 45:

$$M_n = 247 \text{ kip-ft} - (247 \text{ kip-ft} - 158 \text{ kip-ft}) \times \left(\frac{14 \text{ ft} - 6.54 \text{ ft}}{20.2 \text{ ft} - 6.54 \text{ ft}}\right)$$

$$= 198 \text{ kip-ft} > 196 \text{ kip-ft}$$

A W 16 × 45 appears to work, but we need to check to see if the weight of the beam increases the required bending capacity. The weight of the beam is 45 lbs/ft.

$$\begin{aligned} w_u &= 1.2 \times (\text{dead load}) + 1.6 \times (\text{live load}) \\ &= 1.2 \times (45\ \text{lb/ft}) + 1.6 \times (0\ \text{lb/ft}) = 54\ \text{lb/ft}\ [0.79\ \text{kN/m}] \end{aligned}$$

The bending moment due to the maximum factored self-weight of the beam is

$$M_u = \frac{wL^2}{8} = \frac{54\ \text{lb/ft} \times (14\ \text{ft})^2}{8} = 1323\ \text{lb-ft} = 1.32\ \text{kip-ft}\ [1.79\ \text{kNm}]$$

The total bending moment due to the maximum factored self-weight of the beam and the factored superimposed load is

$$M_u = 1.32\ \text{kip-ft} + 176\ \text{kip-ft} = 177\ \text{kip-ft}\ [240\ \text{kNm}]$$

The required bending resistance of the member is

$$M_u = \phi_b \times M_n$$

$$M_n = \frac{M_u}{\phi_b} = \frac{177\ \text{kip-ft}}{0.9} = 197\ \text{kip-ft}\ [267\ \text{kNm}]$$

Because the bending capacity of the selected beam is still greater than the required bending capacity, a W 16 × 45 is selected as the most economical section.

The following problems involve the use of Table 3.1 to choose the most economical beams when lateral bracing is a concern. All beams use A36 steel.

Problem 3.5.A. A W shape steel is to be used for a uniformly loaded simple beam carrying a total dead load of 27 kip [120 kN] and a total live load of 50 kip [222 kN] on a 45-ft [13.7-m] span. Select the lightest weight shape for unbraced lengths of (a) 10 ft [3.05 m]; (b) 15 ft [4.57 m]; (c) 22.5 ft [6.90 m].

Problem 3.5.B. A W shape is to be used for a uniformly loaded simple beam carrying a total dead load of 30 kip [133 kN] and a total live load of 40 kip [178 kN] on a 24-ft [7.32-m] span. Select the lightest-weight shape for unbraced lengths of (a) 6 ft [1.83 m]; (b) 8 ft [2.44 m]; (c) 12 ft [3.66 m].

Problem 3.5.C. A W shape is to be used for a uniformly loaded simple beam carrying a total dead load of 22 kip [98 kN] and a total live load of 50 kip [222 kN] on a 30-ft [9.15-m] span. Select the lightest-weight shape for unbraced lengths of (a) 6 ft [1.83 m]; (b) 10 ft [3.05 m]; (c) 15 ft [4.57 m].

Problem 3.5.D. A W shape is to be used for a uniformly loaded simple beam carrying a total dead load of 26 kip [116 kN] and a total live load of 26 kip [116 kN] on a 36-ft [11-m] span. Select the lightest-weight shape for unbraced lengths of (a) 9 ft [2.74 m]; (b) 12 ft [3.66 m]; (c) 18 ft [5.49 m].

3.6 SHEAR IN STEEL BEAMS

Investigation and design for shear forces in the LRFD method is similar to that for bending moment in that the maximum factored shear force must be less than the factored shear capacity of the beam chosen. This is expressed as:

$$\phi_v V_n \geq V_u$$

where ϕ_v = 0.90
V_n = nominal shear capacity of the beam
V_u = required (factored) shear strength of the beam.

Shear in beams consists of the vertical slicing effect produced by the opposition of the vertical loads on the beams (downwards) and the reactive forces at the beam supports (upwards). The internal shear force mechanism is visualized in the form of the shear diagram for the beam. With a uniformly distributed load on a simply supported beam, this diagram takes the form of that shown in Figure 3.7*a*.

As the shear diagram for the uniformly loaded beam shows, this load condition results in an internal shear force that peaks to a maximum value at the beam supports and steadily decreases in magnitude to zero at the center of the beam span. With a beam having a constant cross section throughout the span, the critical location for shear is thus at the supports, and—if conditions there are adequate—there is no concern for shear at other locations along the beam. Because this is the common condition of loading for many beams, we only need to investigate the support conditions for such beams.

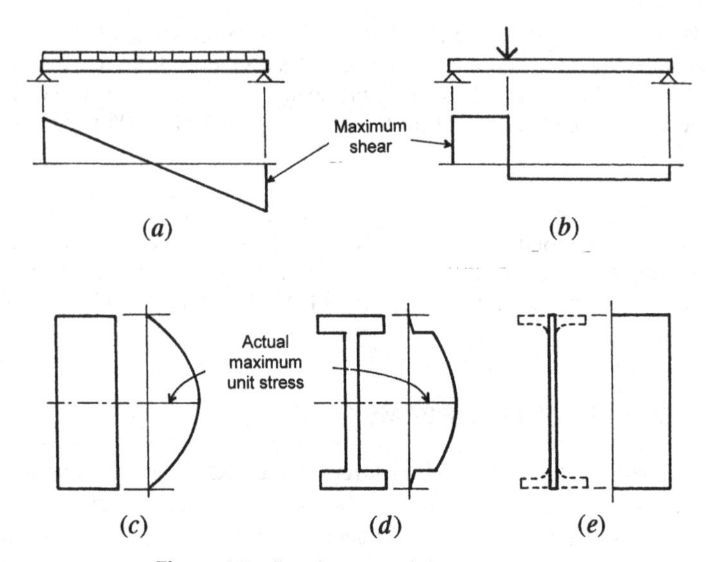

Figure 3.7 Development of shear in beams.

Figure 3.7*b* shows another loading condition: a major concentrated load within the beam span. Framing arrangements for roof and floor systems frequently employ beams that carry the end reactions of other beams, so this is also a common condition. In this case, a major internal shear force is generated over some length of the beam. If the concentrated load is close to one support, a critical internal shear force is created in the shorter portion of the beam length between the load and the closer support.

Internal shear force develops shear stresses in a beam. The distribution of these stresses over the cross section depends on the geometric properties of the beam cross section and significantly depends on the general form of the cross section. For a simple rectangular cross section, such as that of a wood beam, the distribution of beam shear stress is as shown in Figure 3.7*c,* taking the form of a parabola with a maximum shear stress value at the beam neutral axis and a decrease to zero stress at the extreme fiber distances (top and bottom edges).

For the I-shaped cross section of the typical W-shape rolled steel beam, the beam shear stress distribution takes the form shown in Figure 3.7*d* (referred to as the "derby hat" form). Again, the shear stress is a maximum at the beam neutral axis, but the falloff is less rapid between the neutral axis and the inside of the beam flanges. Although the flanges indeed take some shear force, the sudden increase in beam width results in an abrupt drop in the beam unit shear stress. A traditional shear stress investigation for the W shape, therefore, is based on ignoring the flanges and assuming the shear-resisting portion of the beam to be an equivalent vertical plate (Figure 3.7*e*) with a width equal to the beam web thickness and a height equal to the full beam depth. For the ASD method, an allowable value is established for a unit shear stress on this basis, and the computation is performed as:

$$f_v = \frac{V}{t_w d_b} = \frac{V}{A_w}$$

where f_v = the average unit shear stress, based on an assumed distribution as shown in Figure 3.7*e*

V = value for the internal shear force at the cross section

t_w = beam thickness

d_b = overall beam depth

A_w = area of the web

Shear forces internal to a W shape beam may be problematic in terms of compression forces on the section's web. Figure 3.8*d* illustrates this behavior at a beam support consisting of a bearing of the beam end on top of the support, usually in this case the top of a wall or a wall ledge. The potential problem here has more to do with vertical compression force, which results in a squeezing of the beam end and a columnlike action in the thin beam web. This may actually produce a columnlike form of failure. As with a column, the range of possibilities for this form of failure relate to the relative slenderness of the web. Three cases are defined by the slenderness ratio of the web (h/t_w) and control the shear capacity (V_n) of a beam:

1. A very stiff (thick) web that may actually reach something close to the full yield stress limit of the material where $h/t_w \leq 418/\sqrt{F_y}$

$$V_n = (0.6 \times F_y) \times A_w$$

2. A somewhat slender web that responds with some combined yield stress and buckling effect (called an *inelastic buckling* response) where $418/\sqrt{F_y} \leq h/t_w \leq 523/\sqrt{F_y}$.

$$V_n = (0.6 \times F_y) \times A_w \times \left(\frac{418/\sqrt{F_y}}{h/t_w}\right)$$

3. A very slender web that fails essentially in *elastic buckling* in the classic Euler formula manner; basically a deflection failure rather than a stress failure where $h/t_w > 523/\sqrt{F_y}$.

$$V_n = 132{,}000 \times \frac{A_w}{(h/t_w)^2}$$

The dimension h (see Figure 3.8*d*) is the height of the beam web, determined as the beam depth minus two times the flange thickness.

Example 10. A simple beam of A36 steel is 6 ft [1.83 m] long and has a concentrated live load of 36 kips [160 kN] applied 1 ft [0.3 m] from one end. It is found that a W 10 × 19 is adequate for the bending moment. Investigate the beam to determine if the shear capacity is adequate for the required shear.

Solution: The load must first be factored:

$$\begin{aligned} P_u &= 1.2(\text{Dead load}) + 1.6(\text{Live load}) \\ &= 1.2(0 \text{ kips}) + 1.6(36 \text{ kips}) \\ &= 57.6 \text{ kips [256 kN]} \end{aligned}$$

The two reactions for this loading are 48 kips [214 kN] and 9.6 kips [43.2 kN]. The maximum shear (V_u) in the beam is equal to the larger reaction force.

From Table A.3, for the given shape, $d = 10.24$ in., $t_w = 0.250$ in., and $t_f = 0.395$ in. Then:

$$h = d - 2(t_f) = 10.24 - 2(0.395) = 9.45 \text{ in.}$$

$$\frac{h}{t_w} = \frac{9.45}{0.25} = 37.8$$

$$418/\sqrt{F_y} = 418/\sqrt{36 \text{ ksi}} = 69.7 > h/t_w$$

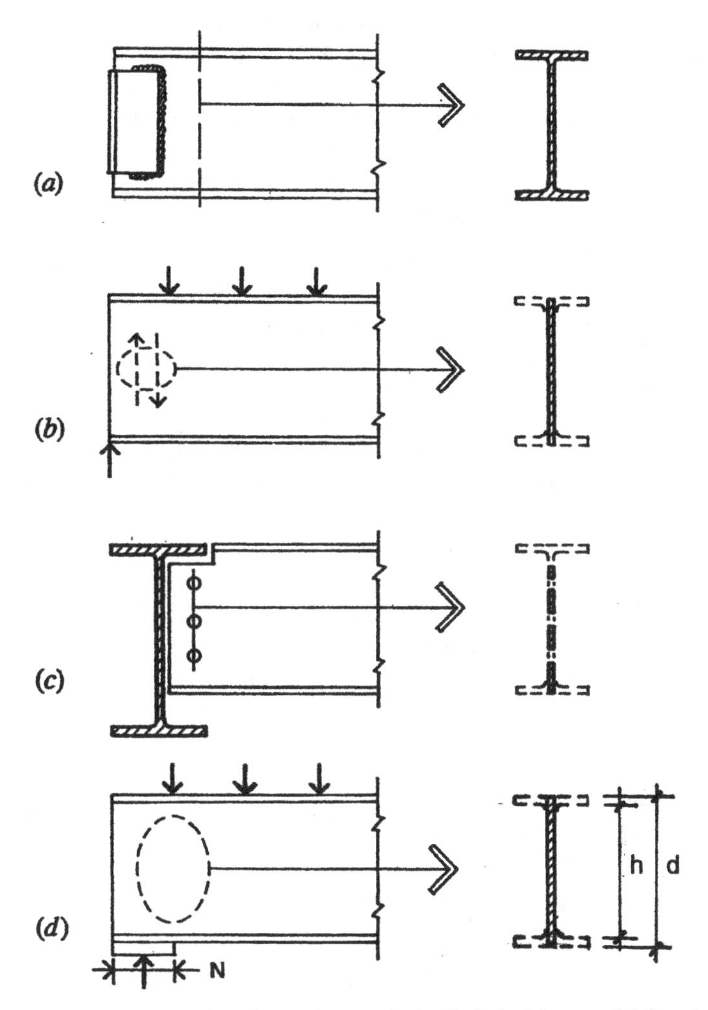

Figure 3.8 Considerations for end support of rolled steel beams: (a) Development of shear by the full beam section with a framed connection. (b) Design assumption for resistance of the web only for shear. (c) Development of shear on a reduced section. (d) Development of compression in the beam web at a bearing support.

Therefore, the shear capacity will be determined using the equation associated with the full yield stress limit of the material:

$$A_w = d \times t_w = 10.24 \text{ in.} \times 0.250 \text{ in.} = 2.56 \text{ in.}^2 \text{ [1652 mm}^2\text{]}$$

$$V_n = (0.6 \times F_y) \times A_w$$

$$V_n = (0.6 \times 36 \text{ ksi}) \times 2.56 \text{ in.}^2 = 55.3 \text{ kips [246 kN]}$$

$$\phi_v V_n = 0.9 \times 55.3 \text{ kips} = 49.8 \text{ kips} > V_u = 48 \text{ kips [214 kN]}$$

Because the factored capacity of the beam is greater than the factored shear, the shape is acceptable.

The net effect of investigations of all the situations so far described, relating to end shear in beams, may be to influence a choice of beam shape with a web that is sufficient. However, other criteria for selection (flexure, deflection, framing details, and so on) may indicate an ideal choice that has a vulnerable web. In the latter case, it is sometimes decided to *reinforce* the web; the usual means is to insert vertical plates on either side of the web and to fasten them to the web as well as to the beam flanges. These plates then both brace the slender web (column) and absorb some of the vertical compression stress in the beam.

For practical design purposes, the beam end shear and end support limitations of unreduced webs can be handled by data supplied in AISC tables.

Problems 3.6.A–C. Compute the shear capacity ($\phi_v V_n$) for the following beams of A36 steel: A, W 24 × 84; B, W 12 × 45; C, W 10 × 33.

3.7 DEFLECTION OF BEAMS

Deformations of structures must often be controlled for various reasons. These reasons may relate to the proper functioning of the structure, but more often relate to effects on the supported construction or to the overall purpose of the structure.

To steel's advantage is the relative stiffness of the material itself. With a modulus of elasticity of 29,000 ksi, it is 8 to 10 times as stiff as average structural concrete and 15 to 20 times as stiff as structural lumber. However, it is often the overall deformation of whole structural assemblages that must be controlled; in this regard, steel structures are frequently quite deformable and flexible. Because of its cost, steel is usually formed into elements with thin parts (beam flanges and webs, for example), and because of its high strength, it is frequently formed into relatively slender elements (beams and columns, for example).

For a beam in a horizontal position, the critical deformation is usually the maximum sag, called the beam's *deflection*. For most beams, this deflection will be too small in magnitude to be detected by eye. However, any load on a beam, such as that in Figure 3.9, will cause some amount of deflection, beginning with the beam's own weight. In the case of a simply supported, symmetrical, single-span beam, the maximum deflection will occur at midspan and it usually is the only deformation value of concern for design. However, as the beam deflects, its ends rotate unless restrained, and this twisting deformation may also be of concern in some situations.

If deflection is determined to be excessive, the usual remedy is to select a deeper beam. Actually, the critical geometric property of the beam cross section is its *moment of inertia* (I) about its major axis (I_x for a W shape), which is typically affected significantly by increases in depth of the beam. Formulas for deflection of beams take a typical form that involves variables as follows:

$$\Delta = C\frac{WL^3}{EI}$$

where Δ = deflection, measured vertically (usually in in. or mm)

C = a constant related to the form of the load and support conditions

W = total load on the beam

L = span of the beam

E = modulus of elasticity of the material of the beam

I = moment of inertia of the beam cross section for the axis about which bending occurs

(Note: The letter D is also used as the symbol for deflection.)

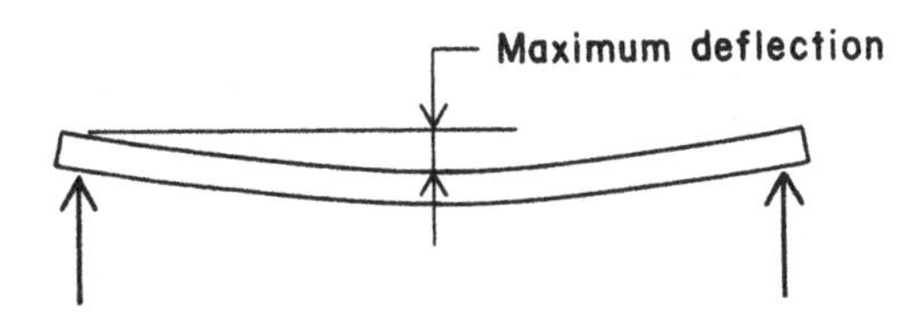

Figure 3.9 Deflection of a simple beam under symmetrical loading.

Note that the magnitude of the deflection is directly proportional to the magnitude of the load; that is, if you double the load, you will double the deflection. However, the deflection is proportional to the third power of the span; double the span and you get 2^3, or eight times, the deflection. For resistance to deflection, increases in either the material's stiffness or the beam's geometric form (I) will cause direct proportional reduction of the deflection. Because E is constant for all steel, design modulation of deflections must deal only with the beam's shape.

Excessive deflection may cause various problems. For roofs, an excessive sag may disrupt the intended drainage patterns for generally flat surfaces. For floors, a common problem is the development of some perceivable bounciness. The form of the beam and its supports may also be a consideration. For the simple-span beam in Figure 3.9, the usual concern is simply for the maximum sag at the beam midspan. For a beam with a projected (cantilevered) end, however, a problem may be created at the unsupported cantilevered end; depending on the extent of the cantilever, this may involve downward deflection (as shown in Figure 3.10*a)* or upward deflection (as shown in Figure 3.10*b)*.

With continuous beams, a potential problem derives from the fact that load in any span causes some deflection in all the spans. This is most critical when loads vary in different spans or the length of the spans differ significantly (see Figure 3.10*c)*.

Most deflection problems in buildings stem from effects of the structural deformation on adjacent or supported elements of the building. When beams are supported by other beams (usually referred to as girders), excessive rotation caused by deflection occurring at the ends of the supported beams can result in cracking or other separation of the floor deck that is continuous over the girders, as shown in Figure 3.10*d*. For such a system, there is also an accumulative deflection caused by the independent deflections of the deck, the beams, and the girders, which can cause problems for maintaining a flat floor surface or a desired roof surface profile for drainage.

An especially difficult problem related to deflections is the effect of beam deflections on nonstructural elements of the construction. Figure 3.10*e* shows the case of a beam occurring directly over a solid wall. If the wall is made to fit tightly beneath the beam, any deflection of the beam will cause it to bear on top of the wall—not an acceptable situation if the wall is relatively fragile (a metal and glass curtain wall, for example). A different sort of problem occurs when relatively rigid walls (plastered, for example) are supported by spanning beams, as shown in Figure 3.10*f*.

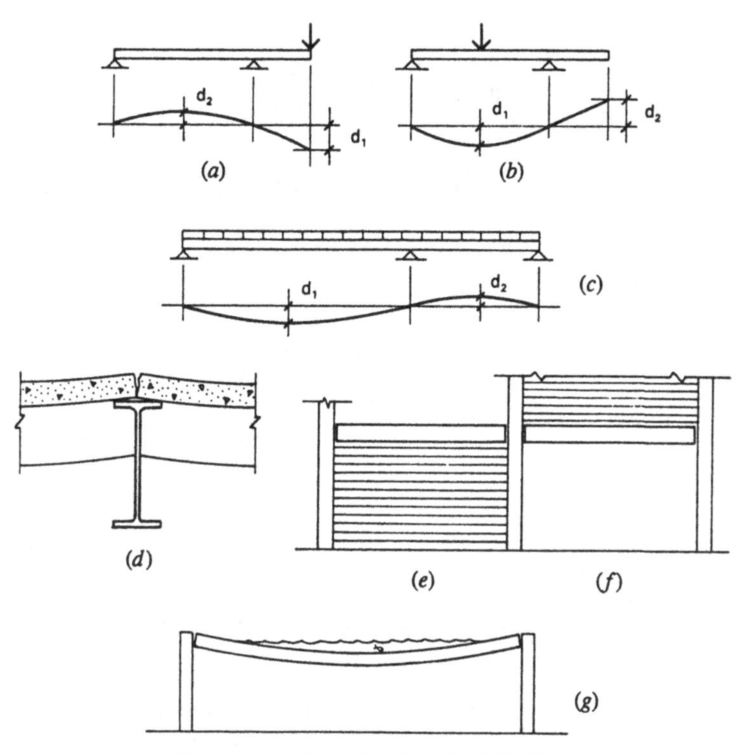

Figure 3.10 Considerations for deflection.

In this case, the wall is relatively intolerant of *any* deformation, so anything significant in the form of sag of the beam is really critical.

For long-span structures (an ambiguous class, usually meaning 100 ft or more span), a special problem is the relatively flat roof surface. In spite of provisions for code-mandated minimum drainage, heavy rain will run slowly off of the surface and linger to cause some considerable loading. Because this results in deflection of the spanning structure, the sag may form a depressed area that the rain can swiftly turn into a pond (see Figure 3.10*g*). The pond then constitutes an additional load that causes more sag—and a resulting deeper pond. This progression can quickly pyramid into a failure of the structure, so that codes (including the AISC Specification) now provide design requirements for investigation of potential ponding.

Standard Equations for Deflection

Determining deflection on a beam is usually done using a series of standard equations. These equations are listed in the AISC Manual (Ref. 3) and in Figure B.1 in Appendix B. The purpose of these equations is to determine the actual deflection of a beam for a given loading, and therefore when using them in steel design, use nonfactored loading. It is important to stay consistent with units. In steel this usually means working in kips and inches. The most used of these equations for simply supported beams are shown in the following table.

Loading Condition	*Maximum Deflection*
Uniform load over entire span	$\Delta = \dfrac{5wL^4}{384EI}, \text{ or } \dfrac{5WL^3}{384EI}$
Point load at midspan	$\Delta = \dfrac{PL^3}{48EI}$

Example 11. A simple beam has a span of 20 ft [6.10 m] with a uniformly distributed load of 1.95 kip/ft [28.5 kN/M]. The beam is a steel W 14 × 34. Find the maximum deflection.

Solution. We can use the standard equation for a uniform load over the entire span. Moment of inertia for the W 14 × 34 is 340 in.^{4}.

$$\Delta = \frac{5wL^4}{384EI}$$

$$= \frac{5\left[1.95\text{ kip/ft} \times \left(\dfrac{1\text{ ft}}{12\text{ in.}}\right)\right]\left[20\text{ ft} \times \left(\dfrac{12\text{ in.}}{1\text{ ft}}\right)\right]^4}{384(29{,}000\text{ ksi})(340\text{ in.}^4)}$$

$$= 0.712\text{ in. [18.1 mm]}$$

Allowable Deflections

What is permissible for beam deflection is mostly a matter of judgment by experienced designers. It is difficult to provide any useful guidance for specific limitations to avoid the various problems described in Figure

3.10. Each situation must be investigated individually and the designers of the structure and those who develop the rest of the building construction must make some cooperative decisions about the necessary design controls.

For spanning beams in ordinary situations, some rules of thumb have been derived over many years of experience. These usually consist of establishing some maximum degree of beam curvature described in the form of a limiting ratio of the deflection to the beam span (*L*), expressed as a fraction of the span. These are sometimes, although not always, specified in design codes or legally enacted building codes. Some typical limitations recognized by designers are the following:

To avoid visible sag under total load on short to medium spans, $L/150$
For total load deflection of a roof structure, $L/180$
For deflection under live load only for a roof structure, $L/240$
For total load deflection of a floor structure, $L/240$
For deflection under live load only for a floor structure, $L/360$

Deflection of Uniformly Loaded Simple Beams

The most frequently used beam in flat roof and floor systems is the uniformly loaded beam with a single, simple span (no end restraint). This situation is shown in Figure B.1 as Case 2. For this case, the following values may be obtained for the beam behavior:

Maximum bending moment:

$$M = \frac{wL^2}{8}$$

Maximum stress on the beam cross section:

$$f = \frac{Mc}{I}$$

Maximum midspan deflection:

$$\Delta = \frac{5}{384} \times \frac{wL^4}{EI}$$

Using these relationships, together with the case of a known modulus of elasticity ($E = 29{,}000$ ksi for steel), a convenient formula can be derived for deflection. Noting that the dimension c in the bending stress formula is $d/2$ for symmetrical shapes, and substituting the expression for M, we can say

$$f = \frac{Mc}{I} = \left(\frac{wL^2}{8}\right)\left(\frac{d/2}{I}\right) = \frac{wL^2 d}{16I}$$

Then:

$$\Delta = \frac{5wL^4}{384EI} = \left(\frac{wL^2 d}{16I}\right)\left(\frac{5L^2}{24Ed}\right) = (f)\left(\frac{5L^2}{24Ed}\right) = \frac{5fL^2}{24Ed}$$

This is a basic formula for any beam symmetrical about its bending axis. Because these deflection equations are only appropriate within the elastic range, we will set the bending stress (f) to the allowable bending stress in allowable stress design ($F_b = 0.67 \times F_y$). Also, for convenience, spans are usually measured in feet, not inches, so a factor of 12 is added. Thus, for A36 steel:

$$\Delta = \frac{5fL^2}{24Ed} = \left(\frac{5}{24}\right)\left[\frac{(0.67 \times 36)}{29{,}000}\right]\left(\frac{(12L)^2}{d}\right) = \frac{0.02483L^2}{d}$$

In metric units, with $f = 165$ MPa, $E = 200$ GPa, and the span in meters

$$\Delta = \frac{0.0001719L^2}{d}$$

For $Fy = 50$ ksi:

$$\Delta = \frac{5fL^2}{24Ed} = \left(\frac{5}{24}\right)\left[\frac{(0.67 \times 50)}{29{,}000}\right] = \left(\frac{(12L)^2}{d}\right) = \frac{0.03449L^2}{d}$$

In metric units, with $f = 230$ MPa, $E = 200$ GPa, and the span in meters:

$$\Delta = \frac{0.0002396L^2}{d}$$

The derived deflection formula involving only span and beam depth can be used to plot a graph that displays the deflection of a beam of a constant depth for a variety of spans. Figure 3.11 consists of a series of such graphs for beams from 6 to 36 in. in depth and a yield stress of 36 ksi. Figure 3.12 is for steel with a yield stress of 50 ksi. Use of these graphs

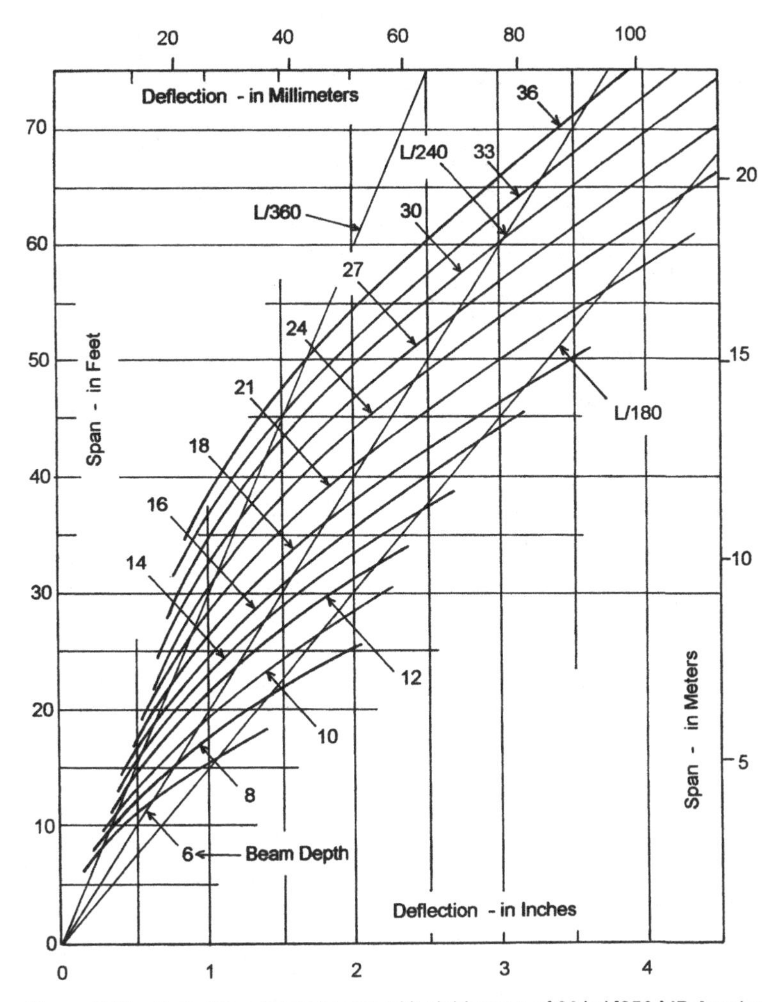

Figure 3.11 Deflection of steel beams with yield stress of 36 ksi [250 MPa] under a constant maximum bending stress of 24 ksi [165 MPa].

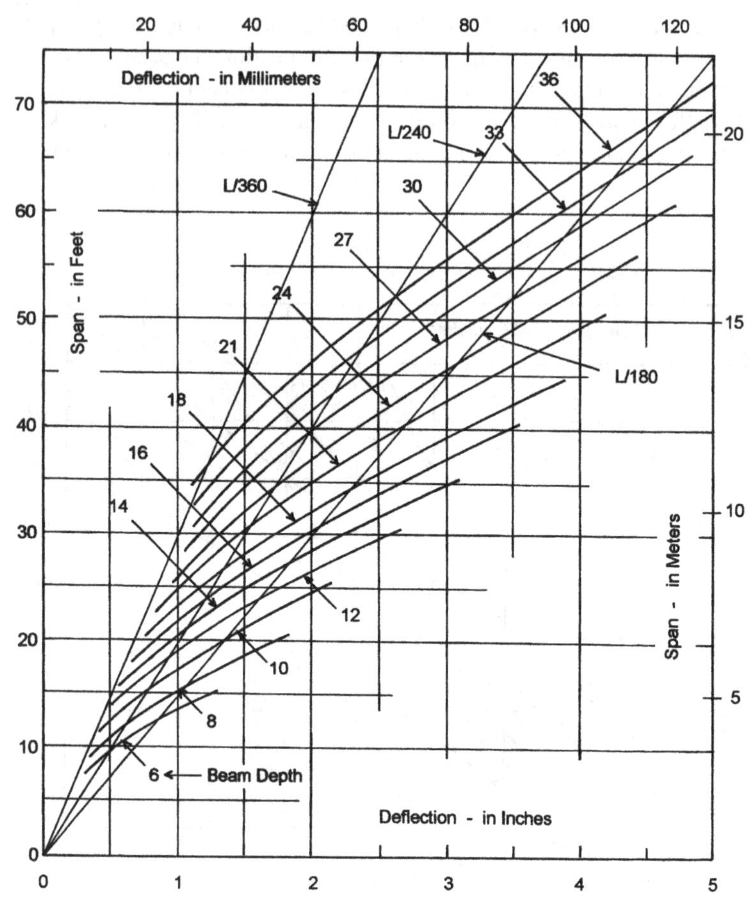

Figure 3.12 Deflection of steel beams with yield stress of 50 ksi [345 MPa] under a constant maximum bending stress of 33 ksi [228 MPa].

presents yet another means for determining beam deflections. An answer within about 5 percent should be considered reasonable from the graphs.

The real value of the graphs in Figure 3.11 and Figure 3.12, however, is in the design process. Once the span is known, it may be initially determined from the graphs what beam depth is required for a given deflection. The limiting deflection may be given in an actual dimension, or more commonly, as a limiting percentage of the span (1/240, 1/360, etc.), as previously discussed. To aid in the latter situation, lines are drawn on the graph representing the usual percentage limits of 1/360, 1/240, and 1/180. Thus, if a beam is to be used for a span of 36 ft, and the total load deflection limit is $L/240$, it may be observed on Figure 3.11 that the lines for a span of 36 ft and a ratio of 1/240 intersect almost precisely on the curve for an 18 in. deep beam. This means that an 18-in.-deep beam will deflect almost precisely 1/240th of the span if stressed in bending to 24 ksi. Thus, any beam chosen with less depth will be inadequate for deflection, and any beam greater in depth will be conservative in regard to deflection.

The minimum depth of a beam required by deflection criteria can be approximated from these equations. They are derived by placing the allowable deflection criteria into the equation derived earlier. It is important to remember in these equations that the beam length (L) is in feet, whereas the beam depth (d) is in inches. The equations are shown in the table that follows.

Yield Stress	$\frac{L}{180}$	$\frac{L}{240}$
36 ksi	$d_{min} = 0.372 \times L \approx \frac{L}{3}$	$d_{min} = 0.497 \times L \approx \frac{L}{2}$
50 ksi	$d_{min} = 0.517 \times L \approx \frac{L}{2}$	$d_{min} = 0.690 \times L \approx \frac{2L}{3}$

Problems 3.7.A–D. Find the maximum deflection in inches for the following simple beams of A36 steel with uniformly distributed load. Find the values using: (a) the equation for deflection of a uniformly distributed load; and (b) the curves in Figure 3.11.

(A) W 10 × 33, span = 18 ft, total service load = 1.67 kips/ft [5.5 m, 24.2 kN/m].

(B) W 16 × 36, span = 20 ft, total service load = 2.5 kips/ft [6 m, 37 kN/m].

(C) W 18 × 46, span = 24 ft, total service load = 2.29 kips/ft [7.3 m, 33.6 kN/m].

(D) W 21 × 57, span = 27 ft, total service load = 2.5 kips/ft [8.2 m, 36.5 kN/m].

3.8 SAFE LOAD TABLES

The simple beam with uniformly distributed load occurs so frequently that it is useful to have a rapid design method for quick selection of shapes based on knowing only the beam load and span. The AISC Manual (Ref. 3) provides a series of such tables with data for the W, M, S, and C shapes most often used as beams.

Tables 3.2 and 3.3 present data for selected beams for $F_y = 36$ ksi and $F_y = 50$ ksi, respectively. Table values are the total factored load for a simple span beam with uniformly distributed loading and lateral bracing at points not farther apart than the limiting dimension of L_p (see discussion in Section 3.5). The table values are determined from the maximum bending moment capacity of the shapes ($M_p = Z_x \times F_y$).

For very short spans, loads are often limited by beam shear or end support conditions, rather than by bending or deflection limits. For this reason, table values are not shown for spans of less than 12 times the beam depth.

For long spans, loads are often limited by deflection, rather than by bending. Thus, table values are not shown for spans exceeding a limit of 24 times the beam depth.

When the distance between points of lateral support exceeds the limiting value of L_p, the values in Tables 3.2 and 3.3 should not be used. See the discussion regarding buckling in Section 3.5.

The self-weight of the beam is also included in the weight given in these tables. It has not been removed from consideration because the factoring will vary depending on the given loading condition. Once a beam is selected, the self-weight of the beam must be factored and added to the total load to determine if the beam is still acceptable.

The following example illustrates the use of Table 3.2 for some common design situations.

Example 12. Design a simply supported beam to carry a uniformly distributed live load of 1.33 kips/ft [19.4 kN/m] and a superimposed uniformly distributed dead load of 0.66 kips/ft [9.6 kN/m] on a span of 24 ft [7.32 m]. The yield stress is 36 ksi [250 Mpa]. Find (1) the lightest shape permitted and (2) the shallowest (least-depth) shape permitted.

Solution: First the loading must be factored and totaled for the entire beam.

$$w_u = 1.4 \times (\text{Dead load})$$

$$= 1.4 \times (0.66 \text{ kip/ft}) = 0.924 \text{ kip/ft } [13.5 \text{ kN/m}]$$

or

$$w_u = 1.2 \times (\text{Dead load}) + 1.6 \times (\text{Live load})$$

$$= 1.2 \times (0.66\ \text{kip/ft}) + 1.6 \times (1.33\ \text{kip/ft}) = 2.92\ \text{kip/ft}\ [42.6\ \text{kN/m}]$$

$$\text{Total superimposed load} = 2.92\ \text{kip/ft} \times 24\ \text{ft} = 70.1\ \text{kips}\ [312\ \text{kN}]$$

From Table 3.2 we find the following:

For Shape	*Allowable Load*
W 12 × 53	70.1 kips [312 kN]
W 18 × 40	70.6 kips [314 kN]
W 16 × 45	74.1 kips [330 kN]
W 14 × 53	78.4 kips [349 kN]
W 18 × 46	81.6 kips [363 kN]
W 21 × 44	85.9 kips [382 kN]
W 12 × 79	107 kips [476 kN]

The W 12 × 53 should be checked to see if it is still appropriate after the self-weight of the beam is added to the superimposed factored load.

$$1.2 \times (\text{Dead load}) + 1.6 \times (\text{Live load})$$

$$= 1.2 \times (53\ \text{lb/ft}) + 1.6 \times (0\ \text{lb/ft}) = 63.6\ \text{lb/ft}\ [0.93\ \text{kN/m}]$$

$$\text{Total self-weight} = 63.6\ \text{lb/ft} \times 24\ \text{ft} \times \left(\frac{1\ \text{kips}}{1000\ \text{lb}}\right) = 1.53\ \text{kips}\ [6.8\ \text{kN}]$$

$$\text{Total load} = 1.53\ \text{kips} + 70.1\ \text{kips} = 71.6\ \text{kips}\ [318\ \text{kN}]$$

Thus, the W 12 × 53 is not viable for this given loading. Similarly the W 18 × 40 does not work. The lightest shape is the W 21 × 44, and the shallowest is the W 12 × 79. It should be noted that although the W 12 × 79 is the shallowest section, it comes at a cost of considerably more steel. If a W 14 × 53 were substituted for the W 12 × 79 it would come with a 45 percent weight savings.

TABLE 3.2 Factored Load-Span Values for 36 ksi Beams

	F_y = 36 ksi				Span (ft)									
Designation	L_p ft	L_r ft	M_p kip-ft	M_r kip-ft	12	14	16	18	20	22	24	26	28	30
W 10 × 17	3.52	11.1	56	35	33.7	28.9	25.2	22.4	20.2					
W 12 × 16	3.22	9.58	60	37	36.2	31.0	27.1	24.1	21.7	19.7	18.1			
W 10 × 19	3.64	12.0	65	41	38.9	33.3	29.2	25.9	23.3					
W 12 × 22	3.53	11.2	88	55	52.7	45.2	39.6	35.2	31.6	28.8	26.4			
W 10 × 26	5.67	18.6	94	60	56.3	48.3	42.3	37.6	33.8					
W 12 × 26	6.29	18.1	112	72	67.0	57.4	50.2	44.6	40.2	36.5	33.5			
W 10 × 33	8.08	27.5	116	76	69.8	59.9	52.4	46.6	41.9					
W 14 × 26	4.50	13.4	121	76		62.0	54.3	48.2	43.4	39.5	36.2	33.4	31.0	
W 16 × 26	4.67	13.4	133	83			59.7	53.0	47.7	43.4	39.8	36.7	34.1	31.8
W 12 × 35	6.42	20.7	154	99	92.2	79.0	69.1	61.4	55.3	50.3	46.1			
W 14 × 34	6.38	19.0	164	105		84.2	73.7	65.5	59.0	53.6	49.1	45.4	42.1	
W 10 × 45	8.38	35.1	165	106	98.8	84.7	74.1	65.9	59.3					
W 16 × 36	6.33	18.2	192	122			86.4	76.8	69.1	62.8	57.6	53.2	49.4	46.1
W 12 × 45	8.13	28.3	193	125	116	99.1	86.7	77.0	69.3	63.0	57.8			
W 10 × 54	10.7	43.9	200	130	120	103	89.9	79.9	71.9					
W 14 × 43	7.88	24.7	209	136		107	94.0	83.5	75.2	68.3	62.6	57.8	53.7	
W 12 × 53	10.3	35.8	234	153	140	120	105	93.5	84.1	76.5	70.1			

W 18 × 40	5.29	15.7	235	148				94.1	84.7	77.0	70.6	65.1	60.5	56.4
W 16 × 45	6.54	20.2	247	158			111	98.8	88.9	80.8	74.1	68.4	63.5	59.3
W 10 × 68	10.8	53.7	256	164	154	132	115	102	92.1					
W 14 × 53	8.00	28.0	261	169		134	118	105	94.1	85.5	78.4	72.4	67.2	
W 18 × 46	5.38	16.6	272	171				109	98.0	89.1	81.6	75.4	70.0	65.3
W 18 × 50	6.88	20.5	303	193				121	109	99.2	90.9	83.9	77.9	72.7
W 16 × 57	6.67	22.8	315	200			142	126	113	103	94.5	87.2	81.0	75.6
W 10 × 88	11.0	68.4	339	213	203	174	153	136	122					
W 14 × 68	10.3	37.3	345	223		177	155	138	124	113	104	95.5	88.7	
W 12 × 79	12.7	51.8	357	232	214	184	161	143	129	117	107			
W 18 × 60	7.00	22.3	369	234				148	133	121	111	102	94.9	88.6
W 14 × 82	10.3	42.8	417	267		214	188	167	150	136	125	115	107	
W 18 × 71	7.08	24.5	438	275				175	158	143	131	121	113	105
W 10 × 112	11.2	86.4	441	273	265	227	198	176	159					
W 12 × 96	12.9	61.4	441	284	265	227	198	176	159	144	132			
W 16 × 77	10.3	35.4	456	295			205	182	164	149	137	126	117	109
W 12 × 120	13.0	75.5	558	353	335	287	251	223	201	183	167			
W 16 × 100	10.4	42.7	600	384			270	240	216	196	180	166	154	144
W 14 × 120	15.6	67.9	636	412		327	286	254	229	208	191	176	164	
W 12 × 152	13.3	94.8	729	453	437	375	328	292	262	239	219			
W 14 × 145	16.6	81.6	780	503		401	351	312	281	255	234	216	201	

(continued)

TABLE 3.2 (*Continued*)

	F_y = 36 ksi				Span (ft)									
Designation	L_p ft	L_r ft	M_p kip-ft	M_r kip-ft	24	26	28	30	32	34	36	38	40	42
W 21 × 44	5.25	15.4	286	177	85.9	79.3	73.6	68.7	64.4	60.6	57.2	54.2	51.5	49.1
W 21 × 50	5.42	16.2	330	205	99.0	91.4	84.9	79.2	74.3	69.9	66.0	62.5	59.4	56.6
W 21 × 57	5.63	17.3	387	241	116	107	99.5	92.9	87.1	82.0	77.4	73.3	69.7	66.3
W 24 × 55	5.58	16.6	405	249	122	112	104	97.2	91.1	85.8	81.0	76.7	72.9	69.4
W 24 × 62	5.71	17.2	462	286	139	128	119	111	104	97.8	92.4	87.5	83.2	79.2
W 21 × 68	7.50	22.8	480	303	144	133	123	115	108	101.6	96.0	90.9	86.4	82.3
W 18 × 86	11.0	35.5	558	360	167	155	143	134	126	118	112			
W 21 × 83	7.63	24.9	588	371	176	163	151	141	132	125	118	111	106	101
W 24 × 76	8.00	23.4	600	381	180	166	154	144	135	127	120	114	108	103
W 18 × 106	11.1	40.4	690	442	207	191	177	166	155	146	138			
W 21 × 101	12.0	37.1	759	492	228	210	195	182	171	161	152	144	137	130
W 24 × 94	8.25	25.9	762	481	229	211	196	183	171	161	152	144	137	131
W 27 × 94	8.83	25.9	834	527			214	200	188	177	167	158	150	143

W 24 × 104	12.1	35.2	867	559	260	240	223	208	195	184	173	164	156	149
W 21 × 122	12.2	41.0	921	592	276	255	237	221	207	195	184	175	166	158
W 24 × 117	12.3	37.1	981	631	294	272	252	235	221	208	196	186	177	168
W 27 × 114	9.08	28.2	1,029	648			265	247	232	218	206	195	185	176
W 30 × 108	8.96	26.3	1,038	648				249	234	220	208	197	187	178
W 24 × 131	12.4	39.3	1,110	713	333	307	285	266	250	235	222	210	200	190
W 21 × 147	12.3	46.4	1,119	713	336	310	288	269	252	237	224	212	201	192
W 30 × 124	9.29	28.2	1,224	769				294	275	259	245	232	220	210
W 33 × 118	9.67	27.8	1,245	778						264	249	236	224	213
W 24 × 146	12.5	42.0	1,254	804	376	347	322	301	282	266	251	238	226	215
W 24 × 162	12.7	45.2	1,404	897	421	389	361	337	316	297	281	266	253	241
W 33 × 141	10.1	30.1	1,542	971						327	308	292	278	264

Source: Compiled from data in the *Manual of Steel Construction* with permission of the publishers, American Institute of Steel Construction.

TABLE 3.3 Factored Load-Span Values for 50 ksi Beams

Designation	$F_y = 50$ ksi				Span (ft)									
	L_p (ft)	L_r ft	M_p kip-ft	M_r kip-ft	12	14	16	18	20	22	24	26	28	30
W 10 × 17	2.99	8.37	78	54	46.8	40.1	35.1	31.2	28.1					
W 12 × 16	2.73	7.44	84	57	50.3	43.1	37.7	33.5	30.2	27.4	25.1			
W 10 × 19	3.09	8.89	90	63	54.0	46.3	40.5	36.0	32.4					
W 12 × 22	3.00	8.41	122	85	73.3	62.8	54.9	48.8	44.0	40.0	36.6			
W 10 × 26	4.81	13.6	130	93	78.3	67.1	58.7	52.2	47.0					
W 12 × 26	5.34	13.8	155	111	93.0	79.7	69.8	62.0	55.8	50.7	46.5			
W 10 × 33	6.86	19.8	162	117	97.0	83.1	72.8	64.7	58.2					
W 14 × 26	3.82	10.2	168	118		86.1	75.4	67.0	60.3	54.8	50.3	46.4	43.1	
W 16 × 26	3.96	10.4	184	128			82.9	73.7	66.3	60.3	55.3	51.0	47.4	44.2
W 12 × 35	5.44	15.2	213	152	128	110	96.0	85.3	76.8	69.8	64.0			
W 14 × 34	5.41	14.4	228	162		117	102	91.0	81.9	74.5	68.3	63.0	58.5	
W 10 × 45	7.11	24.1	229	164	137	118	103	91.5	82.4					
W 16 × 36	5.37	14.0	267	188			120	107	96.0	87.3	80.0	73.8	68.6	64.0
W 12 × 45	6.89	20.3	268	192	161	138	120	107	96.3	87.5	80.3			
W 10 × 54	9.05	30.2	278	200	167	143	125	111	100					
W 14 × 43	6.68	18.3	290	209		149	131	116	104	94.9	87.0	80.3	74.6	
W 12 × 53	8.77	25.6	325	235	195	167	146	130	117	106	97.4			

W 18 × 40	4.49	12.1	327	228				131	118	107	98.0	90.5	84.0	78.4
W 16 × 45	5.55	15.2	343	242			154	137	123	112	103	95.0	88.2	82.3
W 10 × 68	9.16	36.0	355	252	213	183	160	142	128					
W 14 × 53	6.79	20.1	363	259		187	163	145	131	119	109	101	93.3	
W 18 × 46	4.56	12.6	378	263				151	136	124	113	105	97.2	90.7
W 18 × 50	5.83	15.6	421	296				168	152	138	126	117	108	101
W 16 × 57	5.66	16.6	438	307			197	175	158	143	131	121	113	105
W 10 × 88	9.30	45.1	471	328	283	242	212	188	170					
W 14 × 68	8.70	26.4	479	343		246	216	192	173	157	144	133	123	
W 12 × 79	10.8	35.7	496	357	298	255	223	198	179	162	149			
W 18 × 60	5.94	16.6	513	360				205	185	168	154	142	132	123
W 14 × 82	8.77	29.5	579	410		298	261	232	209	190	174	160	149	
W 18 × 71	6.01	17.8	608	423				243	219	199	183	168	156	146
W 10 × 112	9.48	56.5	613	420	368	315	276	245	221					
W 12 × 96	10.9	41.4	613	437	368	315	276	245	221	200	184			
W 16 × 77	8.70	25.5	633	453			285	253	228	207	190	175	163	152
W 12 × 120	11.1	50.0	775	543	465	399	349	310	279	254	233			
W 16 × 100	8.84	29.6	833	590			375	333	300	273	250	231	214	200
W 14 × 120	13.2	46.2	883	633		454	398	353	318	289	265	245	227	
W 12 × 152	11.3	62.1	1,013	697	608	521	456	405	365	331	304			
W 14 × 145	14.1	54.7	1,083	773		557	488	433	390	355	325	300	279	

(continued)

TABLE 3.3 ***(Continued)***

Designation	$F_y = 50$ ksi				Span (ft)									
	L_p (ft)	L_r ft	M_p kip-ft	M_r kip-ft	24	26	28	30	32	34	36	38	40	42
W 21 × 44	4.45	12.0	398	272	119	110	102	95.4	89.4	84.2	79.5	75.3	71.6	68.1
W 21 × 50	4.60	12.5	458	315	138	127	118	110	103	97.1	91.7	86.8	82.5	78.6
W 21 × 57	4.77	13.1	538	370	161	149	138	129	121	114	108	102	96.8	92.1
W 24 × 55	4.74	12.9	563	383	169	156	145	135	127	119	113	107	101	96.4
W 24 × 62	4.84	13.3	642	440	193	178	165	154	144	136	128	122	116	110
W 21 × 68	6.36	17.3	667	467	200	185	171	160	150	141	133	126	120	114
W 18 × 86	9.30	26.1	775	553	233	215	199	186	174	164	155			
W 21 × 83	6.47	18.5	817	570	245	226	210	196	184	173	163	155	147	140
W 24 × 76	6.79	18.0	833	587	250	231	214	200	188	176	167	158	150	143
W 18 × 106	9.40	28.7	958	680	288	265	246	230	216	203	192			
W 21 × 101	10.2	27.6	1,054	757	316	292	271	253	237	223	211	200	190	181
W 24 × 94	7.00	19.4	1,058	740	318	293	272	254	238	224	212	201	191	181
W 27 × 94	7.50	19.9	1,158	810			298	278	261	245	232	219	209	199

W 24 × 104	10.3	26.8	1,204	860	361	333	310	289	271	255	241	228	217	206
W 21 × 122	10.3	29.8	1,279	910	384	354	329	307	288	271	256	242	230	219
W 24 × 117	10.4	27.9	1,363	970	409	377	350	327	307	289	273	258	245	234
W 27 × 114	7.71	21.3	1,429	997			368	343	322	303	286	271	257	245
W 30 × 108	7.60	20.3	1,442	997				346	324	305	288	273	260	247
W 24 × 131	10.5	29.1	1,542	1,097	463	427	396	370	347	326	308	292	278	264
W 21 × 147	10.4	32.8	1,554	1,097	466	430	400	373	350	329	311	294	280	266
W 30 × 124	7.88	21.5	1,700	1,183				408	383	360	340	322	306	291
W 33 × 118	8.20	21.7	1,729	1,197						366	346	328	311	296
W 24 × 146	10.6	30.6	1,742	1,237	523	482	448	418	392	369	348	330	314	299
W 24 × 162	10.8	32.4	1,950	1,380	585	540	501	468	439	413	390	369	351	334
W 33 × 141	8.59	23.1	2,142	1,493						454	428	406	386	367

Source: Compiled from data in the *Manual of Steel Construction* with permission of the publishers, American Institute of Steel Construction.

Problems 3.8.A–H. For each of the following conditions find (a) the lightest permitted shape and (b) the shallowest permitted shape of A36 steel:

	Span	*Live Load*	*Superimposed Dead Load*
A	16 ft	3 kips/ft [43.8 kN/m]	3 kips/ft [43.8 kN/m]
B	20 ft	1 kips/ft [14.6 kN/m]	0.5 kips/ft [7.30 kN/m]
C	36 ft	0.833 kips/ft [12.2 kN/m]	0.278 kips/ft [4.06 kN/m]
D	40 ft	1.25 kips/ft [18.2 kN/m]	1.25 kips/ft [18.2 kN/m]
E	18 ft	0.333 kips/ft [4.86 kN/m]	0.625 kips/ft [9.12 kN/m]
F	32 ft	1.167 kips/ft [17.0 kN/m]	3.5 kips/ft [51.0 kN/m]
G	42 ft	1 kips/ft [14.6 kN/m]	0.238 kips/ft [3.47 kN/m]
H	28 ft	0.5 kips/ft [7.3 kN/m]	0.5 kips/ft [7.3 kN/m]

Equivalent Load Techniques

The safe service loads in Table 3.2 are uniformly distributed loads on simple beams. Actually, the table values are determined on the basis of bending moment and limiting bending stress, so that it is possible to use the tables for other loading conditions for some purposes. Because framing systems always include some beams with other than simple uniformly distributed loadings, this is sometimes a useful process for design.

Consider the following situation: a beam with a load consisting of two equal concentrated loads placed at the beam third points—in other words, Case 3 in Figure B.1. For this condition, the figure yields a maximum moment value expressed as $PL/3$. By equating this to the moment value for a uniformly distributed load, a relationship between the two loads can be derived. Thus:

$$\frac{WL}{8} = \frac{PL}{3} \quad \text{or} \quad W = 2.67P$$

which shows that if the value of one of the concentrated loads in Case 3 of Figure B.1 is multiplied by 2.67, the result would be an *equivalent uniform load* or *equivalent tabular load* (called EUL or ETL) that would produce the same magnitude of maximum moment as the true loading condition.

Although the expression "equivalent uniform load" is the general name for this converted loading, when derived to facilitate the use of tabular materials, it is referred to as the "equivalent tabular load," which is the designation used in this book. Figure B.1 yields the ETL factors for several common loading conditions.

It is important to remember that the EUL or ETL is based only on consideration of flexure (that is, on limiting bending stress), so that investigations for deflection, shear, or bearing must use the true loading conditions for the beam.

This method may also be used for any loading condition, not just the simple, symmetrical conditions shown in Figure B.1. The process consists of first finding the true maximum moment due to the actual loading; then this is equated to the expression for the maximum moment for a theoretical uniform load, and the EUL is determined. Thus:

$$M = \frac{WL}{8} \qquad \text{or} \qquad W = \frac{8M}{L}$$

The expression $W = 8M/L$ is the general expression for an equivalent uniform load (or ETL) for any loading condition.

3.9 MANUFACTURED TRUSSES FOR FLAT SPANS

Factory-fabricated, parallel-chord trusses are produced in a wide range of sizes by a number of manufacturers. Most producers comply with the regulations of industry-wide organizations; for light steel trusses, the principal such organization is the Steel Joist Institute, called the SJI. Publications of the SJI are a chief source of general information (see Ref. 4), although the products of individual manufacturers vary, so that much valuable design information is available directly from the suppliers of a specific product. Final design and development of construction details for a particular project must be done in cooperation with the supplier of the products.

Light steel parallel-chord trusses, called *open-web joists,* have been in use for many years. Early versions used all-steel bars for the chords and the continuously bent web members (see Figure 3.13), so that they were also referred to as *bar joists.* Although other elements are now used for the chords, the bent steel rod is still used for the diagonal web members for some of the smaller-sized joists. The range of size of this basic element has now been stretched considerably, resulting in members as long as 150 ft [46 m] and depths of 7 ft [2.14 m] and more. At the larger size range the members are usually more common forms for steel trusses—double angles, structural tees, and so on. Still, a considerable usage is made of the smaller sizes for both floor joists and roof rafters.

Table 3.4 is adapted from a standard table in a publication of the SJI (Ref. 4). This table lists a number of sizes available in the K series, which is the lightest group of joists. Joists are identified by a three-unit designation. The first number indicates the overall nominal depth of the joist, the letter indicates the series, and the second number indicates the class of size of the members—the higher the number, the heavier and stronger the joist.

Table 3.4 can be used to select the proper joist for a determined load and span situation. Figure 3.14 shows the basis for determination of the span for a joist. There are two entries in the table for each span. The first number represents the total factored load capacity of the joist in pounds per foot of the joist length (lb/ft), and the number in parentheses is the load that will produce a deflection of 1/360 of the span. The following examples illustrate the use of the table data for some common design situations. For the purpose of illustration the examples use data from Table 3.4. However, more joists sizes are available and their capacities are given in the reference for Table 3.4.

Example 13. Open-web steel joists are to be used to support a roof with a unit live load of 20 psf and a unit dead load of 15 psf (not including the weight of the joists) on a span of 40 ft. Joists are spaced at 6 ft center to center. Select the lightest joist if deflection under live load is limited to 1/360 of the span.

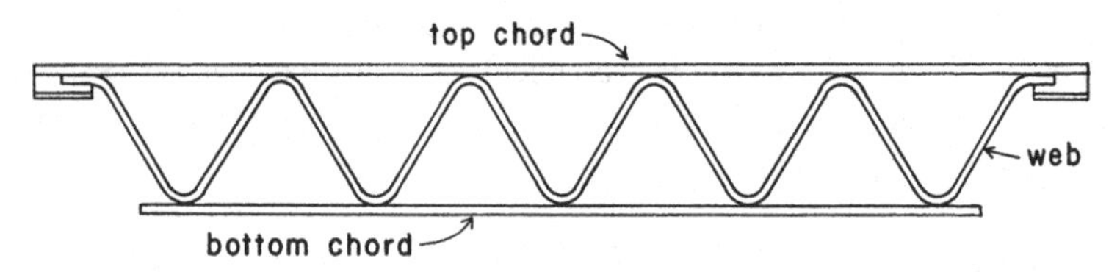

Figure 3.13 Form of a short-span, open-web steel joist.

TABLE 3.4 Safe Factored Loads for K-Series Open-Web Joists[a]

Joist Designation:	12K1	12K3	12K5	14K1	14K3	14K6	16K2	16K4	16K6	18K3	18K5	18K7	20K3	20K5	20K7
Weight (lb/ft):	5.0	5.7	7.1	5.2	6.0	7.7	5.5	7.0	8.1	6.6	7.7	9.0	6.7	8.2	9.3
Span (ft)															
20	357	448	607	421	528	729	545	732	816	687	816	816	767	816	816
	(142)	(177)	(230)	(197)	(246)	(347)	(297)	(386)	(426)	(423)	(490)	(490)	(517)	(550)	(550)
22	295	369	500	347	435	641	449	602	739	567	769	816	632	816	816
	(106)	(132)	(172)	(147)	(184)	(259)	(222)	(289)	(351)	(316)	(414)	(438)	(393)	(490)	(490)
24	246	308	418	291	363	537	377	504	620	475	644	781	530	720	816
	(81)	(101)	(132)	(113)	(141)	(199)	(170)	(221)	(269)	(242)	(318)	(382)	(302)	(396)	(448)
26				246	310	457	320	429	527	403	547	665	451	611	742
				(88)	(110)	(156)	(133)	(173)	(211)	(190)	(249)	(299)	(236)	(310)	(373)
28				212	267	393	276	369	454	347	472	571	387	527	638
				(70)	(88)	(124)	(106)	(138)	(168)	(151)	(199)	(239)	(189)	(248)	(298)
30							239	320	395	301	409	497	337	457	555
							(86)	(112)	(137)	(123)	(161)	(194)	(153)	(201)	(242)
32							210	282	346	264	359	436	295	402	487
							(71)	(92)	(112)	(101)	(132)	(159)	(126)	(165)	(199)
36										209	283	344	233	316	384
										(70)	(92)	(111)	(88)	(115)	(139)
40													188	255	310
													(64)	(84)	(101)

(continued)

TABLE 3.4 (*Continued*)

Joist Designation:	22K4	22K6	22K9	24K4	24K6	24K9	26K5	26K7	26K9	28K6	28K8	28K10	30K7	30K9	30K12
Weight (lb/ft):	8.0	8.8	11.3	8.4	9.7	12.0	9.8	10.9	12.2	11.4	12.7	14.3	12.3	13.4	17.6
Span (ft)															
28	516	634	816	565	693	816	692	816	816	813	816	816			
	(270)	(328)	(413)	(323)	(393)	(456)	(427)	(501)	(501)	(541)	(543)	(543)			
30	448	550	738	491	602	807	601	730	816	708	816	816	816	816	816
	(219)	(266)	(349)	(262)	(319)	(419)	(346)	(417)	(459)	(439)	(500)	(500)	(543)	(543)	(543)
32	393	484	647	430	530	709	528	641	770	620	764	815	743	815	815
	(180)	(219)	(287)	(215)	(262)	(344)	(285)	(343)	(407)	(361)	(438)	(463)	(461)	(500)	(500)
36	310	381	510	340	417	559	415	504	607	490	602	723	586	705	723
	(126)	(153)	(201)	(150)	(183)	(241)	(199)	(240)	(284)	(252)	(306)	(366)	(323)	(383)	(392)
40	250	307	412	274	337	451	337	408	491	395	487	629	473	570	650
	(91)	(111)	(146)	(109)	(133)	(175)	(145)	(174)	(207)	(183)	(222)	(284)	(234)	(278)	(315)
44	206	253	340	227	277	372	277	337	405	326	402	519	390	470	591
	(68)	(83)	(109	(82)	(100)	(131)	(108)	(131)	(155)	(137)	(167)	(212)	(176)	(208)	(258)
48				190	233	313	233	282	340	273	337	436	328	395	542
				(63)	(77)	(101)	(83)	(100)	(119)	(105)	(128)	(163)	(135)	(160)	(216)
52							197	240	289	233	286	371	279	335	498
							(65)	(79)	(102)	(83)	(100)	(128)	(106)	(126)	(184)
56										200	246	319	240	289	446
										(66)	(80)	(102)	(84)	(100)	(153)
60													209	250	389
													(69)	(81)	(124)

Source: Data adapted from more extensive tables in the *Standard Specifications, Load Tables, and Weight Tables for Steel Joists and Joist Girders* (Ref. 4), with permission of the publishers, Steel Joist Institute. The Steel Joist Institute publishes both specifications and load tables; each of these contains standards that are to be used in conjunction with one another.

[a]Loads in pounds per foot of joist span. First entry represents the total factored joist capacity; entry in parentheses is the load that produces a deflection of 1/360 of the span. See Fig. 9.16 for definition of span.

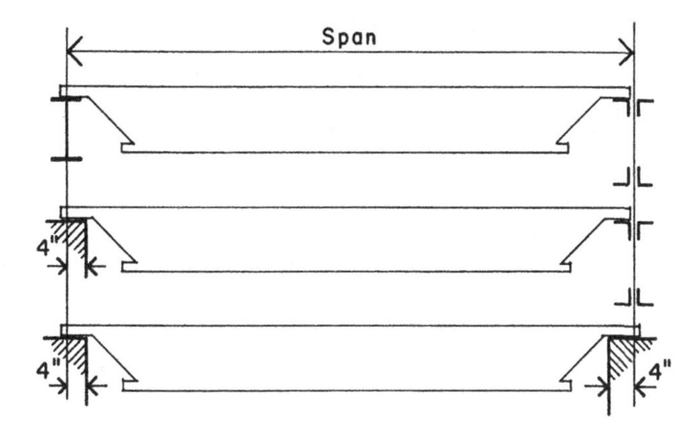

Figure 3.14 Definition of span for open-web steel joists, as given in Ref. 4 by the Steel Joist Institute.

Solution. The first step is to determine the unit load per ft on the joists, thus:

Live load: 6(20) = 120 lb/ft [1.8 kN/m]

Dead load: 6(15) = 90 lb/ft (not including joist weight) [1.3 kN/m]

Total factored load: 1.2(90) + 1.6(120) = 108 + 192 = 300 lb/ft [4.4 kN/m]

This yields the two numbers (total factored load and live load only) that can be used to scan the data for the given span in Table 3.4. Note that the joist weight—so far excluded in the computation—is included in the total load entries in the table. Once a joist is selected, therefore, the actual joist weight (given in the table) must be deducted from the table entry for comparison with the computed values. We thus note from the table the possible choices listed in Table 3.5. Although the joists' weights are all very close, the 24K6 is the lightest choice.

Example 14. Open-web steel joists are to be used for a floor with a unit live load of 75 psf [3.59 kN/m^2] and a unit dead load of 40 psf [1.91 kN/m^2] (not including the joist weight) on a span of 30 ft [9.15 m]. Joists are 2 ft [0.61 m] on center, and deflection is limited to 1/240 of the span under total load and 1/360 of the span under live load only. Determine the lightest possible joist and the lightest joist of least depth possible.

TABLE 3.5 Possible Choices for the Roof Joist

Load Condition	Required Capacity (lb/ft)	Capacity of the Indicated Joists (lb/ft) 22K9	24K6	26K5
Factored total capacity		412	337	337
Joist weight from Table 3.4		11.3	9.7	9.8
Factored joist weight		14	12	12
Net usable capacity	300	398	325	325
Load for deflection of 1/360	120	146	133	157

Solution. As in the previous example, the unit loads are first determined, thus:

Live load: 2(75) = 150 lb/ft (for limiting deflection of L/360) [2.2 kN/m]

Dead load: 2(40) = 80 lb/ft (not including joist weight) [1.2 kN/m]

Total service load: 150 + 80 = 230 lb/ft [3.36 kN/m]

Total factored load: 1.2(80) + 1.6(150) = 96 + 240 = 336 lb/ft [4.9 kN/m]

To satisfy the deflection criteria for total load, the limiting value for deflection in parentheses in the table should be not less than (240/360)(230) = 153 lb/ft [2.2 kN/m]. Because this is slightly larger than the live load, it becomes the value to look for in the table. Possible choices obtained from Table 3.4 are listed in Table 3.6, from which the following may be observed:

The lightest joist is the 18K5.

The shallowest depth joist is the 18K5.

In some situations, it may be desirable to select a deeper joist, even though its load capacity may be somewhat redundant. Total sag, rather than an abstract curvature limit, may be of more significance for a flat roof structure. For example, for the 40-ft [12.2-m] span in Example 13, a sag of 1/360 of the span = (1/360)(40 × 12) = 1.33 in. [338 mm].

The actual effect of this dimension on roof drainage or in relation to interior partition walls must be considered. For floors, a major concern is for bounciness, and this very light structure is highly vulnerable in this regard. Designers therefore sometimes deliberately choose the deepest

TABLE 3.6 Possible Choices for the Floor Joist

Load Condition	Required Capacity (lb/ft)	Capacity of the Indicated Joists (lb/ft)		
		18K5	20K5	22K4
Factored total capacity		409	457	448
Joist weight from Table 3.4		7.7	8.2	8.0
Factored joist weight		10	10	10
Net usable capacity	300	399	447	438
Load for deflection	153	161	201	219

feasible joist for floor structures, in order to get all the help possible to reduce deflection as a means of stiffening the structure in general against bouncing effects.

As mentioned previously, joists are available in other series for heavier loads and longer spans. The SJI, as well as individual suppliers, also have considerably more information regarding installation details, suggested specifications, bracing, and safety during erection for these products.

Stability is a major concern for these elements, because they have very little lateral or torsional resistance. Other construction elements, such as decks and ceiling framing, may help, but the whole bracing situation must be carefully studied. Lateral bracing in the form of x-braces or horizontal ties is generally required for all steel joist construction, and the reference source for Table 3.4 (Ref. 4) has considerable information on this topic.

One means of assisting stability has to do with the typical end support detail, as shown in Figure 3.13. The common method of support consists of hanging the trusses by the ends of their top chords, which is a general means of avoiding the roll-over type of rotational buckling at the supports that is illustrated in Figure 3.26. For construction detailing, however, this adds a dimension to the overall depth of the construction, in comparison to an all-beam system with the joist/beams and supporting girders all having their tops level. This added dimension (the depth of the end of the joist) is typically 2.5 in. [63.5 mm] for small joists and 4 in. [101 mm] for larger joists.

For development of a complete truss system, a special type of prefabricated truss available is that described as a *joist girder*. This truss is specifically designed to carry the regularly spaced, concentrated loads consisting of the end support reactions of joists. A common form of joist girder is shown in Figure 3.15. Also shown in the figure is the form of standard designation for a joist girder, which includes indications of the

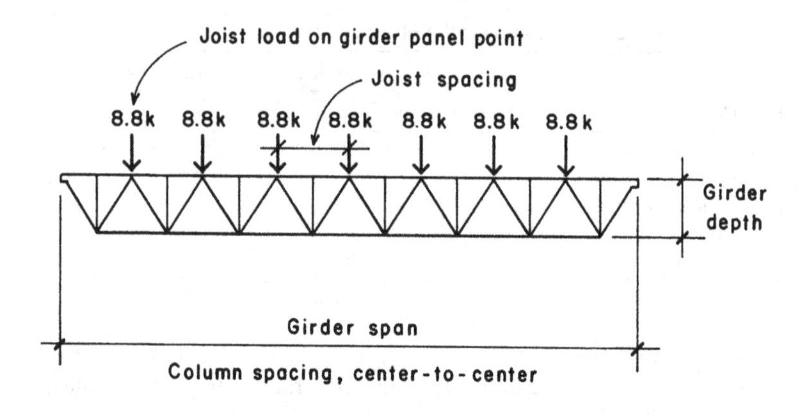

Standard Designation:

48G	8N	8.8 K
Depth in inches	Number of joist spaces	Load on each panel point in kips

Specify: 48G8N8.8K

Figure 3.15 Considerations for layout and designation of joist girders.

nominal girder depth, the number of spaces between joists (called the girder *panel unit*), and the end reaction force from the joists—which is the unit concentrated load on the girder.

Predesigned joist girders (that is, girders actually designed for fabrication by the suppliers) may be selected from catalogs in a manner similar to that for open-web joists. The procedure is usually as follows:

1. The designer determines the joist spacing, joist load, and girder span. (The joist spacing should be a full number division of the girder span.)
2. The designer uses this information to specify the girder by the standard designation.
3. The designer chooses the girder from a particular manufacturer's catalog, or simply specifies it to the supplier.

Illustrations of the use of joists and complete truss systems are given in the building design examples in Chapter 10.

Problem 3.9.A. Open-web steel joists are to be used for a roof with a live load of 25 psf [1.2 kN/m^2] and a dead load of 20 psf [957 N/m^2] (not including the joist weight) on a span of 48 ft [14.6 m]. Joists are 4 ft [1.22 m] on center, and deflection under live load is limited to 1/360 of the span. Select the lightest joist.

Problem 3.9.B. Open-web steel joists are to be used for a roof with a live load of 30 psf [1.44 kN/m^2] and a dead load of 18 psf [862 N/m^2] (not including the joist weight) on a span of 44 ft [13.42 m]. Joists are 5 ft [1.53 m] on center, and deflection is limited to 1/360 of the span due to live load. Select the lightest joist.

Problem 3.9.C. Open-web steel joists are to be used for a floor with a live load of 50 psf [2.39 kN/m^2] and a dead load of 45 psf [2.15 kN/m^2] (not including the joist weight) on a span of 36 ft [11 m]. Joists are 2 ft [0.61 m] on center, and deflection is limited to 1/360 of the span under live load only and to 1/240 of the span under total load. Select (a) the lightest possible joist and (b) the shallowest depth possible joist.

Problem 3.9.D. Repeat Problem 3.9.C, except that the live load is 100 psf [4.79 kN/m^2], the dead load is 35 psf [1.67 kN/m^2], and the span is 26 ft [7.93 m].

3.10 DECKS WITH STEEL FRAMING

Figure 3.16 shows four possibilities for a floor deck used in conjunction with a framing system of rolled steel beams. When a wood deck is used (Figure 3.16*a*), it is usually supported by and nailed to a series of wood joists, which are in turn supported by the steel beams. However, in some cases the deck may be nailed to wood members that are bolted to the tops of the steel beams, as shown in the figure. For floor construction, it is now also common to use a concrete fill on top of the wood deck, for added stiffness, fire protection, and improved acoustic behavior.

A site-cast concrete deck (Figure 3.16*b*) is typically formed with plywood panels placed against the bottoms of the top flanges of the beams. This helps to lock the slab and beams together for lateral effects, although steel lugs are also typically welded to the tops of the beams for composite construction.

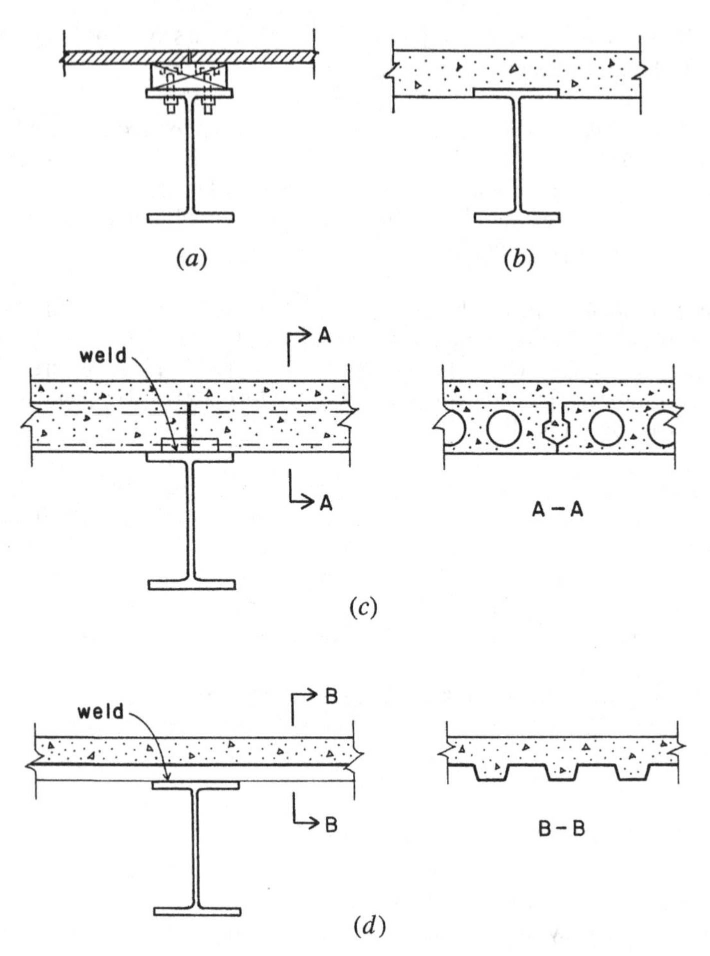

Figure 3.16 Typical forms of floor deck construction used with steel framing.

Concrete may also be used in the form of precast deck units. In this case, steel elements are imbedded in the ends of the precast units and are welded to the beams. A site-poured concrete fill is typically used to provide a smooth top surface, and is bonded to the precast units for added structural performance.

Formed sheet steel units may be used in one of three ways: as the primary structure, as strictly forming for the concrete deck, or as a composite element in conjunction with the concrete. Attachment of this type of deck to the steel beams is usually achieved by welding the steel units to the beams before the concrete is placed.

Three possibilities for roof decks using steel elements are shown in Figure 3.17. Roof loads are typically lighter than floor loads and bounciness of the deck is usually not a major concern. (A possible exception is where suspended elements may be hung from the deck and can create a problem with vertical movements during an earthquake.)

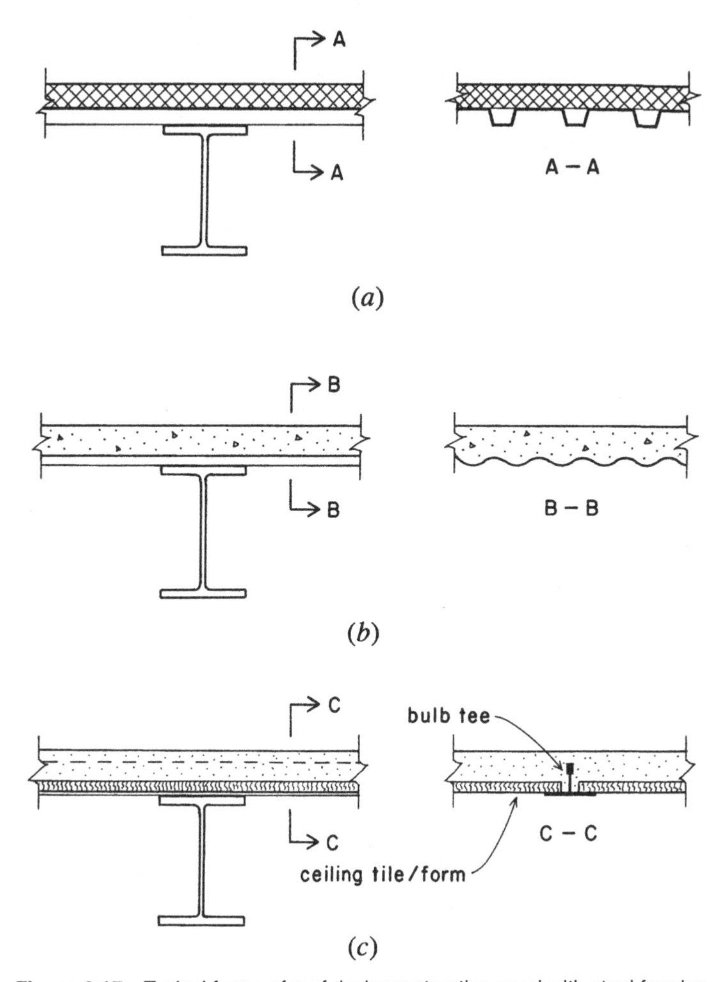

Figure 3.17 Typical forms of roof deck construction used with steel framing.

A fourth possibility for the roof is the plywood deck shown in Figure 3.17*a*. This is, in fact, probably a wider use of this form of construction.

Light-Gage Formed Steel Decks

Steel decks consisting of formed sheet steel are produced in a variety of configurations, as shown in Figure 3.18. The simplest is the corrugated sheet, shown in Figure 3.18*a*. This may be used as the single, total surface for walls and roofs of utilitarian buildings (tin shacks). For more demanding usage, it is used mostly as the surfacing of a built-up panel or general sandwich-type construction. As a structural deck, the simple corrugated sheet is used for very short spans, typically with a structural-grade concrete fill that effectively serves as the spanning deck—the steel sheet serving primarily as forming for the concrete.

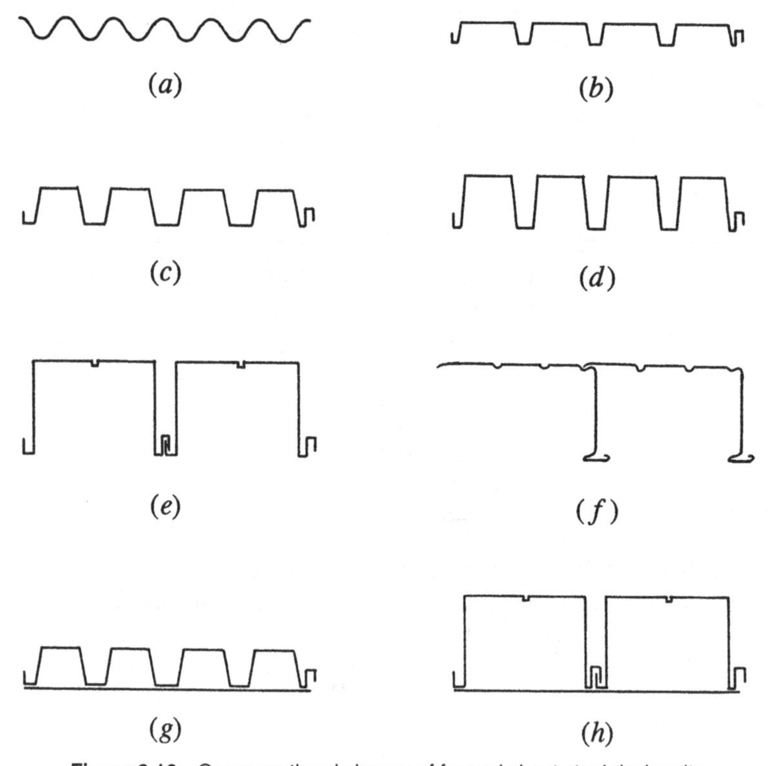

Figure 3.18 Cross-sectional shapes of formed sheet steel deck units.

A widely used product is that shown in three variations in Figure 3.18*b* through *d*. When used for roof deck, where loads are light, a flat-top surface is formed with a very lightweight fill of foamed concrete or gypsum concrete or a rigid sheet material that also serves as insulation. For floors—with heavier loads, a need for a relatively hard surface, and concern for bouncing—a structural-grade concrete fill is used, and the deeper ribs of the units shown in Figure 3.18*c* and *d* may be selected to achieve greater spans with wide-spaced beams. Common overall deck heights are 1.5, 3, and 4.5 in.

There are also formed sheet steel decks produced with greater depth, such as those shown in Figure 3.18*e* and *f*. These can achieve considerable span, generally combining the functions of joist and deck in a single unit.

Although used somewhat less now, with the advent of different wiring products and techniques for facilitating the need for frequent and rapid change of building wiring, a possible use for steel deck units is as a conduit for power, signal, or communication wiring. This can be accomplished by closing the deck cells with a flat sheet of steel, as shown in Figures 3.18*g* and *h*. This usually provides for wiring in one direction in a grid; the perpendicular wiring is achieved in conduits buried in the concrete fill.

Decks vary in form (shape of the cross section) and in the thickness (gage) of the steel sheet used to form them. Design choices relate to the desire for a particular form and to the load and span conditions. Units are typically available in lengths of 30 ft or more, which usually permits design for a multiple-span condition; this reduces bending effects only slightly but has a considerable influence on reduction of deflection and bouncing.

Fire protection for floor decks is provided partly by the concrete fill on top of the deck. Protection of the underside is achieved by sprayed-on materials (as also used on beams) or by use of a permanent fire-rated ceiling construction. The latter is no longer favored, however, because many disastrous fires have occurred in the void space between ceilings and overhead floors or roofs.

Other structural uses of the deck must also be considered. The most common use is as a horizontal diaphragm for distribution of lateral forces from wind and earthquakes. Lateral bracing of beams and columns is also often assisted or completely achieved by structural decks.

When structural-grade concrete is used as a fill, there are three possibilities for its relationship to a forming steel deck:

1. The concrete serves strictly as a structurally inert fill, providing a flat surface, fire protection, added acoustic separation, and so on, but no significant structural contribution.
2. The steel deck functions essentially only as a forming system for the concrete fill; the concrete is reinforced and designed as a spanning structural deck.
3. The concrete and sheet steel work together in what is described as *composite structural action.* In effect, the sheet steel on the bottom serves as the reinforcement for midspan bending stresses, leaving a need only for top reinforcement for negative bending moments over the deck supports.

Table 3.7 presents data relating to the use of the type of deck unit shown in Figure 3.18*b* for roof structures. The three forms of deck patterns, referred to in the table, are shown in Figure 3.19. This data is adapted from a publication that is distributed by an industry-wide organization referred to in the table footnotes and is adequate for preliminary design work. The reference publication also provides considerable information and standard specifications for usage of the deck. For any final design work for actual construction, structural data for any manufactured products should be obtained directly from the suppliers of the products.

The common usage for roof decks with units as shown in Table 3.7 is that described as structural dependence strictly on the steel deck units. That is the basis for the data in the table given here.

Three different rib configurations are shown for the deck units in Table 3.7, described as *narrow, intermediate,* and *wide* rib deck. This has some effect on the properties of the deck cross section and thus produces three separate sections in the table. Although structural performance may be a factor in choosing the rib width, there are usually other predominating reasons. If the deck is to be welded to its supports (usually required for good diaphragm action), this is done at the bottom of the ribs and the wide rib is required. If a relatively thin topping material is used, the narrow rib is favored.

Rusting is a critical problem for the very thin sheet steel deck. With its top usually protected by other construction, the main problem is the treatment of the underside of the deck. A common practice is to provide the appropriate surfacing of the deck units in the factory. The deck weights in Table 3.7 are based on simple painted surfaces, which is usually the least expensive surface. Surfaces consisting of bonded enamel or galvanizing are also available, adding somewhat to the deck weight.

TABLE 3.7 Safe Service Load Capacity of Formed Steel Roof Deck

Deck Type[a]	Span Condition	Weight[b] (lb/sq ft)	Total Safe Service Load (Dead + Live)[c] for spans indicated in ft												
			4.0	4.5	5.0	5.5	6.0	6.5	7.0	7.5	8.0	8.5	9.0	9.5	10.0
NR22	Simple	1.6	73	58	47										
NR20		2.0	91	72	58	48	40								
NR18		2.7	125	99	80	66	55	47							
NR22	Two	1.6	80	63	51	42									
NR20		2.0	97	76	62	51	43								
NR18		2.7	128	101	82	68	57	48	42						
NR22	Three and +	1.6	100	79	64	53	44								
NR20		2.0	121	96	77	64	54	46							
NR18		2.7	160	126	102	85	71	61	52	45					
IR22	Simple	1.6	84	66	54	44									
IR20		2.0	104	82	67	55	46								
IR18		2.7	142	112	91	75	63	54	46	40					
IR22	Two	1.6	90	71	58	48	40								
IR20		2.0	110	87	70	58	49	41							
IR18		2.7	145	114	93	77	64	55	47	40					
IR22	Three and +	1.6	113	89	72	60	50	43							
IR20		2.0	137	108	88	72	61	52	45						
IR18		2.7	181	143	116	96	81	69	59	52	45	40			
WR22	Simple	1.6			90	70	56	46							
WR20		2.0			113	88	70	57	48	40					
WR18		2.7			159	122	96	77	64	54	46	40			
WR22	Two	1.6			96	79	67	57	49	43					
WR20		2.0			123	102	86	73	63	55	48	43			
WR18		2.7			164	136	114	98	84	73	64	57	51	46	41
WR22	Three and +	1.6			119	99	83	71	61	53	47	41	36		
WR20		2.0			153	127	107	91	79	68	58	50	43		
WR18		2.7			204	169	142	121	105	91	79	67	58	51	43

[a] Letters refer to rib type (see Fig. 3.19). Numbers indicate gage (thickness) of deck sheet steel.

[b] Approximate weight with paint finish; other finishes available.

[c] Total safe allowable service load in lb/sq ft.

Source: Adapted from the *Steel Deck Institute Design Manual for Composite Decks, Form Decks, and Roof Decks* (Ref. 5), with permission of the publishers, the Steel Deck Institute.

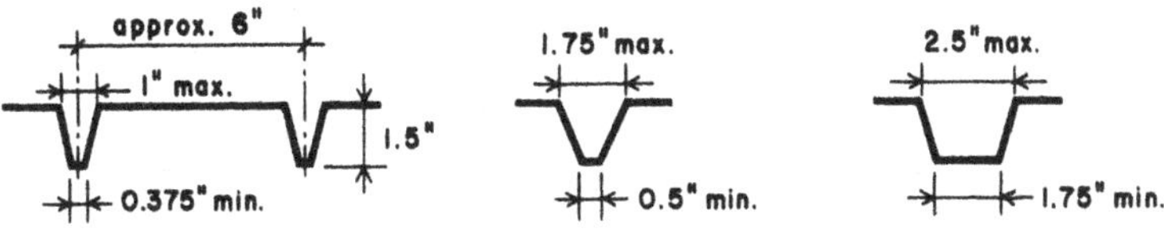

Figure 3.19 Reference for Table 3.7.

As described previously, these products are typically available in lengths up to 30 ft or more. Depending on the spacing of supports, therefore, various conditions of continuity of the deck may occur. Because of this condition, the table provides three cases for continuity: simple span (one span), two-span, and three or more spans.

Problem 3.10.A–F. Using data from Table 3.7, select the lightest steel deck for the following:

A. Simple span of 7 ft, total load of 45 psf
B. Simple span of 5 ft, total load of 50 psf
C. Two-span condition, span of 8.5 ft, total load of 45 psf
D. Two-span condition, span of 6 ft, total load of 50 psf
E. Three-span condition, span of 6 ft, total load of 50 psf
F. Three-span condition, span of 8 ft, total load of 50 psf

3.11 CONCENTRATED LOAD EFFECTS IN BEAMS

An excessive bearing reaction on a beam, or an excessive concentrated load at some point in the beam span, may cause either localized yielding or *web crippling* (i.e., buckling of the thin beam web). The AISC Specification requires that beam webs be investigated for these effects and that web stiffeners be used if the concentrated load exceeds limiting values.

The three common situations for this effect occur as shown in Figure 3.20. Figure 3.20*a* shows the beam end bearing on a support (commonly a masonry or concrete wall), with the reaction force transferred to the beam bottom flange through a steel bearing plate. Figure 3.20*b* shows a column load applied to the top of the beam at some point within the beam span. Figure 3.20*c* shows what may be the most frequent occurrence of this condition: a beam supported in bearing on top of a column with the beam continuous through the joint.

Figure 3.20*d* shows the development of the effective portion of the web length (along the beam span) that is assumed to resist bearing forces. For yield resistance, the maximum design end reaction R_u, and the maximum design load P_u within the beam span, are defined as follows (see Figure 3.20*d*):

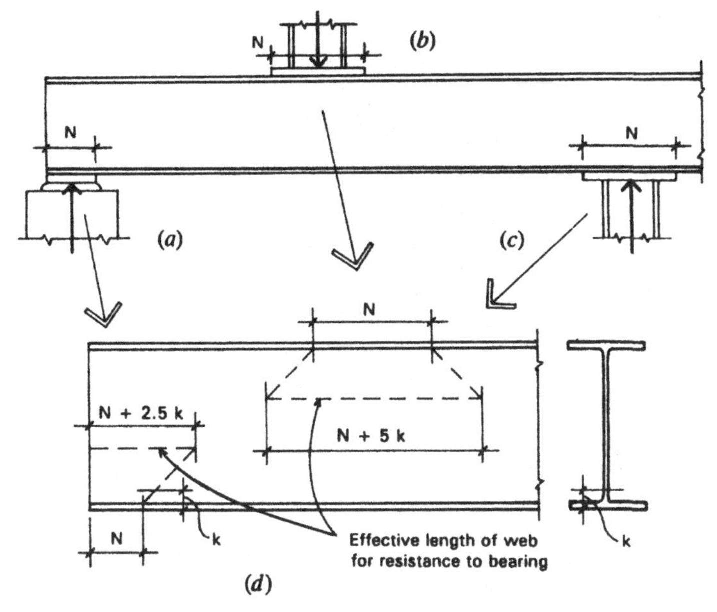

Figure 3.20 Considerations for bearing in steel beams with thin webs as related to web crippling (buckling of the thin web in diagonal compression).

$$R_u = (\phi F_y)(t_w)\{N + 2.5(k)\}$$

$$P_u = (\phi F_y)(t_w)\{N + 5(k)\}$$

where t_w = thickness of the beam web
N = length of the bearing
k = distance from the outer face of the beam flange to the web toe of the fillet (radius) of the corner between the web and the flange
ϕ = 1.0

For W shapes, the dimensions t_w and k are provided in the AISC Manual (Ref. 3) tables of properties for rolled shapes.

When these values are exceeded, web stiffeners—usually consisting of steel plates welded into the channel-shaped sides of the beam as shown in Figure 3.21—should be provided at the location of the concentrated load. These stiffeners may indeed add to bearing resistance, but the usual reason for using them is to alleviate the other potential form of failure:

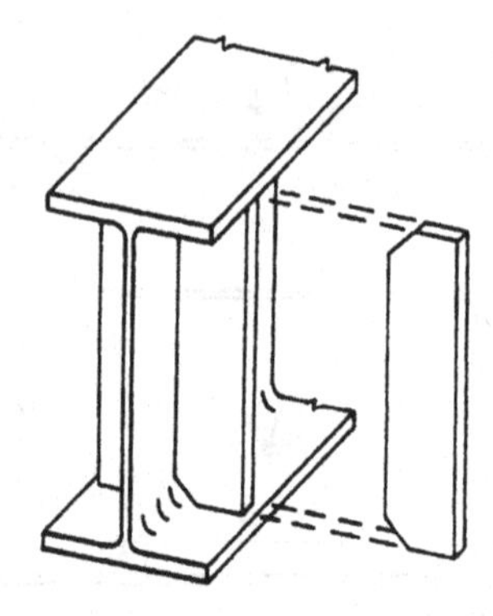

Figure 3.21 Use of stiffeners to prevent buckling of a thin web in a beam.

web crippling due to an excessively slender web; hence, their usual description as *web stiffeners.*

The AISC provides additional formulas for computation of limiting loads due to web crippling. However, the AISC Manual (Ref. 3) also provides data for shortcut methods.

Example 15. A W 18 × 50 beam of 50 ksi steel has an end reaction that is developed in bearing over a length $N = 10$ in. [254 mm]. Investigate the beam for yielding and web crippling if the reaction force is 65 kips [289 kN].

Solution: Table A.3 yields values of 1.25 in. [32 mm] for k and 0.355 in. [9 mm] for t_w. Determine the end bearing limit:

$$R_u = (F_y)(t_w)\{N + 2.5(k)\} = (50)(0.355)\{10 + 2.5(1.25)\}$$
$$= 232 \text{ kips } [1030 \text{ kN}]$$

To investigate for buckling (web crippling), the following data from the AISC Manual (Ref. 3) is used:

$\phi R_1 = 55.5$ kips
$\phi R_2 = 17.8$ kips/in.
$\phi_r R_3 = 57.7$ kips
$\phi_r R_4 = 4.73$ kips/in.
$\phi_r R_5 = 52$ kips
$\phi_r R_6 = 6.30$ kips/in.

The maximum end reaction is the least of two values, one based on yield and the other on buckling. Thus, for yield:

$$R_u = \phi R_1 + N(\phi R_2) = 55.5 + 10(17.8) = 233 \text{ kips}$$

For buckling, if $N/d \leq 0.2$

$$R_u = \phi_r R_3 + N(\phi_r R_4)$$

And if $N/d > 0.2$

$$R_u = \phi_r R_5 + N(\phi_r R_6)$$

Because $N/d = 10/18 = 0.556 > 0.2$

$$R_u = \phi_r R_5 + N(\phi_r R_6) = 52 + 10(6.30) = 115 \text{ kips}$$

The buckling limit prevails. So the maximum design end reaction is 115 kips. Because this value is greater than the required force of 65 kips, the beam and its support detail are adequate as is.

In Figures 3.20*b* and *c*, the beam is continuous through the bearing condition. In such cases, web crippling must be investigated using the AISC formulas. However, as with yielding, the crippling limit is slightly higher than that for end bearing. As a result, the web crippling may be determined as for end bearing and if the required load is close, the situation is not critical.

Example 16. The beam in Example 15 carries a factored column load of 100 kips [445 kN] within the beam span. The bearing length of the column on the beam is 12 in. [305 mm]. Investigate the beam for yield and web crippling.

Solution: First investigate for yield.

$$P_u = (\phi F_y)(t_w)\{N + 5(k)\} = (50)(0.355)\{12 + 5(1.25)\}$$
$$= 324 \text{ kips } [1440 \text{ kN}]$$

For web crippling, consider the limit for end bearing as follows:

Since $N/d = 12/18 = 0.67 > 0.2$:

$$P_u = \phi_r R_5 + N(\phi_r R_6) = 52 + 12(6.3) = 128 \text{ kips } [569 \text{ kN}]$$

Because both values exceed the required force, the beam is sufficient.

Problem 3.11.A. Find the maximum allowable reaction force for a W 18 × 50 beam of 50 ksi steel with an end bearing plate 8 in. [203 mm] long.

Problem 3.11.B. A column load of 120 kips [534 kN] with a bearing length of 11 in. [279 mm] is placed atop the beam defined in Problem 3.11.A. Are web stiffeners required?

3.12 TORSIONAL EFFECTS

In various situations, steel beams may be subjected to torsional twisting effects in addition to the primary conditions of shear and bending. These effects may occur when the beam is loaded in a plane that does not coincide with the *shear center* of the member cross section. For doubly symmetrical forms such as the W, M, and S shapes, the shear center coincides with the member centroid (the intersection of its principal axes). Off-center loadings such as those shown in Figure 3.22 will develop twisting of the member.

As shown in Figure 3.23, a loading exactly coinciding with the plane of the minor axis (the *y-y* axis) will produce a pure bending about the major axis (the *x-x* axis). The opposite is also true for bending about the *y-y* axis. However, any misalignment of the loading will produce a twisting, torsional effect, which becomes additive to the usual effects of shear and bending. For torsionally weak members—such as open I-shaped sections with thin parts and low bending resistance on the *y-y* axis—the

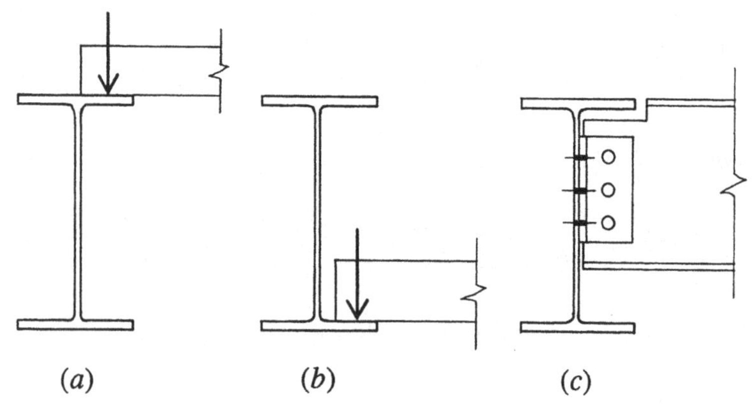

Figure 3.22 Torsional moment in a beam produced by off-center loading.

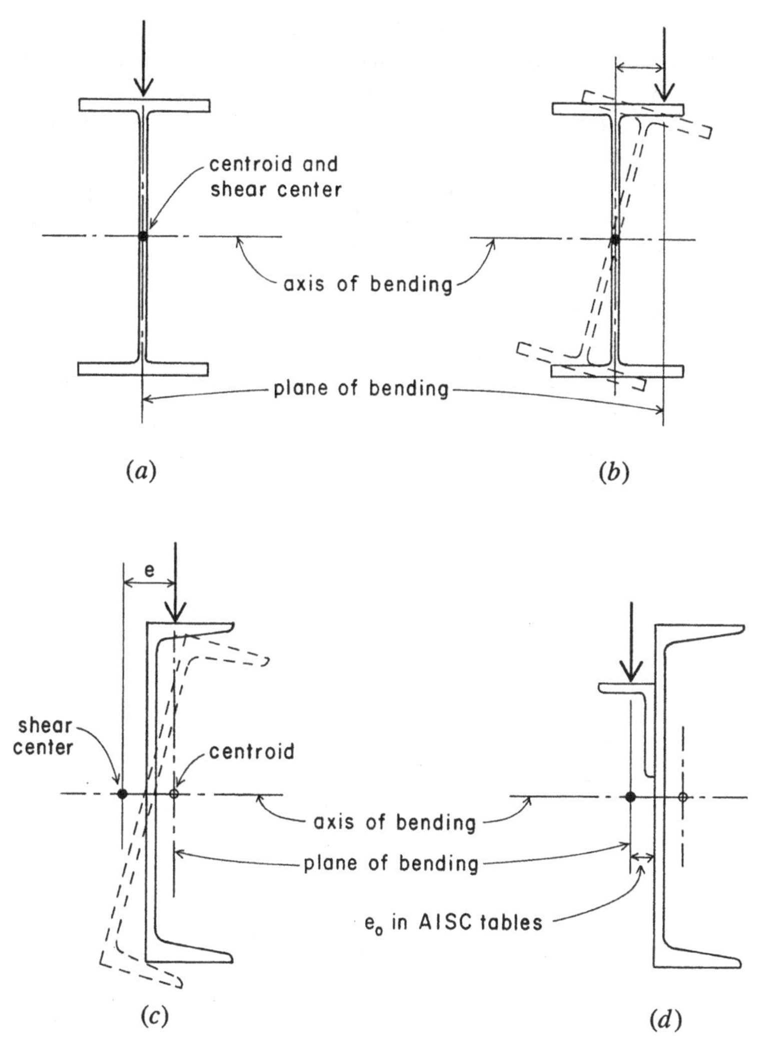

Figure 3.23 Torsion in beams developed by loading not aligned with the beam's shear center.

torsional action may quite easily become the predominant mode of ultimate failure of the beam.

For shapes that are not symmetrical about both axes—such as C and angle shapes—the location of the shear center does not coincide with the centroid of the section. For the channel section, for example, it is located some distance behind the back of the section, as shown in Figure 3.23*c*. Thus, loading the channel through its centroid or loading it through its vertical web portion can produce twisting. Where possible, one way to avoid this is to attach an angle to the channel to permit the load to be applied close to the channel's shear center (see Figure 3.23*d*).

The locations of shear centers and centroids for various shapes and combinations of shapes are shown in Figure 3.24. For the members for which these two centers coincide (see Figure 3.24*a*), the concern is limited to ensuring that loadings remain in the plain of the member's cen-

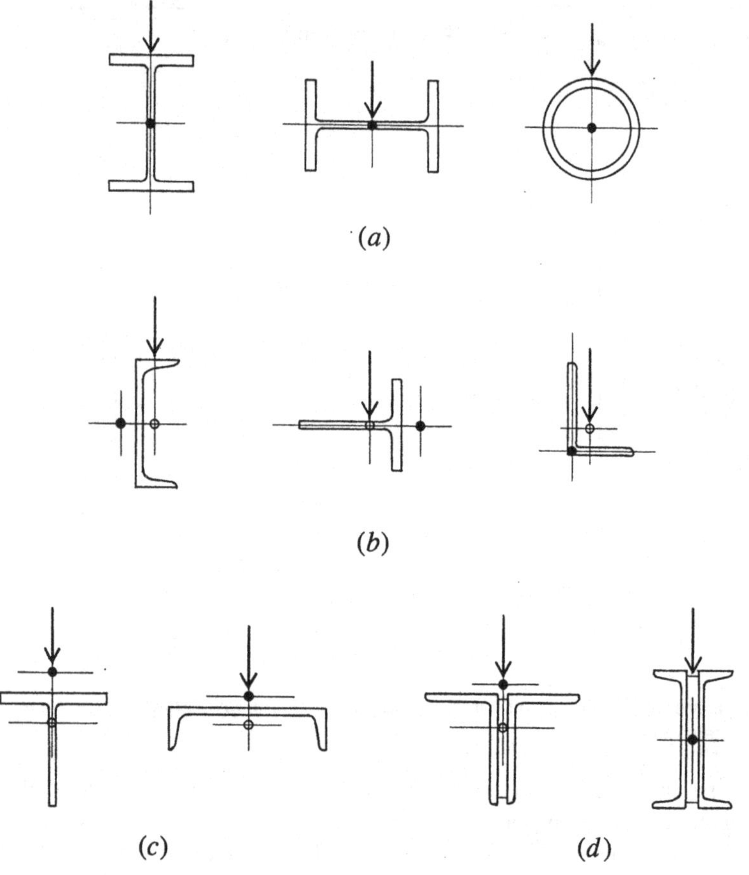

Figure 3.24 Shear centers and centroids for various beam sections. For all cases, the loading is shown passing through the centroid.

troidal axis. For the members for which the two centers are separated (see Figure 3.24*b*), centroidal loading will produce twisting. However, even if the two centers are separated, a loading through the centroid may not produce twisting if the plane of loading also passes through the shear center, as shown in Figure 3.24*c* for the single tee shape and the channel and in Figure 3.24*d* for the built-up shapes of double angles and double channels.

There are means for computation of torsional effects and torsional resistance of structural members. Indeed, some members may truly need to resist torsion and can be designed for the effects. For beams in structural frameworks, however, a more common technique is to try to avoid the torsional effects as much as possible. One way to do this is to choose members appropriate to the situation involving the manner of application of loads and the details of the frame assembly. Another way is to provide bracing for members to essentially prevent the possibility of their being twisted.

Torsion is an issue that should be considered for beams, in addition to all the other basic concerns. Assurance of safety in this case must be established by following through to the development of the details for the construction.

3.13 BUCKLING OF BEAMS

Buckling of beams, in one form or another, is mostly a problem with beams that are relatively weak on their transverse axes—that is, the axis of the beam cross section at right angles to the axis of bending. This is not a frequent condition in concrete beams, but it is a common one with beams of wood or steel or with trusses that perform beam functions. The cross sections shown in Figure 3.25 illustrate members that are relatively susceptible to buckling in beam action.

When buckling is a problem, one solution is to redesign the beam for more resistance to lateral movement. Another possibility is to analyze for the lateral buckling effect and reduce the usable bending capacity as appropriate. However, the solution most often used is to brace the beam against the movement developed by the buckling effect. To visualize where and how such bracing should be done, we must first consider the various possibilities for buckling. The three main forms of beam buckling are shown in Figure 3.26.

Figure 3.26*b* shows the response described as *lateral* (that is *sideways*) *buckling*. This action is caused by the compressive stresses in the top of the beam that make it act like a long column, which is thus subject

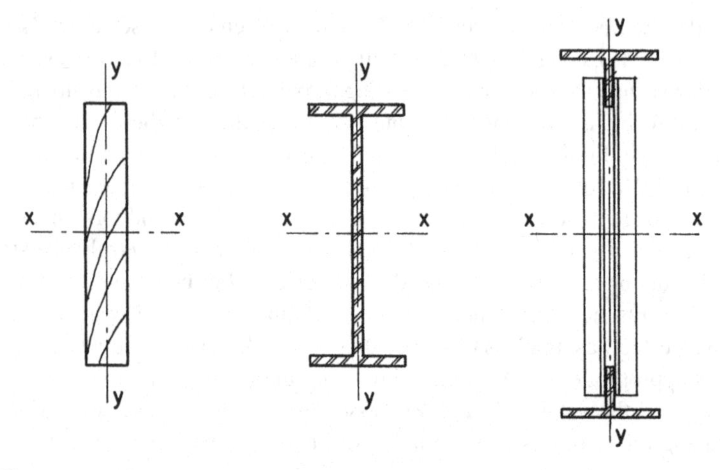

Figure 3.25 Shapes of beams with low resistance to lateral bending and torsion.

to a sideways movement as with any slender column. Bracing the beam for this action means simply preventing its sideways movement at the beam edge where compression exists. For simple span beams, this edge is the top of the beam. For beams, joists, rafters, or trusses that directly support roof or floor decks, the supported deck may provide this bracing if it is adequately attached to the supporting members. For beams that support other beams in a framing system, the supported beams at right angles to the supporting member may provide lateral bracing. In the latter case, the unsupported length of the buckling member becomes the distance between the supported beams, rather than its entire span length.

Another form of buckling for beams is that described as *torsional buckling,* as shown in Figure 3.26*d.* This action may be caused by tension stress, resulting in a rotational, or twisting, effect. This action can occur even when the top of the beam is braced against lateral movement and is often due to a lack of alignment of the plane of the loading and the vertical axis of the beam. Thus, a beam that is slightly tilted is predisposed to a torsional response. An analogy for this is shown in Figure 3.26*e,* which shows a trussed beam with a vertical post at the center of the span. Unless this post is perfectly vertical, a sideways motion at the bottom end of the post is highly likely.

To most positively prevent both lateral and torsional buckling, the beam must be braced sideways at both its top and bottom. If the roof or floor deck is capable of bracing the top of the beam, the only extra brac-

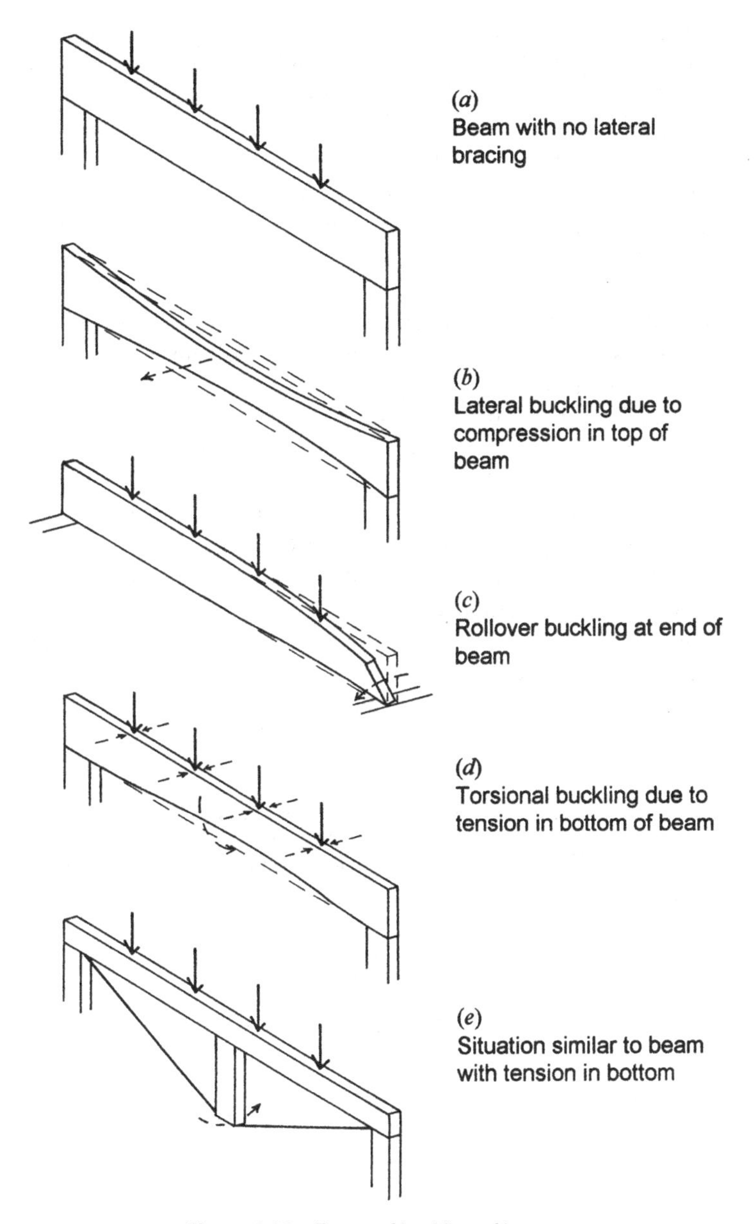

Figure 3.26 Forms of buckling of beams.

ing required is that for the bottom. For closely spaced trusses, this bracing is usually provided by simple horizontal ties between adjacent trusses. For beams in wood or steel framing systems, lateral bracing may be provided as shown in Figure 3.27. The beam shown is braced for both lateral and torsional buckling.

It is not always easy to decide whether a beam is sufficiently braced against buckling. Figure 3.28 shows a number of common situations in which construction is attached to beams, but not all represent sufficient bracing.

In times past, to provide fire protection, concrete was cast around steel members as shown in Figure 3.28*a*. Combining this with a site-cast concrete slab provides a sturdy bracing for the steel beam, although this form of construction is no longer common.

Steel beams are sometimes used for direct support of wood joists, as shown in Figure 3.28*b*. With no apparent attachment between the joist and the supporting beam, this should not be considered to provide adequate lateral bracing for the beam. With a wood nailer attached to the beam, and the joists attached to the nailer—as shown in Figure 3.28*d*—the bracing should be adequate.

Figure 3.28*c* shows a common detail for support of open-web steel joists, with the joist ends welded to the supporting beam. This is also an adequate bracing system, leaving an unsupported length equal to the joist spacing.

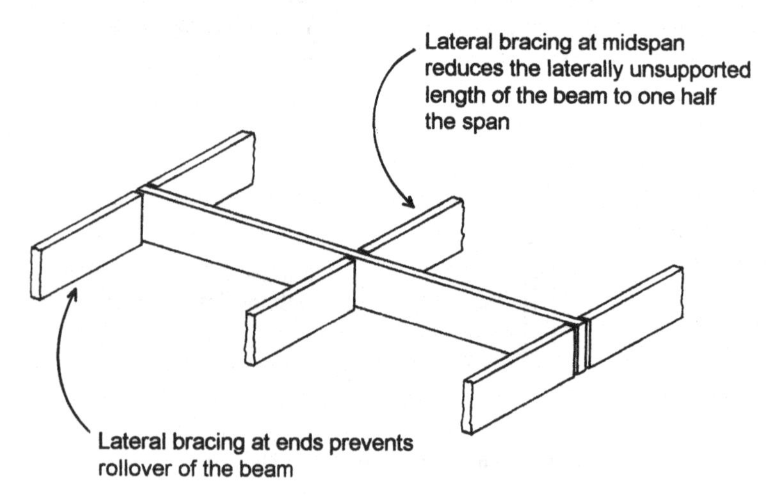

Figure 3.27 Lateral bracing for beams.

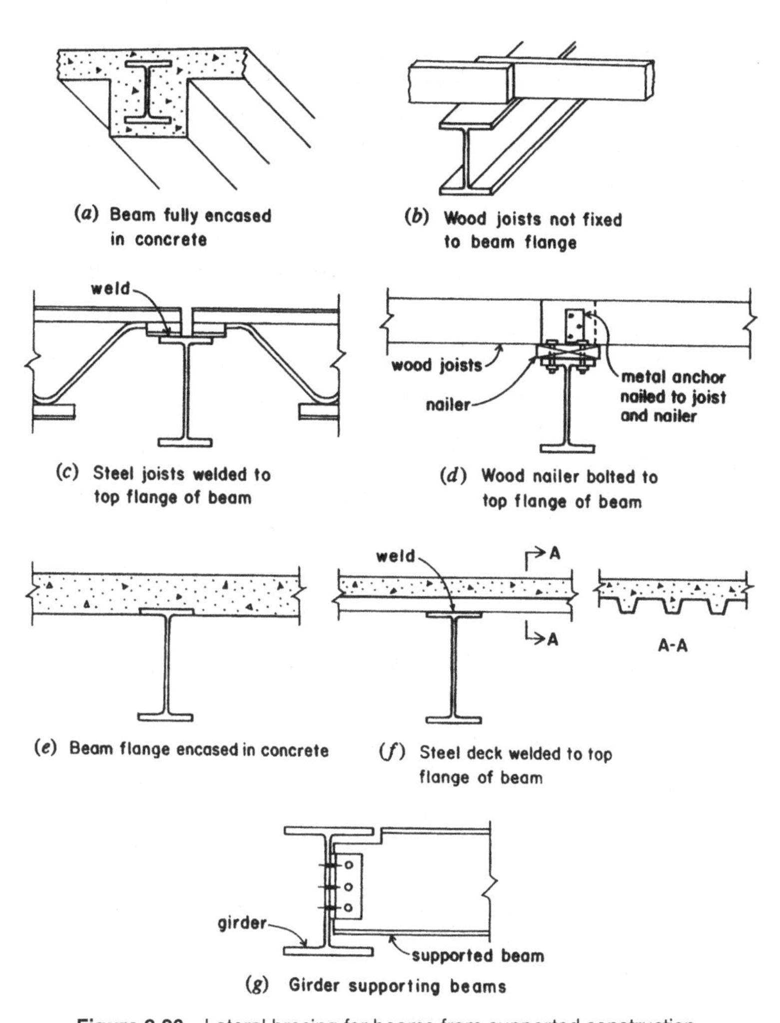

Figure 3.28 Lateral bracing for beams from supported construction.

For beams that support concrete slab decks, a detail sometimes used is that shown in Figure 3.28*e*. In this case, the deck is formed with plywood sheets placed on the underside of the beam flange, which provides a secure lateral bracing by the deck. When composite construction is used, with steel lugs welded to the top of the beam, a secure engagement of the beam and deck is also ensured.

A very common deck construction is that using formed sheet steel units that are attached by welding to the supporting beams, as shown in Figure 3.28*f*. This bracing is effective for the beams that support the deck directly, but it does not work for any beams parallel to the fluted ribs of the deck.

For beams that support other beams, the connection shown in Figure 3.28*g* is common. Although the beam flanges are not directly braced, this connection is usually considered to be adequate for both lateral and torsional bracing.

Lateral support for the top flange of a beam is sufficient for simple beams. However, when bending is reversed—as in the case of cantilever beams, beams with overhanging ends, or beams continuous through multiple spans—some bracing must be provided for the bottom flange of the beam.

Current design specifications are not clear about what is precisely required in the form of construction to provide adequate bracing against buckling of beams. This leaves designers to exercise their own judgment in this regard.

3.14 BEAM BEARING PLATES

Steel beams that are supported on masonry or concrete usually rest directly on a steel bearing plate. The plates have a primary purpose of spreading the bearing compression force over a larger area of contact with the supporting material than that offered by the width of the beam flange. However, even when the flange width is actually adequate for the bearing load, a plate is usually used to facilitate the connection and anchoring of the beam on its support. Figure 3.29 illustrates the form of a beam bearing plate and contains some reference dimensions used in investigation and design of the plate.

Generally, three separate situations must be investigated for this form of connection:

1. The bearing pressure of the plate on the support. This is essentially a matter of concrete or masonry design and is a part of the design of the supporting structure. For the plate design, the only critical concern is for the strength of the supporting material and the ratio of the area of the plate and the effective cross-sectional area of the supporting structure (column, wall, or pier).
2. The bending of the plate.

3. The potential failure of the beam web in yielding or crippling (see Section 3.11). The critical factor in this regard is the dimension *N* in Figure 3.29.

For bearing on a wall, the dimension *N* is usually controlled by the thickness of the wall and is typically limited to about 2 in. less than the wall thickness. If the wall is thin and the bearing load is large, a considerable value for the dimension *B* may be needed. As *B* increases, the bending of the plate requires an increase in the plate thickness, *t*. At some point, increase in *B* may cause the general form of the connection to become unfeasible, and some modification of the supporting structure may be required.

The required thickness for the bearing plate is determined by considering the bending generated by the cantilever distance labeled *n* in Figure 3.29. The load on this cantilever is the uniformly distributed pressure

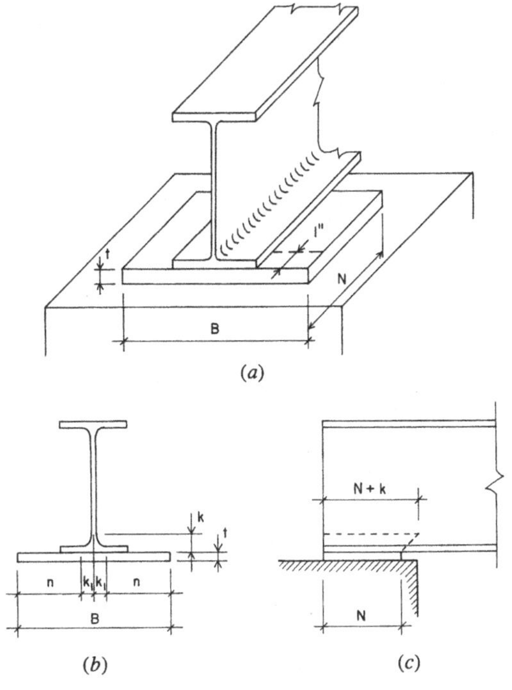

Figure 3.29 Reference dimensions for bearing plates for beams.

on the bottom of the plate. Based on this bending moment, a derived formula for the required plate thickness takes the form

$$t \times \sqrt{\frac{2.22 R_u n^2}{A_1 F_y}}$$

where t is the plate thickness
R_u is the factored end reaction force on the beam
n is the cantilever distance
A_1 is the contact bearing area of the plate ($B \times N$)
F_y is the yield strength of the plate

Using current standards for concrete design, the LRFD method permits a maximum bearing stress on concrete of $0.85\,f'_c$. Thus, with the contact area of A_1, the total bearing load on the concrete is expressed as:

$$\phi_c P_p = 0.6(0.85 f'_c A_1)$$

Equating this to the required bearing load of R_u, an expression for the required area A_1 becomes

$$\phi P_p = 0.6(0.85 f'_c A_1) = R_u$$

$$A_1 = \frac{R_u}{0.6(0.85 f'_c)} = \frac{R_u}{0.51 f'_c}$$

This equation is valid when the supporting member has the same area as the bearing plate. When the supporting member is larger, P_p may be increased by a factor of $\sqrt{A_2/A_1}$ with a maximum value of 2. A_2 is the maximum portion of the supporting surface that is geometrically similar to the bearing plate and concentric to it.

The following example illustrates the design procedure, assuming that $A_1 = A_2$.

Example 17. The end of a steel beam consisting of a W 27 $\times$ 94 rests on top of a concrete wall with $f'_c = 3$ ksi. The factored reaction force for the beam is 180 kips. A bearing plate should be used, for which the dimension N (see Figure 3.29) is limited to 8 in. From the AISC Manual (Ref. 3) the dimension k_1 is 1.0625 in. Find the required dimensions for a steel bearing plate with $F_y = 36$ ksi.

Solution: For the required plate area, assuming full bearing on the support:

$$A_1 = \frac{R_u}{0.51f'_c} = \frac{180}{0.51(3)} = 117.6 \text{ in.}^2$$

Using the full limit of 8 in. for *N*, the required dimension for *B* is

$$B = \frac{117.6}{8} = 14.7 \text{ in.}$$

Rounding this up to 15 in., the actual value for A_1 becomes

$$A_1 = 15 \times 8 = 120 \text{ in.}^2$$

For the cantilever distance *n*

$$n = \frac{B}{2} - k_1 = 7.5 - 1.0625 = 6.44 \text{ in.}$$

The required thickness is thus:

$$t = \sqrt{\frac{2.22R_u n^2}{A_1 F_y}} = \sqrt{\frac{2.22(180)(6.44)^2}{120(36)}} = \sqrt{3.836} = 1.96 \text{ in.}$$

The selected size of the plate is thus: PL $2 \times 8 \times 15$.

Selection of the actual plate size—especially the dimension *B*—would be done as part of the complete design of the beam-to-support connection. It is likely that some anchor bolts would be used to secure the steel beam to the concrete support. If so, there must be room between the edge of the beam flange and the end of the plate to accommodate the bolts. In this case, this dimension is only about 2.5 in., which is probably a little tight. However, increase in the length of the plate produces more bending and requires a thicker plate, so the design must be carefully considered.

If the area of the support is considerably greater than the area of the plate, an increased value can be used for the bearing resistance of the concrete. This can result in a reduction of the plate size.

For a complete design of the beam-to-support connection, the following must also be considered:

1. Vertical compression yielding of the web
2. Web crippling of the beam
3. Compression load capacity of the concrete support (as a column, pilaster, pier, pedestal, and so on)

Beam web yielding and crippling are discussed in Section 3.11. Design of the concrete support is not within the scope of this book.

Problem 3.14.A. The end of a steel beam consisting of a W 24 × 84 rests on top of a concrete wall with $f'_c = 3$ ksi. The factored reaction force for the beam is 200 kips [890 kN]. A bearing plate should be used, for which the dimension N (see Figure 3.29) is limited to 10 in. [254 mm]. From the AISC Manual (Ref. 3) the dimension k_1 is 1.25 in. [32 mm]. Find the required dimensions for an A36 steel plate.

Problem 3.14.B. Same data as Problem 3.14.A, except the beam is a W 21 × 73, the end reaction is 94 kips [418 kN], N is limited to 6 in. [150 mm], and k_1 is 1.44 in. [37 mm].

4

STEEL COLUMNS

Steel compression members range from small, single-piece columns and truss members to huge, built-up sections for high-rise buildings and large tower structures. The basic column function is one of simple compressive force resistance, but it is often complicated by the effects of buckling and the possible presence of bending actions. This chapter deals with various issues relating to the design of individual compression members and with the development of building structural frameworks.

4.1 COLUMN SHAPES

For modest load conditions, the most frequently used shapes are the round pipe, the rectangular tube, and the W shapes with wide flanges (see Figure 4.1). Accommodation of beams for framing is most easily achieved with W shapes of 10-in. or larger nominal depth.

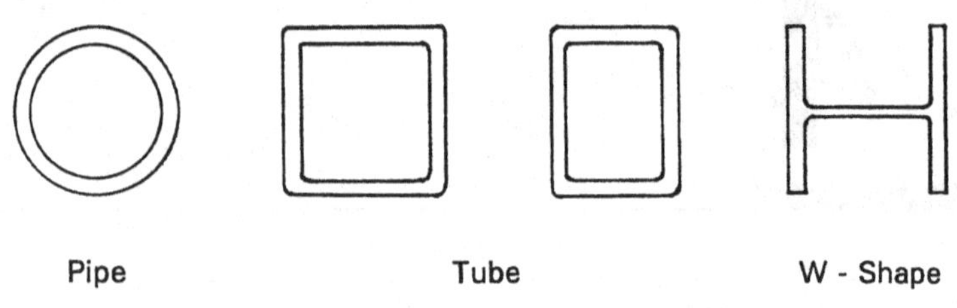

Figure 4.1 Common cross-sectional shapes for steel columns.

For various reasons, it is sometimes necessary to make up a column section by assembling two or more individual steel elements. Figure 4.2 shows some such shapes that are used for special purposes. The customized assemblage of built-up sections is usually costly, so a single piece is typically favored if one is available.

One widely used built-up section is the double angle, shown in Figure 4.2*f*. This occurs most often as a member of a truss or as a bracing member in a frame, the general stability of the paired members being much better than that of a single angle. This section is seldom used as a building column, however.

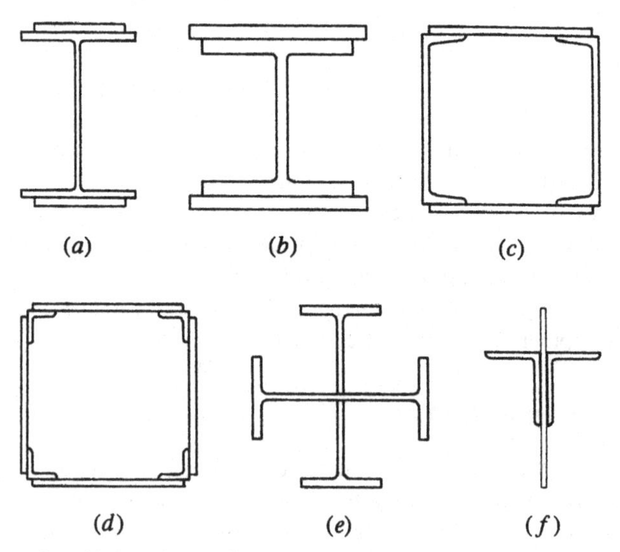

Figure 4.2 Various forms of combined, built-up shapes for steel columns.

4.2 SLENDERNESS AND END CONDITIONS

For steel columns, the value of the critical stress (F_c) in compression is determined from formulas in the AISC Specification; it includes variables of the steel yield stress and modulus of elasticity, the relative slenderness of the column, and special considerations for the restraint at the column ends.

Column slenderness is determined as the ratio of the column unbraced length to the radius of gyration of the column section: L/r. Effects of end restraint are considered by use of a modifying factor (K). (See Figure 4.3.) The modified slenderness is thus expressed as KL/r.

Figure 4.4 is a graph of the critical axial compressive stress for a column for two grades of steel with F_y of 36 ksi and 50 ksi. Values for full-number increments of KL/r are also given in Table 4.1.

For practical reasons, most building columns tend to have relative slenderness between about 50 and 100, with only very heavily loaded

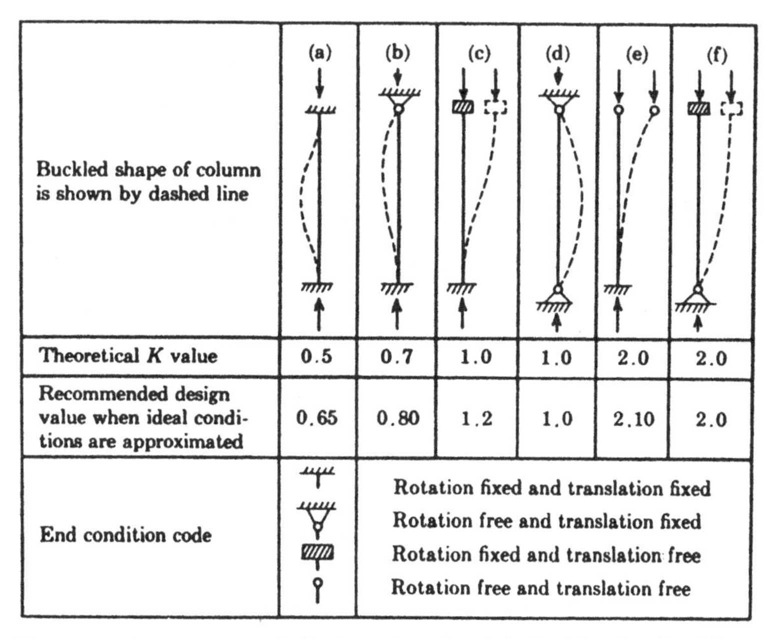

	(a)	(b)	(c)	(d)	(e)	(f)
Buckled shape of column is shown by dashed line						
Theoretical K value	0.5	0.7	1.0	1.0	2.0	2.0
Recommended design value when ideal conditions are approximated	0.65	0.80	1.2	1.0	2.10	2.0

End condition code:
- Rotation fixed and translation fixed
- Rotation free and translation fixed
- Rotation fixed and translation free
- Rotation free and translation free

Figure 4.3 Determination of effective column length for buckling. Reprinted from *Steel Construction Manual,* with permission of the publisher, the American Institute of Steel Construction.

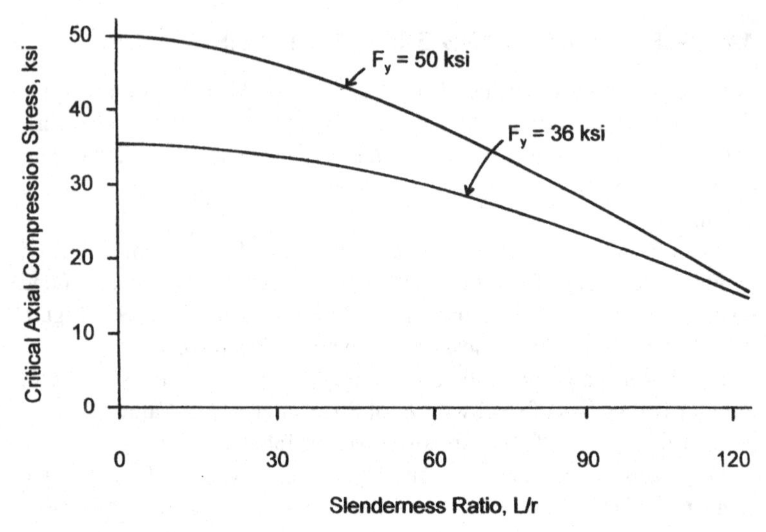

Figure 4.4 Critical unfactored compressive stress for steel columns F_c, as a function of yield stress and column slenderness.

columns falling below this and most designers avoiding use of extremely slender columns. The AISC Specification for steel discourages use of any compression member with a slenderness ratio greater than 200.

4.3 SAFE AXIAL LOADS FOR STEEL COLUMNS

The design strength in axial compression for a column is computed by multiplying the design stress ($\phi_c F_c$) by the cross-sectional area of the column, where $\phi_c = 0.85$. Thus:

$$P_u = \phi_c \times P_n = \phi_c \times F_c \times A$$

where P_u = Maximum factored load
ϕ_c = Resistance factor, 0.85 for columns
P_n = Nominal load resistance (unfactored) of the column
F_c = Critical compressive stress for the column, based on *KL*/*r*
A = Area of the column cross section

The following examples demonstrate the process. For single-piece columns, a more direct process consists of using column load tables. For built-up sections, however, it is necessary to compute the properties of the section.

Example 1. A W 12 × 53 of A36 steel is used as a column with an unbraced length of 16 ft [4.88 m]. Compute the maximum factored load (P_u).

Solution: Referring to Table A.3, A = 15.6 in.^{2}, r_x = 5.23 in., and r_y = 2.48 in. If the column is unbraced on both axes, it is limited by the lower r value for the weak axis. With no stated end conditions, Case (d) in Figure 4.3 is assumed, for which K = 1.0; that is, no modification is made. (This is the unmodified condition.) Thus, the relative stiffness is computed as:

$$\frac{KL}{r} = \frac{1 \times 16 \times 12}{2.48} = 77.4$$

In design work, it is usually considered acceptable to round the slenderness ratio off to the nearest whole number. Thus, with a KL/r value of 77, Table 4.1 yields a value for F_c of 26.3 ksi. The maximum factored load for the column is then:

$$P_u = \phi_c \times F_c \times A = 0.85 \times 26.3 \text{ ksi} \times 15.6 \text{ in.}^2$$
$$= 349 \text{ kips } [1.55 \times 10^3 \text{ kN}]$$

Example 2. Compute the maximum factored load for the column in Example 1 if the top is pinned but prevented from lateral movement and the bottom is totally fixed.

Solution: Referring to Figure 4.3, note that the case for this is (b) and the modifying factor is 0.8. Then:

$$\frac{KL}{r} = \frac{0.8 \times 16 \times 12}{2.48} = 62$$

From Table 4.1, F_c = 29.4 ksi. Then:

$$P_u = \phi_c \times F_c \times A = 0.85 \times 29.4 \text{ ksi} \times 15.6 \text{ in.}^2$$
$$= 390 \text{ kips } [1.73 \times 10^3 \text{ kN}]$$

TABLE 4.1 Critical Unfactored Compressive Stress for Columns, F_c [a]

KL/r	Critical Stress, F_c		KL/r	Critical Stress, F_c		KL/r	Critical Stress, F_c	
	F_y = 36 ksi	F_y = 50 ksi		F_y = 36 ksi	F_y = 50 ksi		F_y = 36 ksi	F_y = 50 ksi
1	36.0	50.0	41	33.0	44.2	81	25.5	30.9
2	36.0	50.0	42	32.8	43.9	82	25.3	30.6
3	36.0	50.0	43	32.7	43.7	83	25.0	30.2
4	36.0	49.9	44	32.5	43.4	84	24.8	29.8
5	36.0	49.9	45	32.4	43.1	85	24.6	29.5
6	35.9	49.9	46	32.2	42.8	86	24.4	29.1
7	35.9	49.8	47	32.0	42.5	87	24.2	28.7
8	35.9	49.8	48	31.9	42.2	88	23.9	28.4
9	35.8	49.7	49	31.7	41.9	89	23.7	28.0
10	35.8	49.6	50	31.6	41.6	90	23.5	27.7
11	35.8	49.6	51	31.4	41.3	91	23.3	27.3
12	35.7	49.5	52	31.2	41.0	92	23.1	26.9
13	35.7	49.4	53	31.1	40.7	93	22.8	26.6
14	35.6	49.3	54	30.9	40.4	94	22.6	26.2
15	35.6	49.2	55	30.7	40.1	95	22.4	25.8
16	35.5	49.1	56	30.5	39.8	96	22.2	25.5
17	35.5	49.0	57	30.3	39.4	97	21.9	25.1
18	35.4	48.8	58	30.2	39.1	98	21.7	24.8
19	35.3	48.7	59	30.0	38.8	99	21.5	24.4

20	35.2	48.6	60	29.8	38.4	100	21.3	24.1
21	35.2	48.4	61	29.6	38.1	101	21.0	23.7
22	35.1	48.3	62	29.4	37.7	102	20.8	23.4
23	35.0	48.1	63	29.2	37.4	103	20.6	23.0
24	34.9	47.9	64	29.0	37.1	104	20.4	22.7
25	34.8	47.8	65	28.8	36.7	105	20.1	22.3
26	34.7	47.6	66	28.6	36.4	106	19.9	22.0
27	34.6	47.4	67	28.4	36.0	107	19.7	21.6
28	34.5	47.2	68	28.2	35.7	108	19.5	21.3
29	34.4	47.0	69	28.0	35.3	109	19.3	21.0
30	34.3	46.8	70	27.8	34.9	110	19.0	20.6
31	34.2	46.6	71	27.6	34.6	111	18.8	20.3
32	34.1	46.4	72	27.4	34.2	112	18.6	20.0
33	34.0	46.2	73	27.2	33.9	113	18.4	19.7
34	33.9	45.9	74	27.0	33.5	114	18.2	19.3
35	33.8	45.7	75	26.8	33.1	115	17.9	19.0
36	33.6	45.5	76	26.6	32.8	116	17.7	18.7
37	33.5	45.2	77	26.3	32.4	117	17.5	18.3
38	33.4	45.0	78	26.1	32.0	118	17.3	18.0
39	33.2	44.7	79	25.9	31.7	119	17.1	17.7
40	33.1	44.5	80	25.7	31.3	120	16.9	17.4

[a] Usable design stress limit in ksi for obtaining of nominal strength (unfactored) of steel columns.

*Source:*Developed from data in the *Manual of Steel Construction* with permission of the publishers, American Institute of Steel Construction.

The following example illustrates the situation where a W shape is braced differently on its two axes.

Example 3. Figure 4.5*a* shows an elevation of the steel framing at the location of an exterior wall. The column is laterally restrained but rotationally free at the top and bottom in both directions. The end condition is as for Case (d) in Figure 4.3. With respect to the *x*-axis of the section, the column is laterally unbraced for its full height. However, the existence of the horizontal framing in the wall plane provides lateral bracing with respect to the *y*-axis of the section; thus, the buckling of the column in this direction takes the form shown in Figure 4.5*b*. If the column is a W 12 × 53 of A36 steel, L_1 is 30 ft [9.15 m], and L_2 is 18 ft [5.49 m], what is the maximum factored compression load?

Solution: The basic procedure here is to investigate both axes separately and to use the highest value for relative slenderness obtained to find the critical stress. (*Note:* This is the same section used in Example 1, for which properties were previously obtained from Table A.3.) For the *x*-axis, the situation is Case (d) from Figure 4.3. Thus:

$$x\text{-axis: } \frac{KL}{r} = \frac{1 \times 30 \times 12}{5.23} = 68.8, \text{ say } 69$$

For the *y*-axis, the situation is also assumed to be Case (d) from Figure 4.3, except that the deformation occurs in two parts (see Figure 4.5*b*). The lower part is used as it has the greater unbraced length. Thus:

$$y\text{-axis: } \frac{KL}{r} = \frac{1 \times 18 \times 12}{2.48} = 87.1, \text{ say } 87$$

Despite the bracing, the column is still critical on its weak axis. From Table 4.1 the value for F_c is 24.2 ksi, and the allowable load is thus:

$$P_u = \phi_c \times F_c \times A = 0.85 \times 24.2 \text{ ksi} \times 15.6 \text{ in.}^2$$
$$= 321 \text{ kips } [1.43 \times 10^3 \text{ kN}]$$

For the following problems, use A36 steel with $F_y = 36$ ksi.

Problem 4.3.A. Determine the maximum factored axial compression load (P_u) for a W 10 × 49 column with an unbraced height of 15 ft [4.57 m]. Assume $K = 1.0$.

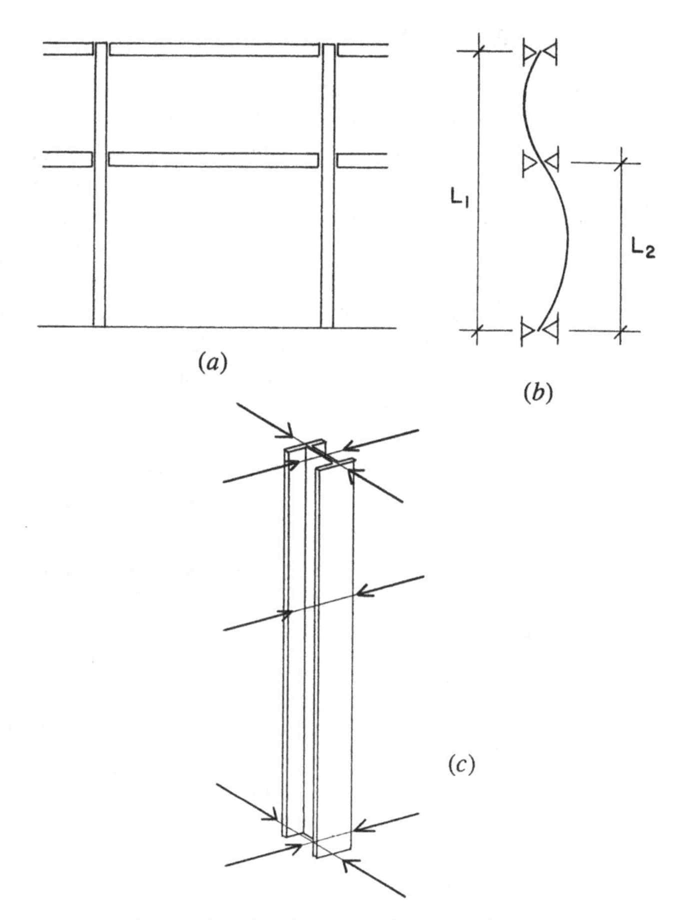

Figure 4.5 Biaxial bracing for steel columns.

Problem 4.3.B. Determine the maximum factored axial compression load (P_u) for a W 12 × 120 column with an unbraced height of 22 ft [6.71 m] if both ends are fixed against rotation and horizontal movement.

Problem 4.3.C. Determine the maximum factored axial compression load (P_u) in Problem 4.3.A if the conditions are as shown in Figure 4.5 with L_1 = 15 ft [4.6 m] and L_2 = 8 ft [2.44 m].

Problem 4.3.D. Determine the maximum factored axial compression load (P_u) in Problem 4.3.B if the conditions are as shown in Figure 4.5 with L_1 = 40 ft [12 m] and L_2 = 22 ft [6.7 m].

4.4 DESIGN OF STEEL COLUMNS

Unless a computer-supported procedure is used, design of steel columns is mostly accomplished through the use of tabulated data. The following discussions in this section consider the latter process, using materials from the AISC Manual (Ref. 3). Design of a column without using load tables is hampered by the fact that the critical stress (F_c) is not able to be precisely determined until *after* the column shape is selected. That is, *KL*/*r* and the associated value for F_c cannot be determined until the critical *r* value for the column is known. This leads to a trial-and-error approach, which can become laborious in even simple circumstances. This process is unavoidable for built-up sections, so their design is unavoidably tedious.

The real value of the safe load tables for single rolled shapes is the ability to use them directly, once only the factored load, column height, and *K* factor are determined. The AISC Manual for LRFD (Ref. 3) provides load tables for W shapes for F_y equal to 50 ksi.

The work that follows in this section provides examples of the use of the AISC Manual tables for selection of columns of a variety of shapes. For the sake of brevity, the examples are limited to a single steel grade, A36, except for the tubular shapes, which are commonly fabricated in a different grade of steel.

In many cases, the simple, axial compression capacity of a column is all that is involved in its design selection. However, columns are also frequently subjected to bending and shear, and in some cases to torsion. In combined actions, however, the axial load capacity is usually included, so its singular determination is still a factor—thus, the safe compression load under simple axial application conditions is included in some way in just about every column design situation.

Single Rolled Shapes as Columns

The single rolled shape most commonly used as a column is the squarish, H-shaped element, with a nominal depth of 8 in. or more. Responding to this, steel rolling mills produce a wide range of sizes in this category. Most are in the W shape series, although some are designated as M shapes.

The AISC Manual (Ref. 3) provides extensive tables with factored loads for W and M shapes. The tables also yield considerable other data for the listed shapes. Table 4.2 summarizes data from the series of AISC tables for shapes ranging from the W 8 × 24 to the W 14 × 211 for steel with yield of 36 ksi. Table 4.3 has similar values for steel with yield of 50 ksi. Table values are based on the r value for the y-axis, with K = 1.0. Also included in this table are values for the bending factor (m), which is used for approximate design for combined compression and bending, as discussed previously in this section.

To illustrate one use of Table 4.2, refer to Example 1 of Section 4.3. For the safe load for the W 12 × 53 with unbraced height of 16 ft, the table yields a value of 348 kips, which agrees closely with the computed load found in the example.

The real value of Tables 4.2 and 4.3, however, is for quick design selections. The basic equation for design of steel columns in LRFD is

$$\phi_c \times P_n \geq P_u$$

where $\phi_c = 0.85$
P_n = nominal column strength
P_u = maximum factored load

Example 4. Using Table 4.2, select a W shape A36 steel column section for an axial load of 100 kips [445 kN] dead load and 150 kips [667 kN] live load if the unbraced length is 24 ft [7.32 m] and the end conditions are pinned at top and bottom.

Solution: The load must be factored using the load combinations to determine the maximum factored load on the column.

$$P_u = 1.4 \times (\text{Dead load})$$
$$= 1.4 \times (100 \text{ kips}) = 140 \text{ kips [623 kN]}$$

or

$$P_u = 1.2 \times (\text{Dead load}) + 1.6 \times (\text{Live load})$$
$$= 1.2 \times (100 \text{ kips}) + 1.6 \times (150 \text{ kips}) = 360 \text{ kips [1600 kN]}$$

From the basic equation for the maximum column load:

$$\phi_c \times P_n \geq P_u = 360 \text{ kips}$$

TABLE 4.2 Safe Factored Loads for Selected A36 W Shapes[a]

	Ratio	Effective Length (KL) in feet																		
$F_y = 36$ ksi	r_x/r_y	0	6	8	10	12	14	16	18	20	22	24	26	28	30	32	34	36	38	40
W 14 × 211	1.61	1897	1866	1842	1812	1776	1734	1687	1636	1580	1520	1458	1392	1325	1257	1187	1118	1048	980	912
W 14 × 176	1.60	1585	1559	1538	1512	1482	1446	1406	1362	1314	1263	1210	1154	1097	1039	980	922	863	805	748
W 14 × 145	1.59	1307	1284	1267	1246	1220	1190	1156	1119	1079	1036	992	945	898	849	800	751	703	655	608
W 14 × 120	1.67	1080	1059	1043	1023	999	971	940	906	870	831	791	749	706	663	620	577	535	494	454
W 14 × 82	2.44	734	703	679	649	615	577	536	493	449	404	361	319	279	243	214	189	169	151	137
W 14 × 68	2.44	612	585	565	540	511	479	444	408	371	334	297	262	229	199	175	155	138	124	112
W 14 × 53	3.07	477	443	418	389	355	319	282	245	210	176	148	126	109	95	83				
W 12 × 336	1.85	3023	2956	2904	2839	2761	2672	2573	2465	2350	2229	2104	1975	1845	1716	1587	1460	1337	1218	1102
W 12 × 279	1.82	2506	2447	2402	2345	2278	2201	2115	2021	1922	1818	1710	1600	1490	1379	1270	1164	1061	960	866
W 12 × 230	1.75	2072	2021	1982	1933	1875	1809	1735	1656	1571	1482	1391	1298	1204	1111	1020	931	845	761	687
W 12 × 190	1.79	1707	1664	1631	1589	1540	1483	1421	1353	1281	1206	1129	1051	973	895	819	745	674	605	546
W 12 × 152	1.77	1368	1332	1304	1270	1229	1182	1130	1074	1015	954	891	827	763	700	638	578	520	467	421
W 12 × 120	1.76	1080	1051	1028	1000	966	928	886	841	793	743	692	640	589	538	489	442	395	355	320
W 12 × 96	1.76	863	839	820	797	770	739	704	667	628	588	546	505	463	422	383	345	308	276	249
W 12 × 79	1.75	710	689	674	654	631	605	576	545	512	479	444	409	375	341	308	277	247	221	200
W 12 × 53	2.11	477	457	441	422	400	375	348	320	292	263	235	207	181	158	139	123	110	98	89
W 12 × 45	2.64	401	373	353	328	301	271	241	210	181	152	128	109	94	82	72				
W 10 × 112	1.74	1007	969	941	906	865	819	768	715	660	604	548	493	440	389	342	303	270	242	219
W 10 × 88	1.73	793	762	739	710	677	639	599	556	511	466	422	378	336	295	259	230	205	184	166
W 10 × 68	1.71	612	588	569	547	520	490	458	424	389	354	319	285	252	221	194	172	153	138	124
W 10 × 54	1.71	483	464	449	431	409	385	360	332	304	276	248	221	195	170	150	133	118	106	96
W 10 × 45	2.15	407	380	361	337	311	282	252	222	192	164	138	118	102	88	78				
W 10 × 33	2.16	297	276	261	243	222	200	177	155	133	112	94	80	69	60	53				
W 8 × 58	1.74	523	492	469	441	409	374	337	300	263	228	194	165	143	124	109	97			
W 8 × 40	1.73	358	335	319	298	275	251	225	198	173	148	125	107	92	80	70	62			
W 8 × 31	1.72	279	261	248	232	214	194	173	153	133	114	96	82	70	61	54				
W 8 × 24	2.12	217	195	180	162	142	122	102	84	68	56	47	40							
Bending Factor (m)			2.2	2.1	2.0	1.9	1.8	1.7	1.6	1.5	1.3	1.3	1.3	1.3	1.3	1.3	1.3	1.3	1.3	1.3

[a] Factored nominal strength of columns in kips.

Source: Developed from data in the *Manual of Steel Construction* with permission of the publishers, American Institute of Steel Construction.

TABLE 4.3 Safe Factored Loads for Selected 50 ksi Yield Stress W Shapes[a]

	Ratio	Effective Length (KL) in feet																		
F_y = 50 ksi	r_x/r_y	0	6	8	10	12	14	16	18	20	22	24	26	28	30	32	34	36	38	40
W 14 × 211	1.61	2635	2575	2530	2473	2405	2326	2239	2145	2043	1937	1827	1715	1601	1487	1374	1264	1156	1052	951
W 14 × 176	1.60	2202	2150	2112	2063	2004	1938	1863	1783	1696	1606	1513	1417	1321	1225	1130	1037	946	859	775
W 14 × 145	1.59	1815	1772	1739	1698	1649	1593	1531	1463	1391	1316	1237	1158	1078	998	919	842	767	694	626
W 14 × 120	1.67	1500	1460	1430	1391	1346	1294	1237	1176	1110	1042	972	902	832	762	694	628	565	507	457
W 14 × 82	2.44	1020	959	914	860	797	729	658	586	514	445	380	324	279	243	214	189	169	151	137
W 14 × 68	2.44	850	798	760	714	662	604	544	484	424	366	311	265	229	199	175	155	138	124	112
W 14 × 53	3.07	663	598	552	498	439	379	319	263	213	176	148	126	109	95	83				
W 12 × 336	1.85	4199	4069	3970	3847	3702	3538	3357	3163	2960	2750	2537	2325	2116	1911	1715	1525	1360	1221	1102
W 12 × 279	1.82	3481	3367	3281	3174	3048	2906	2749	2582	2408	2228	2047	1867	1690	1519	1354	1199	1070	960	866
W 12 × 230	1.75	2877	2779	2706	2614	2505	2383	2250	2107	1959	1807	1654	1503	1354	1212	1073	951	848	761	687
W 12 × 190	1.79	2372	2288	2225	2147	2054	1951	1837	1717	1592	1464	1336	1209	1085	967	853	755	674	605	546
W 12 × 152	1.77	1900	1830	1778	1713	1637	1551	1458	1359	1256	1151	1047	944	844	749	658	583	520	467	421
W 12 × 120	1.76	1500	1443	1401	1347	1285	1215	1139	1059	976	892	808	726	646	569	500	443	395	355	320
W 12 × 96	1.76	1199	1152	1117	1073	1023	966	904	838	771	703	635	569	505	443	390	345	308	276	249
W 12 × 79	1.75	986	947	917	880	838	790	738	683	627	570	514	459	406	355	312	277	247	221	200
W 12 × 53	2.11	663	623	594	559	518	474	428	381	334	290	247	210	181	158	139	123	110	98	89
W 12 × 45	2.64	557	504	466	422	374	324	274	227	185	152	128	109	94	82	72				
W 10 × 112	1.74	1398	1326	1273	1208	1132	1049	961	870	778	688	601	518	447	389	342	303	270	242	219
W 10 × 88	1.73	1101	1042	999	945	884	817	746	672	599	527	458	393	339	295	259	230	205	184	166
W 10 × 68	1.71	850	803	769	727	678	625	569	511	454	398	344	294	254	221	194	172	153	138	124
W 10 × 54	1.71	672	634	606	572	533	490	445	399	353	309	266	227	196	170	150	133	118	106	96
W 10 × 45	2.15	565	515	478	436	388	339	290	243	199	164	138	118	102	88	78				
W 10 × 33	2.16	413	373	345	312	276	238	202	167	135	112	94	80	69	60	53				
W 8 × 58	1.74	727	667	624	572	515	455	394	335	279	231	194	165	143	124	109	97			
W 8 × 40	1.73	497	454	423	386	345	303	260	219	180	149	125	107	92	80	70	62			
W 8 × 31	1.72	388	353	329	299	267	234	200	168	138	114	96	82	70	61	54				
W 8 × 24	2.12	301	260	232	200	168	136	106	84	68	56	47	40							
Bending Factor (m)			2.1	2.0	1.9	1.8	1.7	1.6	1.4	1.3	1.2	1.2	1.2	1.2	1.2	1.2	1.2	1.2	1.2	1.2

[a]Factored nominal strength of columns in kips.

Source: Developed from data in the *Manual of Steel Construction* with permission of the publishers, American Institute of Steel Construction.

From Table 4.2 some possible choices are as follows:

Section	*Design Load* ($\phi_c \times P_n$)
W 10 × 88	422 kips [1880 kN]
W 12 × 79	444 kips [1970 kN]
W 14 × 82	361 kips [1610 kN]

With no additional parameters required by the problem, we will select the W 12 × 79, because it is the most economical section based upon material weight.

Tables 4.2 and 4.3 are set up to work when the y-axis is the most slender—that is, when $K_yL_y/r_y > K_xL_x/r_x$. This is always the case when the end conditions are the same with respect to both axes and when the unbraced lengths are also the same or when the unbraced length is greater about the y-axis than it is about the x-axis. It is a bit trickier if the unbraced length about the x-axis is greater than the unbraced length about the y-axis. For this we need to use another piece of information provided in Tables 4.2 and 4.3—namely, the ratio between r_x/r_y. This ratio varies with each section but is usually within the range of 1.6 to 3.1, with an average value equal to 1.75. If $L_x/L_y \leq r_x/r_y$, then the slenderness ratio about the y-axis still controls and the unbraced length about the y-axis is used. If $L_x/L_y > r_x/r_y$ then the slenderness about the x-axis controls. When this happens, a new equivalent length ($KL_y{}'$) must be used to find the most appropriate steel section for a column. The new equivalent length can be found by using the following equation:

$$KL_y{}' = \frac{KL_x}{r_x/r_y}$$

Example 5. Using Table 4.2, select an A36 steel column section for an axial load of 100 kips [445 kN] dead load and 150 kips [667 kN] live load if the unbraced length is 24 ft [7.32 m] about the x-axis, the unbraced length is 8 ft [2.44 m] about the y-axis, and the end conditions are pinned at top and bottom.

Solution: The load is the same as it is in Example 4; therefore, we already know the maximum factored load is 360 kips. We then determine the ratio of the unbraced lengths about the two axes:

$$\frac{L_x}{L_y} = \frac{24 \text{ ft}}{8 \text{ ft}} = 3.0$$

Looking at Table 4.2, note that only one section has a ratio of r_x/r_y greater than 3.0; therefore, we will assume that the x-axis will control. Next, we determine a new effective length for the column and use it to determine the most appropriate columns. We don't know the exact value for the ratio of r_x/r_y; therefore, we will assume it to be 1.75, because that is an average value from Table 4.2.

$$KL_y{}' = \frac{KL_x}{r_x/r_y} = \frac{1 \times 24 \text{ ft}}{1.75} = 13.7, \text{ say } 14$$

Options from Table 4.2 are as follows:

Section	*Design Load* ($\phi_c \times P_n$)	r_x/r_y	*Actual New Equivalent Length*
W 8 × 58	374 kips [1660 kN]	1.74	13.8 ft [4.21 m]
W 10 × 54	385 kips [1710 kN]	1.71	14.0 ft [4.27 m]
W 12 × 53	375 kips [1670 kN]	2.11	11.4 ft [3.48 m]
W 14 × 68	479 kips [2130 kN]	2.44	9.84 ft [3.0 m]

The actual new equivalent length for each section chosen is less than or equal to the 14 ft [4.27 m] used in the calculation, and each is greater than the unbraced length (KL_y) about the y-axis. Thus, all options are acceptable. Barring other parameters on the design, the most economical section would be the W 12 × 53.

Problem 4.4.A. Using Table 4.2, select a column section for an axial dead load of 60 kips [267 kN] and an axial live load of 88 kips [391 kN] if the unbraced height about both the x- and the y-axes is 12 ft [3.66 m]. A36 steel is to be used and K is assumed as 1.0.

Problem 4.4.B. Select a column section using the same data as in Problem 4.4.A, except the dead load is 103 kips [468 kN] and live load is 155 kips [689 kN]. The unbraced height about the x-axis is 16 ft [4.88 m] and the unbraced height about the y-axis is 12 ft [3.66 m].

Problem 4.4.C. Select a column section using the same data as in Problem 4.4.A, except the dead load is 142 kips [632 kN] and the live load is 213 kips [947 kN]. The unbraced height about the x-axis is 20 ft [6.10 m] and about the y-axis is 10 ft [3.05 m].

Problem 4.4.D. Using Table 4.3, select a column section for an axial dead load of 400 kips [1779 kN] and a live load of 600 kips [2669 kN]. The unbraced height about the x-axis is 16 ft [4.88 m] and about the y-axis is 4 ft [1.22m]. The steel is to have a yield stress of 50 ksi and K is assumed as 1.0.

Steel Pipe Columns

Round steel pipe columns most frequently occur as single-story columns, supporting either wood or steel beams that sit on top of the columns. Pipe is available in three weight categories: *standard* (Std), *extra strong* (XS), and *double-extra strong* (XXS). Pipe is designated with a nominal diameter, which is slightly less than the outside diameter. The outside diameter is the same for all three weight categories, with variation occurring in terms of the pipe wall thickness and inside diameter. See Table A.7 for properties of standard weight pipe. Table 4.4 gives safe loads for pipe columns of steel with a yield stress of 35 ksi.

Example 6. Using Table 4.4, select a standard weight steel pipe to carry a dead load of 15 kips [67 kN] and a live load of 26 kips [116 kN] if the unbraced height is 12 ft [3.66 m].

Solution: The load must be factored using the load combinations to determined the maximum factored load on the column.

$$P_u = 1.4 \times (\text{Dead load})$$
$$= 1.4 \times (15 \text{ kips}) = 21 \text{ kips } [93 \text{ kN}]$$

or

$$P_u = 1.2 \times (\text{Dead load}) + 1.6 \times (\text{Live load})$$
$$= 1.2 \times (15 \text{ kips}) + 1.6 \times (26 \text{ kips}) = 59.6 \text{ kips } [265 \text{ kN}]$$

For the height of 12 ft, the table yields a value of 95 kips as the design load for a 5-in. pipe. A 4-in. pipe is close, but its design strength is 59 kips [262 kN], just short of that required.

TABLE 4.4 Safe Factored Loads for Selected 35 ksi Yield Stress Pipe Columns[a]

			Area	Effective Length (*KL*) in feet																		
$F_y = 35$ ksi			(sq. in.)	0	6	8	10	12	14	16	18	20	22	24	26	28	30	32	34	36	38	40
Pipe	12	XS	19.2	571	563	557	549	540	529	517	503	488	472	455	438	420	401	382	363	343	324	305
Pipe	12	Std.	14.6	434	428	424	418	411	403	394	384	372	361	348	335	321	307	293	279	264	249	235
Pipe	10	XS	16.1	479	469	462	453	442	429	415	400	383	365	347	328	309	290	270	251	232	214	196
Pipe	10	Std.	11.9	354	347	342	335	327	318	308	297	284	272	258	245	231	216	202	188	174	161	148
Pipe	8	XXS	21.3	634	612	596	575	551	524	495	463	430	397	363	329	297	265	235	208	186	166	150
Pipe	8	XS	12.8	381	369	360	348	335	320	303	286	267	248	228	209	190	171	153	136	121	109	98
Pipe	8	Std.	8.4	250	242	237	229	221	211	201	190	178	165	153	140	128	116	104	93	83	75	67
Pipe	6	XXS	15.6	464	436	415	390	361	330	298	264	232	200	170	145	125	109	96	85			
Pipe	6	XS	8.4	250	236	226	214	200	185	169	152	135	119	103	88	76	66	58	52	46		
Pipe	6	Std.	5.6	166	158	151	144	135	125	114	104	93	82	72	62	53	47	41	36	32		
Pipe	5	XXS	11.3	336	307	287	262	235	206	178	150	124	102	86	73	63						
Pipe	5	XS	6.1	182	168	158	146	133	119	104	90	76	63	53	45	39	34					
Pipe	5	Std.	4.3	128	119	112	104	95	85	75	65	56	47	39	33	29	25					
Pipe	4	XXS	8.1	241	209	187	163	137	112	88	70	56	47									
Pipe	4	XS	4.4	131	116	106	94	81	68	55	44	36	30	25								
Pipe	4	Std.	3.2	94	84	77	68	59	50	41	33	27	22	19								
Pipe	3.5	XS	3.7	109	94	83	71	59	47	37	29	23										
Pipe	3.5	Std.	2.7	80	69	61	53	44	36	28	22	18	15									
Pipe	3	XXS	5.5	163	128	106	83	62	46	35												
Pipe	3	XS	3.0	90	73	62	51	40	30	23	18											
Pipe	3	Std.	2.2	66	54	47	38	30	23	17	14											

[a] Factored nominal strength of columns in kips.

Developed from data in the *Manual of Steel Construction* with permission of the publishers, American Institute of Steel Construction.

Problems 4.4.E–H. Select the minimum size standard weight steel pipe for an axial dead load of 20 kips, a live load of 30 kips, and the following unbraced heights: (e) 8 ft; (f) 12 ft; (g) 18 ft; (h) 25 ft.

Structural Tubing Columns

Structural tubing is used for building columns and for members of trusses. Members are available in a range of designated nominal sizes that indicate the actual outer dimensions of the rectangular tube shapes. Within these sizes, various wall thicknesses (the thickness of the steel plates used to make the tubes) are available. For building columns, sizes used range upward from the 3-in. square tube to the largest sizes fabricated (48 in. square at present). Tubing may be specified in various grades of steel. That used for the AISC table, as noted, is 46 ksi.

Table 4.5 yields design strengths for square tubes: 3 in. to 12 in. Use of the tables is similar to that for the other design strength tables.

Problem 4.4.I. A structural tubing column, designated as HHS 4 × 4 × 3/8, of steel with F_y = 46 ksi [317 MPa], is used with an effective unbraced length of 12 ft [3.66 m]. Find the maximum factored axial load.

Problem 4.4.J. A structural tubing column, designated as HHS 3 × 3 × 5/16, of steel with F_y = 46 ksi [317 MPa], is used with an effective unbraced length of 15 ft [4.57 m]. Find the maximum factored axial load.

Problem 4.4.K. Using Table 4.5, select the lightest structural tubing column to carry an axial dead load of 30 kips [133 kN] and a live load of 34 kips [151 kN] if the effective unbraced length is 10 ft [3.05 m].

Problem 4.4.L. Using Table 4.5, select the lightest structural tubing column to carry an axial dead load of 90 kips [400 kN] and a live load of 60 kips [267 kN] if the effective unbraced length is 12 ft [3.66 m].

Double-Angle Compression Members

Matched pairs of angles are frequently used for trusses or for braces in frames. The common form consists of the two angles placed back-to-back but separated a short distance to achieve joints by use of gusset plates or by sandwiching the angles around the web of a structural tee. Compression members that are not columns are frequently called *struts.*

The AISC Manual contains safe load tables for double angles with an assumed average separation distance of ⅜ in. [9.5 mm]. For angles with

TABLE 4.5 Safe Factored Loads for Selected 46 ksi Yield Stress Square Tube Steel Columns[a]

	Area	Effective Length (KL) in feet																		
$F_y = 46$ ksi	(sq. in.)	0	6	8	10	12	14	16	18	20	22	24	26	28	30	32	34	36	38	40
HHS 12 × 12 × 5/8	25.7	1005	989	976	960	941	919	895	867	838	807	774	739	704	668	631	595	558	522	486
HHS 12 × 12 × 3/8	16	626	616	609	599	588	575	560	544	526	507	488	467	446	424	402	379	357	335	313
HHS 12 × 12 × 1/4	10.8	422	416	411	405	397	389	379	368	357	344	331	317	303	289	274	259	244	230	215
HHS 10 × 10 × 1/2	17.2	673	657	645	630	612	592	569	545	519	491	462	433	404	375	346	317	290	263	237
HHS 10 × 10 × 5/16	11.1	434	424	417	408	397	384	370	355	338	321	303	285	266	248	229	211	193	176	160
HHS 10 × 10 × 3/16	6.76	264	259	254	249	242	235	226	217	207	197	186	176	164	153	142	131	121	110	100
HHS 8 × 8 × 1/2	13.5	528	508	494	475	454	430	404	376	347	318	289	260	232	205	181	160	143	128	116
HHS 8 × 8 × 5/16	8.76	343	331	322	310	297	282	266	249	231	212	194	176	158	141	124	110	98	88	79
HHS 8 × 8 × 3/16	5.37	210	203	197	191	183	174	164	154	143	132	121	110	99	89	79	70	62	56	50
HHS 7 × 7 × 5/8	14	547	519	499	473	444	412	377	342	306	271	237	204	176	153	135	119	107	96	86
HHS 7 × 7 × 3/8	8.97	351	334	322	307	289	270	249	227	205	183	162	142	123	107	94	83	74	67	60
HHS 7 × 7 × 1/4	6.17	241	230	222	212	201	188	174	159	145	130	115	101	88	77	68	60	53	48	43
HHS 6 × 6 × 1/2	9.74	381	355	336	313	288	260	231	203	175	148	125	106	92	80	70	62	55		
HHS 6 × 6 × 5/16	6.43	251	236	224	210	194	176	158	140	122	104	88	75	65	56	50	44	39	35	
HHS 6 × 6 × 3/16	3.98	156	146	139	131	121	111	100	89	78	68	58	49	42	37	32	29	26	23	
HHS 5 × 5 × 3/8	6.18	242	219	202	183	162	140	119	98	80	66	56	47	41	36					
HHS 5 × 5 × 1/4	4.3	168	153	142	130	116	101	86	72	59	49	41	35	30	26	23				
HHS 5 × 5 × 1/8	2.23	87	80	75	68	61	54	47	39	33	27	23	19	17	15	13				
HHS 4 × 4 × 3/8	4.78	187	159	140	119	98	78	60	47	38	32	27								
HHS 4 × 4 × 1/4	3.37	132	113	101	87	72	58	45	36	29	24	20								
HHS 4 × 4 × 1/8	1.77	69	60	54	47	40	32	26	20	16	14	11	10							
HHS 3 × 3 × 5/16	3.52	138	116	101	85	69	54	41	32	26	22									
HHS 3 × 3 × 3/16	2.24	88	75	66	57	47	37	29	23	18	15	13								

[a] Factored nominal strength of columns in kips.

Source: Developed from data in the *Manual of Steel Construction* with permission of the publishers, American Institute of Steel Construction.

unequal legs, two back-to-back arrangements are possible, described either as *long legs back-to-back* or as *short legs back-to-back.* Table 4.6 has been adapted from data in the AISC tables for selected pairs of double angles with long legs back-to-back. Note that separate data is provided for the variable situation of either axis being used for the determination of the effective unbraced length. If conditions relating to the unbraced length are the same for both axes, then the lower value for the safe load from Table 4.6 must be used. Properties for selected double angles with long legs back-to-back are given in Table A.6.

Like other members that lack biaxial symmetry, such as the structural tee, there may be some reduction applicable due to slenderness of the thin elements of the cross section. This reduction is incorporated in the values provided in the AISC safe load tables.

Problem 4.4.M. A double-angle compression member 8 ft [2.44 m] long is composed of two A36 steel angles 4 × 3 × ⅜ in. with the long legs back-to-back. Determine the maximum factored axial compression load for the angles.

Problem 4.4.N. A double-angle compression member 12 ft [3.66 m] long is composed of two A36 steel angles 6 × 4 × ½ in. with the long legs back-to-back. Determine the maximum factored axial compression load for the angles.

Problem 4.4.O. Using Table 4.6, select a double-angle compression member for an axial compression dead load of 25 kips [111 kN] and a live load of 25 kips [111 kN] if the effective unbraced length is 10 ft [3.05 m].

Problem 4.4.P. Using Table 4.6, select a double-angle compression member for an axial compression dead load of 75 kips [334 kN] and a live load of 100 kips [445 kN] if the effective unbraced length is 16 ft [4.88 m].

4.5 COLUMNS WITH BENDING

Steel columns must frequently sustain bending in addition to the usual axial compression. Figures 4.6*a* through *c* show three of the most common situations that result in this combined effect. When loads are supported on a bracket at the column face, the eccentricity of the compression adds a bending effect (Figure 4.6*a*). When moment-resistive connections are used to produce a rigid frame, any load on the beams will induce a

TABLE 4.6 Safe Factored Loads for Double-Angle Compression Members of A36 Steel[a]

Size (in.)			4 × 3			3.5 × 2.5			3 × 2		2.5 × 2	
Thickness (In.)			3/8	5/16		5/16	1/4		5/16	1/4	5/16	1/4
Weight (Lb/ft)			16.9	14.2		12.2	9.88		10.1	8.18	8.97	7.30
Area (in.^{2})			4.98	4.19		3.58	2.90		2.96	2.40	2.64	2.14
r_x			1.26	1.27		1.11	1.12		0.945	0.953	0.774	0.782
r_y			1.30	1.29		1.09	1.08		0.897	0.883	0.943	0.930
Effective Buckling Length in Feet (L/r) with Respect to Indicated Axis	X-X Axis	0	152	128	0	110	86	0	90.6	73.4	80.8	65.5
		4	141	119	2	107	84	2	87.6	71.0	76.8	62.3
		6	128	108	4	99	78	3	83.9	68.1	72.1	58.6
		8	112	95	6	88	69	4	79.1	64.3	66.0	53.7
		10	94	80	8	74	59	5	73.3	59.6	58.9	48.0
		12	77	65	10	59	48	6	66.7	54.4	51.2	41.9
		14	60	51	12	45	37	8	52.6	43.0	35.9	29.6
		16	46	39	14	33	27	10	38.8	31.9	23.4	19.4
		18	36	31	16	25	21	12	27.2	22.4	16.3	13.5
		20	29	25	18	20	17	14	20.0	16.5	-	-
	Y-Y Axis	0	152	128	0	110	86	0	90.6	73.4	80.8	65.5
		4	130	104	2	96	71	2	80.7	62.4	74.4	58.6
		6	118	95	4	89	66	3	76.8	59.5	71.0	55.9
		8	104	84	6	78	59	4	71.6	55.6	66.5	52.4
		10	88	72	8	65	50	5	65.4	50.9	61.1	48.2
		12	71	58	10	51	40	6	58.5	45.7	55.1	43.5
		14	55	46	12	38	30	8	43.9	34.4	42.2	33.4
		16	43	35	14	29	22	10	31.9	24.9	31.5	24.8
		18	34	28	16	23	17	12	22.3	17.5	22.0	17.4
		20	28	23	18	18	14	14	16.5	12.9	16.2	12.8

Size (in.)			8 × 6			6 × 4			5 × 3.5		5 × 3	
Thickness (In.)			3/4	1/2		1/2	3/8		1/2	3/8	3/8	5/16
Weight (Lb/ft)			68.0	46.3		32.3	24.6		27.2	20.8	19.5	16.4
Area (in.^{2})			20.0	13.6		9.50	7.22		8.01	6.10	5.73	4.81
r_x			2.52	2.55		1.91	1.93		1.58	1.59	1.60	1.61
r_y			2.47	2.43		1.64	1.61		1.48	1.46	1.22	1.21
Effective Buckling Length in Feet (L/r) with Respect to Indicated Axis	X-X Axis	0	612	379	0	291	201	0	245	183	172	134
		10	543	341	10	236	167	6	220	165	155	122
		12	515	325	12	216	154	8	202	152	143	113
		14	484	308	14	193	140	10	181	137	129	103
		16	451	289	16	171	125	12	158	120	113	91
		20	380	248	18	148	110	14	135	103	97	79
		24	308	206	20	127	96	16	113	86	82	68
		28	240	165	22	106	82	18	91	70	67	57
		32	184	128	26	76	59	22	61	47	45	38
		36	145	101	30	57	44	26	44	34	32	27
	Y-Y Axis	0	612	379	0	291	201	0	245	183	172	134
		10	493	276	6	231	148	4	213	149	135	100
		12	467	264	10	194	128	6	199	140	123	92
		14	438	251	12	172	115	8	180	127	108	82
		16	406	236	14	148	101	10	158	113	90	70
		20	338	203	16	125	87	12	135	97	72	57
		24	269	167	18	102	73	14	111	81	55	45
		28	205	131	20	83	60	16	89	65	43	35
		32	158	102	22	69	50	18	71	52	34	28
		36	126	82	26	50	36	22	48	35		

[a]Factored nominal axial compression strength for members in kips.

Source: Developed from data in the *Manual of Steel Construction* with permission of the publishers, American Institute of Steel Construction.

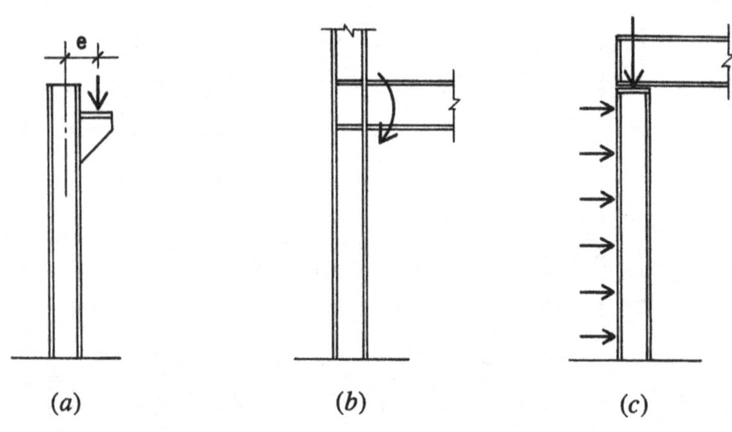

Figure 4.6 Considerations for bending in steel columns.

twisting (bending) effect on the columns (Figure 4.6*b*). Columns built into exterior walls (a common occurrence) may become involved in the spanning effect of the wall in resisting wind forces (Figure 4.6*c*).

Adding bending to a direct compression effect results in a combined stress, or net stress, condition, with something other than an even distribution of stress across the column cross section. The two effects may be investigated separately and the stresses added to determine the net effect. However, the two *actions*—compression and bending—are essentially different, so that a combination of the separate actions, not just the stresses, is more significant. This combination is accomplished with the so-called *interaction* analysis that takes the form of:

$$\frac{P_u}{\phi_c P_n} + \frac{M_{ux}}{\phi_b M_{nx}} + \frac{M_{uy}}{\phi_b M_{ny}} \leq 1$$

On a graph, the interaction formula describes a straight line, which is the classical form of the relationship in elastic theory. However, variations from the straight-line form occur because of special conditions having to do with the nature of the materials, the usual form of columns, and some recognition of usual fabrication and construction practices.

For steel columns, major issues include slenderness of column flanges and webs (for W shapes), ductility of the steel, and overall column slenderness that effects potential buckling in both axial compression and bending. Understandably, the AISC formulas are considerably more complex than the simple interaction formula above, reflecting these as well as other concerns. The AISC interaction formulas for compression and bending are as follows:

If $P_u/\phi_c P_n \geq 0.2$,

$$\frac{P_u}{\phi_c P_n} + \frac{8}{9}\left[\frac{M_{ux}}{\phi_b M_{nx}} + \frac{M_{uy}}{\phi_b M_{ny}}\right] \leq 1.0$$

If $P_u/\phi_c P_n < 0.2$,

$$\frac{P_u}{2 \times \phi_c P_n} + \left[\frac{M_{ux}}{\phi_b M_{nx}} + \frac{M_{uy}}{\phi_b M_{ny}}\right] \leq 1.0$$

Another potential problem with combined compression and bending is that of the *P*-delta effect. This occurs when a relatively slender column that is subjected to compression plus bending is curved significantly by the bending effect. The deflection due to this curvature (called delta, Δ) results in an eccentricity of the compression force, and thus is an added bending moment equal to the product of *P* times delta. Any bending of the column can produce this effect; an especially critical case is a free-standing, towerlike column with no top restraint and a load on its top. Obviously, a very stiff column with little deflection will not suffer much from the *P*-delta effect, whereas a very slender column may be quite vulnerable.

For use in preliminary design work, or to quickly obtain a first trial section for use in a more extensive design investigation, a procedure developed by Uang, Wattar, and Leet (Ref. 8) may be used that involves the determination of an equivalent axial load that incorporates the bending effect. This is accomplished by use of a bending factor *(m)*, which is listed here for the W shapes in Tables 4.2 and 4.3 at the bottom of the tables. Using this factor, the equivalent axial load P' is obtained as:

$$P_u' = P_u + m \times M_{ux} + (2 \times m) \times M_{uy}$$

If $P_u/P_u' < 0.2$ recalculate P_u' using the following equation:

$$P_u' = \frac{P_u}{2} + \frac{9}{8} \times [m \times M_{ux} + (2 \times m) \times M_{uy}]$$

where P_u' = equivalent factored axial compression load for design in kips

P_u = actual factored compression load in kips

m = bending factor

M_x = bending moment about the column x-axis in kip-ft

M_y = bending moment about the column y-axis in kip-ft

The following examples illustrate the use of this approximation method.

Example 7. We want to use a 10-in. W shape for a column in a situation such as that shown in Figure 4.7. The factored axial load from above on the column is 175 kips [778 kN], and the factored beam load at the column face is 35 kips [156 kN]. The column has an unbraced height of 16 ft [4.88 m] and a K factor of 1.0. Select a trial section for the column.

Solution: Because the loading has already been factored, in order not to confuse matters more with this equation, we do not need to factor it as we have in earlier examples. Determining the axial load capacity and bending capacities will also not be illustrated. They were determined using methods shown earlier in this text.

First, we will determine an equivalent axial load. The bending factor (m) is equal to 1.7 and is taken from Table 4.2.

$$
\begin{aligned}
P_u' &= P_u + m \times M_{ux} \\
&= (175 \text{ kips} + 35 \text{ kips}) + (1.7 \times 35 \text{ kips} \times 5 \text{ in.} \times \text{ft}/12 \text{ in.}) \\
&= 210 \text{ kips} + 24.8 \text{ kips} = 235 \text{ kips } [1050 \text{ kN}]
\end{aligned}
$$

Check to see if the correct equation for equivalent axial load was used:

$$
\frac{P_u}{P_u'} = \frac{210 \text{ kips}}{235 \text{ kips}} = 0.9 > 0.2
$$

Therefore, the correct equation was used.

Using $P_u' = 235$ kips, we use Table 4.2 to determine a trial member size of W 10 × 45. Next, check the section for compliance with the AISC interaction formula for axial compression and bending:

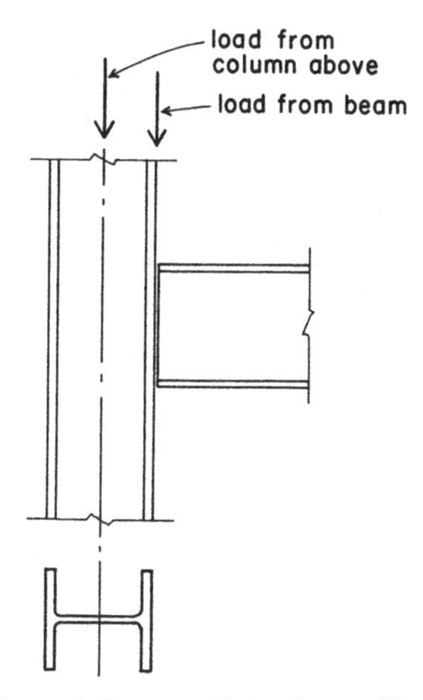

Figure 4.7 Development of an eccentric loading condition with steel framing.

$$\phi_c P_n = 252 \text{ kips (Table 4.2)}$$

$$\phi_b M_{nx} = 175 \text{ kip-ft } (L_p < [L = 16 \text{ ft}] < L_r)$$

$$\frac{P_u}{\phi_c P_n} = \frac{210 \text{ kips}}{252 \text{ kips}} = 0.833 > 0.2$$

$$\frac{P_u}{\phi_c P_n} + \frac{8}{9}\left[\frac{M_{ux}}{\phi_b M_{nx}} + \frac{M_{uy}}{\phi_b M_{ny}}\right]$$

$$= \frac{210 \text{ kips}}{252 \text{ kips}} + \frac{8}{9}\left[\frac{14.6 \text{ kip-ft}}{175 \text{ kip-ft}} + 0\right]$$

$$= 0.907 < 1$$

A W 10 × 45 is an acceptable option for this loading.

Although this process may seem laborious, be assured that direct use of the AISC formulas is considerably more laborious.

When bending occurs about both axes, as it does in full three-dimensional rigid frames, all three parts of the approximation formula must be used. The following example demonstrates the process for this.

Example 8. We want to use a 12-in. [305 mm] W shape for a column that sustains the following factored loading: axial load of 85 kips [378 kN], M_{ux} = 100 kip-ft [136 kN-m], M_{uy} = 75 kip-ft [102 kN-m]. Select a column made from 50-ksi steel for an unbraced height of 12 ft [3.66 m].

Solution: In Table 4.3, observe that the bending factor for a 12-ft length is 1.8. Thus:

$$P_u' = P_u + m \times M_{ux} + (2 \times m) \times M_{uy}$$

$$= (85 \text{ kips}) + (1.8 \times 100 \text{ kip-ft}) + (2 \times 1.8 \times 75 \text{ kip-ft})$$

$$= 85 \text{ kips} + 180 \text{ kips} + 270 \text{ kips} = 535 \text{ kips } [2380 \text{ kN}]$$

Check to see if the correct equation for equivalent axial load was used:

$$\frac{P_u}{P_u'} = \frac{85 \text{ kips}}{535 \text{ kips}} = 0.159 < 0.2$$

Therefore, the incorrect equation was used.

Recalculate P_u':

$$P_u' = \frac{P_u}{2} + \frac{9}{8} \times [m \times M_{ux} + (2 \times m) \times M_{uy}]$$

$$= \frac{85 \text{ kips}}{2} + \frac{9}{8} \times [(1.8 \times 100 \text{ kip-ft}) + (2 \times 1.8 \times 75 \text{ kip-ft})]$$

$$= 42.5 \text{ kips} + \frac{9}{8} \times (180 \text{ kips} + 270 \text{ kips}) = 549 \text{ kips } [2440 \text{ kN}]$$

Using P_u' = 549 kips, use Table 4.3 to determine a trial member size of W 12 × 79. Next, we check the section for compliance with the AISC interaction formula for axial compression and bending:

$$\phi_c P_n = 838 \text{ kips (Table 4.3)}$$

$$\phi_b M_{nx} = 706 \text{ kip-ft } (L_p < [L = 12 \text{ ft}] < L_r)$$

$$\phi_b M_{ny} = 201 \text{ kip-ft } (\Phi_b M_{py})$$

$$\frac{P_u}{\phi_c P_n} = \frac{85 \text{ kips}}{838 \text{ kips}} = 0.101 < 0.2$$

$$\frac{P_u}{2\phi_b P_n} + \left[\frac{M_{ux}}{\phi_b M_{nx}} + \frac{M_{uy}}{\phi_b M_{ny}}\right]$$

$$= \frac{85 \text{ kips}}{2 \times 838 \text{ kips}} + \left[\frac{100 \text{ kip-ft}}{706 \text{ kip-ft}} + \frac{75 \text{ kip-ft}}{201 \text{ kip-ft}}\right]$$

$$= 0.0507 + 0.142 + 0.373$$

$$= 0.565 < 1$$

A W 12 × 79 is an acceptable option for this loading, though this loading is using only 56.5 percent of its capacity. Try a W 12 × 53 and check it against the AISC interaction formulas:

$$\phi_c P_n = 518 \text{ kips (Table 4.3)}$$

$$\phi_b M_{nx} = 287 \text{ kip-ft } (L_p < [L = 12 \text{ ft}] < L_r)$$

$$\phi_b M_{ny} = 108 \text{ kip-ft } (\Phi_b M_{py})$$

$$\frac{P_u}{\phi_c P_n} = \frac{85 \text{ kips}}{518 \text{ kips}} = 0.164 < 0.2$$

$$\frac{P_u}{2\phi_b P_n} + \left[\frac{M_{ux}}{\phi_b M_{nx}} + \frac{M_{uy}}{\phi_b M_{ny}}\right]$$

$$= \frac{85 \text{ kips}}{2 \times 518 \text{ kips}} + \left[\frac{100 \text{ kip-ft}}{287 \text{ kip-ft}} + \frac{75 \text{ kip-ft}}{108 \text{ kip-ft}}\right]$$

$$= 0.0820 + 0.348 + 0.694$$

$$= 1.124 > 1$$

The W 12 × 53 does not work for this given loading. Therefore, the W 12 × 79 is the most appropriate section. (*Note:* We have limited choices for the example to those listed in Tables 4.2 and 4.3. However, there are many more shapes listed in the reference document, the AISC Manual, Ref. 3.)

A major occurrence of the condition of columns with bending is that of columns in rigid frame bents. For steel columns, this typically means the use of a whole steel frame with beams in two directions and with columns having beams framing into both sides—or, indeed, into *all* sides for interior columns.

Problem 4.5.A. We want to use a 12-in. [305-mm] W shape for a column to support a beam as shown in Figure 4.6. Select a trial size for the column for the following data: column factored axial load from above = 200 kips [890 kN], factored beam reaction = 30 kips [133 kN], and unbraced column height is 14 ft [4.27 m].

Problem 4.5.B. Check the section found in Problem 4.5.A to see if it complies with the AISC interaction formulas for axial compression and bending.

Problem 4.5.C. Same as Problem 4.5.A, except factored axial load is 485 kips [2157 kN], factored beam reaction is 100 kips [445 kN], and unbraced height is 18 ft [5.49 m].

Problem 4.5.D. Check the section found in Problem 4.5.C to see if it complies with the AISC interaction formulas for axial compression and bending.

Problem 4.5.E. A 14-in. [356-mm] W shape is to be used for a column that sustains bending on both axes. Select a trial section for the column for the following factored data: total axial load = 80 kips [356 kN], M_x = 85 kip-ft [115 kN-m], M_y = 64 kip-ft [87 kN-m], and unbraced column height is 16 ft [4.88 m].

Problem 4.5.F. Same as Problem 4.5.E, except axial load is 200 kips [890 kN], M_x is 45 kip-ft [61 kN-m], M_y is 30 kip-ft [41 kN-m], and unbraced height is 35 ft [10.7 m].

4.6 COLUMN FRAMING AND CONNECTIONS

Connection details for columns must be developed with considerations of the column shape and size; the shape, size, and orientation of other framing; and the particular structural functions of the joints. Some common forms of simple connections for light frames are shown in Figure 4.8. The usual means for attachment are by welding, by bolting with high-strength steel bolts, or with anchor bolts embedded in concrete or masonry.

When beams sit directly on top of a column (Figure 4.8*a*), the usual solution is to weld a bearing plate on top of the column and bolt the bottom flanges of the beam to the plate. For this, and for all connections, it is necessary to consider what parts of the connection are achieved in the fabrication shop and what is achieved as part of the erection of the frame at the job site (called the *field*). In this case, it is likely that the plate will be attached to the column in the shop (where welding is preferred), and the beam will be attached in the field (where bolting is preferred). In this joint the plate serves no special structural function, because the beam could theoretically bear directly on top of the column. However, field assembly of the frame works better with the plate, and the plate also probably helps to spread the bearing stress more fully over the column cross section.

In many situations, beams must frame into the side of a column. If simple transfer of vertical force is all that is basically required, a common solution is the connection shown in Figure 4.8*b*, in which a pair of steel angles is used to connect the beam web to the column face. With minor

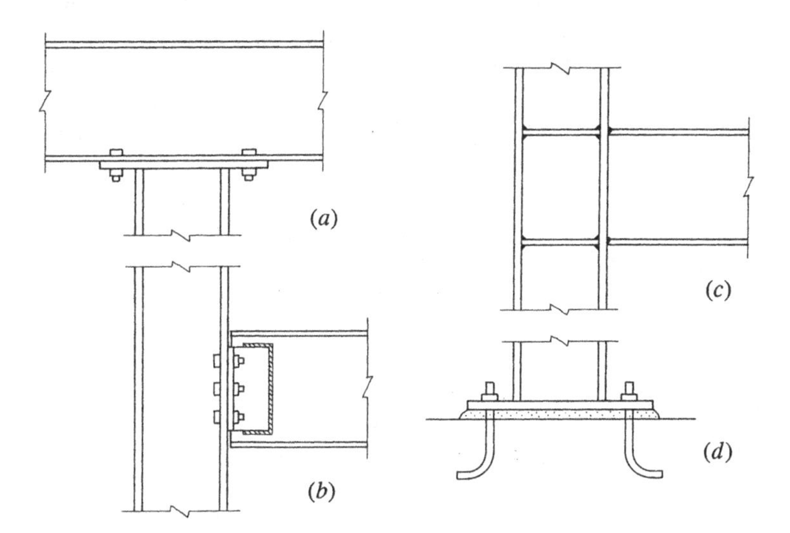

Figure 4.8 Typical fabrication details for steel columns in lightly loaded frames.

variation, this form of connection can also be used to connect a beam to the column web when framing intersects the column differently. When the latter is the case, the outspread legs of the angles must fit between the column flanges, which generally requires at least a 10-in. W shape column—thus, the popularity of the 10-, 12-, and 14-in. W shapes for columns.

If bending moment must be transferred between a beam's end and its supporting column, a common solution is to weld the cut ends of the beam flanges directly to the column face, as shown in Figure 4.8*c*. Because the bending must be developed in both column flanges, and the beam directly grabs only one, the filler plates shown are often used for a more effective transfer of the bending from the beam. This leaves the beam web as yet unconnected, so some attachment must also be made there, because the beam web actually carries most of the beam shear force. Although common for years and still widely used for gravity and wind loads, this form of connection has recently received a lot of scrutiny because of poor performance in earthquakes, and some refinements are surely in order if it is used for this load condition.

At the column bottom, where bearing is commonly on top of a concrete pier or footing, the major concern is for reduction of the bearing pressure on the soft concrete. With upwards of 20 ksi or more of compression in the column steel, and possibly little over 1000 psi of resistance in the concrete, the contact bearing area must be quite spread out. For this reason, as well as the simple practical one of holding the column in place, the common solution is a steel bearing plate attached to the column in the shop and made to bear on a filler material between the smooth plate and the rough concrete surface (see Figure 4.8*d*). This form of connection is adequate for lightly loaded columns, and a design for a column base plate is presented in the next section. For transfer of very high column loads, development of uplift or bending moment resistance, or other special concerns, this joint can receive a lot of special modification. Still, the simple joint shown is the most common form.

For tall steel frames, an issue that must be dealt with is the splicing of columns. All rolling mills have some limitations on how long a single piece can be rolled. This also depends on the practical concerns of handling the finished rolled product. For example, if a very small W shape with a significantly weak *y*-axis is picked up, its own weight can cause it to bend permanently if it is excessively long. In any event, there are practical limitations on length for elements delivered to the job site for erection. On the other hand, *all* joints are relatively expensive with regard to fabrication and erection costs. Thus, the frame with a *minimum* number of joints is likely to be the least expensive. For tall columns, therefore, it is common to use a single piece for as long a vertical distance as possible—surely at least two stories in most multistory construction.

Figure 4.9 shows some means for achieving splices in stacked pieces for a tall column. In these joints, transfer of vertical load is made by having the upper column part bear directly on the end of the lower column part, the bolted plates serving primarily to keep them from separating by slipping sideways. A special problem with this form of joint is the matching of the upper and lower parts to permit a contact bearing. For example, the joint in Figure 4.9 only works if the two column parts have very close to the same overall depth (out-to-out dimension for the flanges) and the web of the upper part is thinner than the web of the lower part.

If the matchup of the two column parts is as shown in Figure 4.9*b*, where the *inside* dimension between flanges is the same, the bearing contact still works, but there is a gap between the splice plates and the face of the flanges in the upper part. This could be solved by using filler plates, as shown in the figure. In fact, this case is the more common one, because this dimension (inside-to-inside of column flanges) tends to remain quite close to the same for groups of W shapes (see Table A.3).

If a major column size change is made—for example, a switch from a 10-in. upper column to a 12-in. lower one—the column flanges may be totally unmatched, as shown in Figure 4.9. In this case, a bearing connection may be achieved through the use of a very thick bearing plate, made to cover the whole top of the lower column.

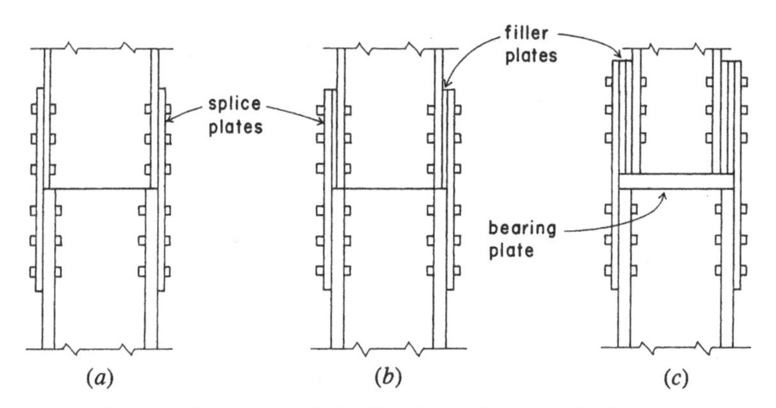

Figure 4.9 Typical bolted splices for multistory steel columns.

It is also possible to develop column splice joints as completely welded ones. This is more likely to be done when the structure is developed as a rigid frame, but it may also simply be a preferred method by individual steel fabricators or erection contractors.

Various considerations for development of both bolted and welded connections for steel framing are treated in Chapter 10. This subject is treated at length in the AISC manuals and in other AISC publications.

4.7 COLUMN BASES

Force transfer to supporting materials at the base of a steel column frequently involves only direct bearing pressure. In this case, a base plate is used as shown in Figure 4.8*d*. For very light column loads, it is not uncommon that the area required for this plate is actually less than that of the column's footprint (product of the shape depth times the flange width); in this event, a larger plate must be used, if for no other reason than to facilitate the placement of anchor bolts and the welding of the column to the base plate. The following procedure may be used to determine the required dimensions for a column base plate, assuming the required plate area to be larger than the column footprint.

The AISC Specification gives the design capacity of a column base plate as

$$\phi_c P_p = \phi_c [0.85 f'_c A_1]$$

or

$$\phi_c P_p = \phi_c [0.85 f'_c A_1] \sqrt{A_2 / A_1}$$

where ϕ_c = 0.60
f'_c = ultimate compressive strength of the supporting concrete
A_1 = area of the steel bearing plate
A_2 = maximum area of the support that is geometrically similar to and concentric with the plate

$$\sqrt{A_2 / A_1} \leq 2$$

For a W-shape column, the basis for determining the thickness of the plate due to bending is shown in Figure 4.10. Once the required value for A_1 is found, the dimensions B and N are selected so that the projections

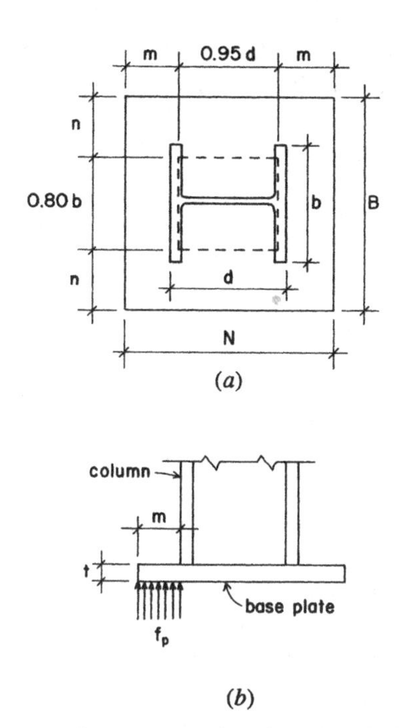

Figure 4.10 Reference dimensions for column base plates.

m and n are approximately equal. Choice of dimensions must also relate to the locations for anchor bolts and for the development of attachment of the plate to the column.

The required thickness, based on flexure, is determined with the following AISC formula:

$$t_{min} = L\sqrt{\frac{2P_u}{0.9F_yBN}}$$

where P_u = required ultimate compression load
L = larger of m or n (see Figure 4.10)
F_y = yield strength of the plate
BN = area of the plate (A_1)

The following example illustrates the process for a column with a relatively light load.

Example 9. Design a base plate of A36 steel for a W 10×49 column with an ultimate load of 350 kips. The column bears on a concrete footing with $f'_c = 3$ ksi.

Solution: Assuming the footing plan area to be considerably larger than the area of the base plate, the maximum value of 2 may be used for the factor $\sqrt{A_2/A_1}$. Then:

$$P_u = 350 \text{ kips} = \phi_c P_p = \phi_c[0.85 f'_c A_1]\sqrt{A_2/A_1} = 0.6[0.85(3)(A_1)]2$$

$$A_1 = \frac{350}{3.06} = 114 \text{ in.}^2$$

For an approximation of the plate dimensions, find

$$\sqrt{114} = 10.7 \text{ in.}$$

This indicates that the required area is somewhat small for the column footprint. The layout shown in Figure 4.11 provides more room for attachment of the column to the plate and development of the anchor bolts. The maximum edge cantilever distance is 2.26 in. and the required thickness is

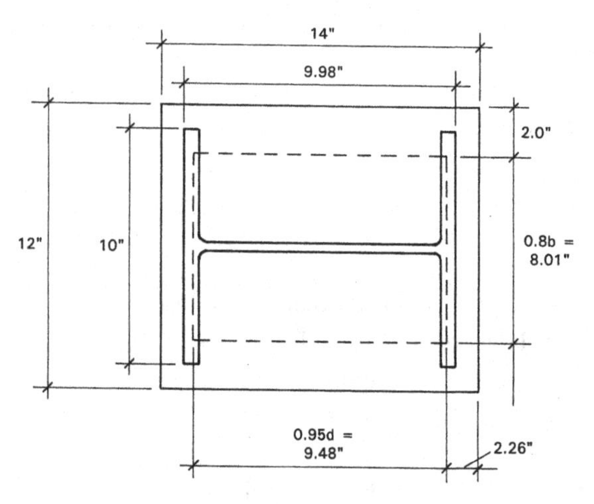

Figure 4.11 Reference for Example 9.

$$t = 2.26\sqrt{\frac{2(350)}{0.9(36)(12)(14)}} = 2.26\sqrt{0.128} = 2.26(0.3586) = 0.810 \text{ in.}$$

An acceptable plate size is ⅞ or 0.875 in.

Problem 4.7.A. Design a column base plate of A36 steel for a W 12 × 120 column that carries a total factored load of 1200 kips [5338 kN] and rests on a concrete pier with $f'_c = 4$ ksi [27.6 MPa]. The side dimensions of the pier are to be approximately the same as the plan size of the column base plate.

Problem 4.7.B. Redesign the base plate for the column in Problem 4.7.A, if bearing is on an 8-ft-square [2.4 m-square] footing of the same concrete strength.

5

FRAME BENTS

A major use of steel for building structures is in frames consisting of vertical columns and horizontal-spanning members. These systems are often constituted as simple post-and-beam arrangements, with columns functioning as simple, axially loaded compression members and with horizontal members functioning as simple beams. In some cases, however, there may be more complex interactions among the frame members. Such is the case of the rigid frame, in which members are connected for moment transfer, and the braced frame, in which diagonal members produce truss actions. In this chapter, we consider some aspects of behavior and problems of design of rigid and braced frames.

5.1 DEVELOPMENT OF BENTS

A *bent* is a planar frame formed to develop resistance to lateral loads, such as those produced by wind or earthquake effects. The simple frame in Figure 5.1*a* consists of three members connected by pinned joints with pinned joints also at the bottom of the columns. In theory, this frame may be stable under vertical load only, if both the load and the frame are perfectly symmetrical. However, any lateral (in this case, horizontal) load, or even a slightly unbalanced vertical load, will topple the frame. One means for restoring stability to such a frame is to connect the tops of the columns to the ends of the beam with moment-resistive connections, as shown in Figure 5.1*b*. If the modification is made, the deformation of the frame under vertical loading will be as shown in Figure 5.2*a*, with moments developed in the beam and columns as indicated. If the frame is subjected to a lateral load, the frame deformation will be of the form shown in Figure 5.2*b*.

Although the transformation of a frame into a rigid frame bent may be done primarily for the purpose of achieving lateral stability, the form of response to vertical load is also unavoidably altered. Thus, in the frame shown in Figure 5.1*a*, the columns (stable or not) would be subject to only vertical axial compression under vertical loading on the beam, whereas they are also subject to bending when connected to the beam to produce rigid frame action. Under the combination of vertical and lateral loads, the bent will function as shown in Figure 5.2*c*.

5.2 MULTIUNIT RIGID FRAMES

Single-unit rigid frames, such as that shown in Figure 5.1, are used frequently for single-space, one-story buildings. However, the greater use for rigid frame bents is in buildings with multiple horizontal bays and multiple levels (multistoried). Figure 5.3*a* shows the response of a two-story, two-bay rigid frame to lateral loading. Note that all members of the

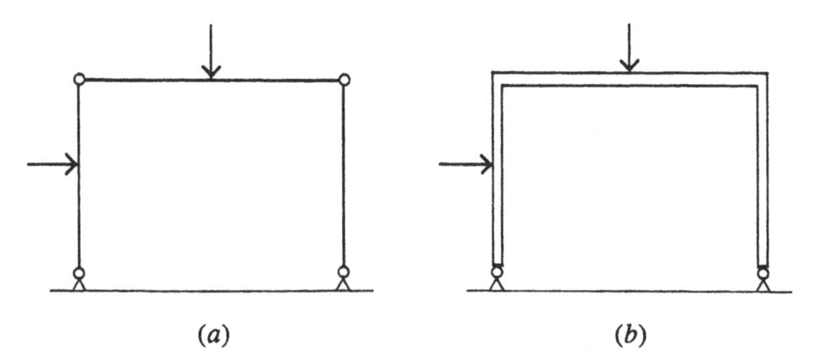

Figure 5.1 Behavior of a single-unit frame.

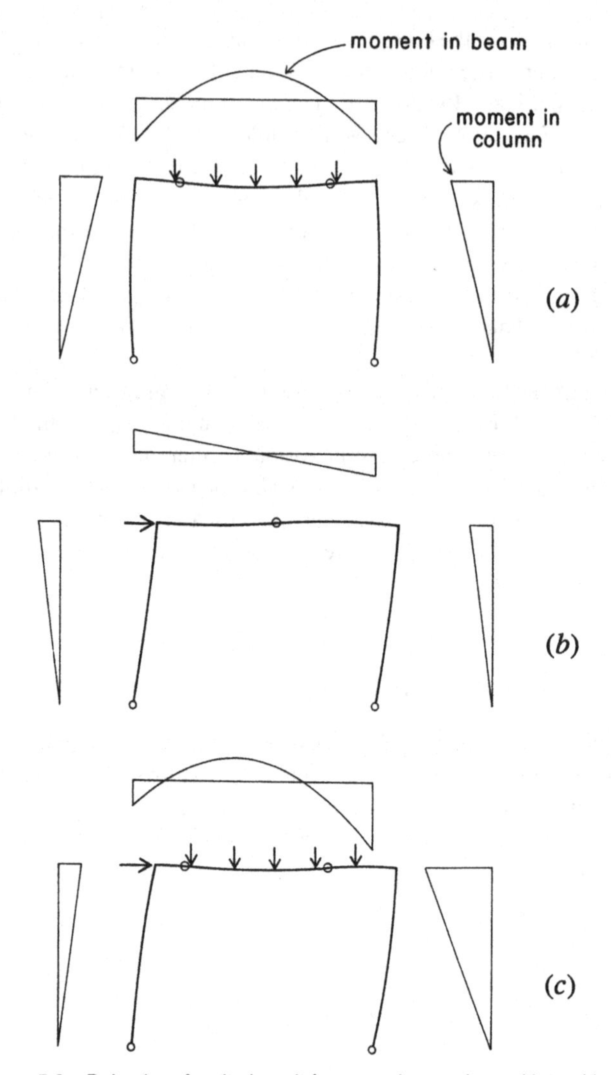

Figure 5.2 Behavior of a single-unit frame under gravity and lateral loads.

frame are bent, indicating that they all contribute to the development of resistance to the loading. Even when a single member is loaded, such as the single beam in Figure 5.3*b*, some response is developed by all the members in the frame. This is a major aspect of the nature of such frames.

In steel frames moment-resistive connections are most often achieved with welding. The term "rigid" as used in referring to a frame actually ap-

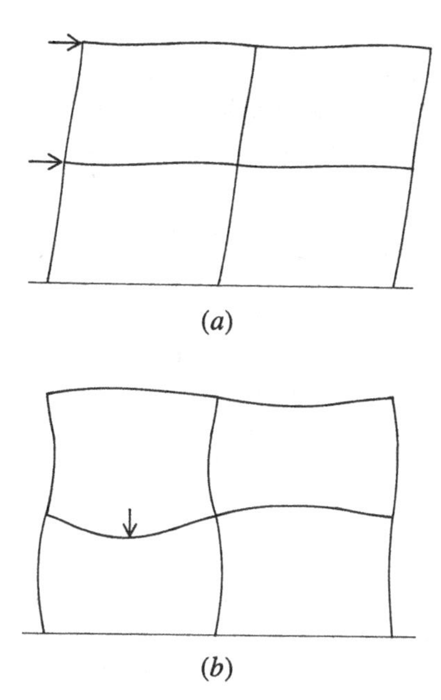

Figure 5.3. Behavior of a multistory, multiple-span rigid frame under (a) lateral loads and (b) gravity load on a single beam.

plies essentially to the connections; implying that the joint resists deformation sufficiently to prevent any significant rotation of one connected member with respect to the other. The term is actually not such a good description of the general nature of the frame with regard to lateral resistance, because the other means for bracing frames against lateral loads (by shear panels or trussing) typically produce more rigid (stiff, deformation-resistive) structures.

Rigid frames are generally statically indeterminate, and their investigation and design is beyond the scope of the work in this book. Some of the problems of designing for combined compression and bending are discussed in Chapter 4, and some detailing for moment-resistive connections is discussed in Chapter 8. Approximate design of rigid bents is illustrated in the design examples in Chapter 10.

5.3 THREE-DIMENSIONAL FRAMES

In most building structures that use column and beam frames, there is some reasonable order to the layout of the system. Columns ordinarily occur in rows with even spacing in a row and some regular spacing of rows; thus, the beams that connect columns are also in regularly spaced order. In this situation, what occurs typically are vertical planes of columns and beams that define individually a series of two-dimensional bents. (See Figure 5.4.) These do not exist as total, freestanding entities but form a subset within the whole structure. Actions of these individual bents may be investigated for effects of both gravity and lateral loads.

Multiunit frames are typically also three-dimensional with regard to the total building framework. Rigid frame actions may be three-dimensional or may be limited to the actions of selected planar (two-dimensional) bents. With site-cast concrete frames, three-dimensional frame action is natural and generally unavoidable. With frames of wood or steel, however, the normal nature of the framework is that of a simple post-and-beam system. Something special must usually be done to produce rigid frame action for multiunit steel frames—that is, to produce moment-resistive joints between the beams and the columns.

In any rigid frame, in addition to the actions of individual columns and beams, there is the action of the whole frame. This involves the collective *interactions* of all the columns and beams. One aspect of this is the highly statically indeterminate nature of the investigation for internal forces. In large frames, the full consideration for this behavior, including that for

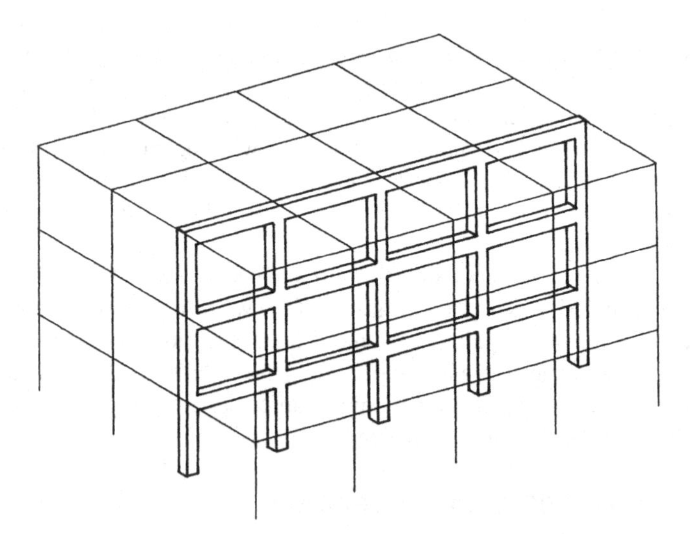

Figure 5.4 Planar (i.e., two-dimensional) column-beam frame bent as a subset in a three-dimensional framing system.

possible variations of loading, is itself a formidable computational problem, preceding any design efforts.

A major use of the vertical rigid frame bent is bracing for lateral forces due to wind and earthquakes. Once constituted as a rigid (moment-resisting) frame, however, its continuous, indeterminate responses will occur for all loadings. Thus, both the individual responses to gravity and lateral loadings must be considered, as well as their potential combined effects.

The complete engineering investigation and design of multiunit rigid frames is well beyond the scope of this book. Some of the general considerations for design of multistory frames and examples of methods for approximate design are treated in the discussion for example Building Three in Chapter 10.

5.4 MIXED FRAME AND WALL SYSTEMS

Most buildings consist of a mixture of framed systems and walls. For structural use, walls may vary in potential. Metal and glass skins on buildings are typically not components of the general building structure, even though they must have some structural character to resist gravity and wind effects. Walls of cast concrete or concrete masonry construction are frequently used as parts of the building structure, which brings the necessity to analyze the relationships between the walls and the framed structure. This section treats some of the issues involved in this analysis.

Coexisting, Independent Elements

Frames and walls may act independently for some functions, even though they interact for other purposes. For low-rise buildings, walls are often used to brace the building for lateral forces, even when a complete gravity-load-carrying frame structure exists. Such is the typical case with light wood frame construction using plywood shear walls. It may also be the case for a concrete frame structure with cast concrete or concrete masonry walls.

Attachment of walls and frames must be done so as to ensure the actions desired. Walls are typically very stiff, whereas frames often have significant deformation due to bending in the frame members. If interactions of the walls and frames are desired, they may be rigidly attached to achieve the necessary load transfers. However, if independent actions are desired, it may be necessary to develop special attachments that achieve load transfer for some purposes while allowing for independent movements due to other effects. In some cases, total separation may be desired.

A frame may be designed for gravity load resistance only, with lateral load resistance developed by walls acting as shear walls (see Figure 5.5). This method usually requires that some elements of the frame function as collectors, stiffeners, shear wall end members, or chords for horizontal diaphragms. If the walls are intended to be used strictly for lateral bracing, care must be exercised in developing attachment of wall tops to overhead beams; deflection of beams must be permitted to occur without transferring loads to the walls below.

Load Sharing

When walls are rigidly attached to columns, they usually provide continuous lateral bracing in the plane of the wall. This permits the column to be designed only for the relative slenderness in the direction perpendicular to the wall. This is more often useful for wood and steel columns (for example, wood stud 2 × 4s and steel W shapes with narrow flanges), but it may be significant for a concrete or masonry column with a cross section other than square or round.

In some buildings, both walls and frames may be used for lateral load resistance at different locations or in different directions. Figure 5.6 shows four such situations. In Figure 5.6*a*, a shear wall is used at one end

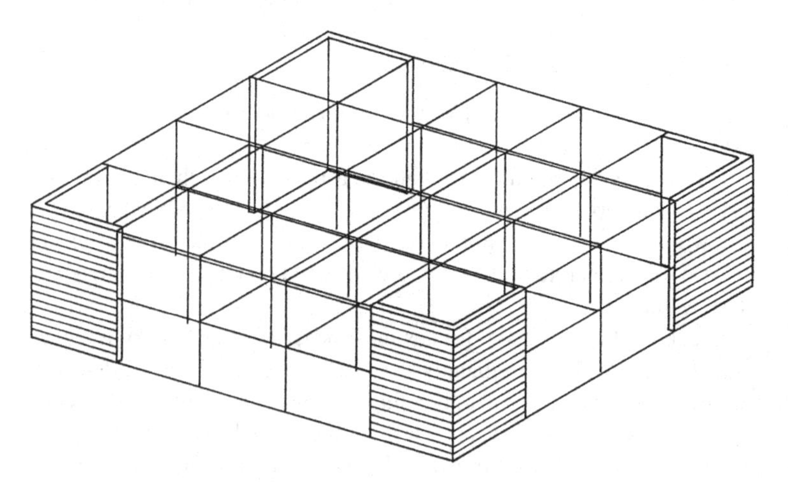

Figure 5.5 Framed structure braced by walls.

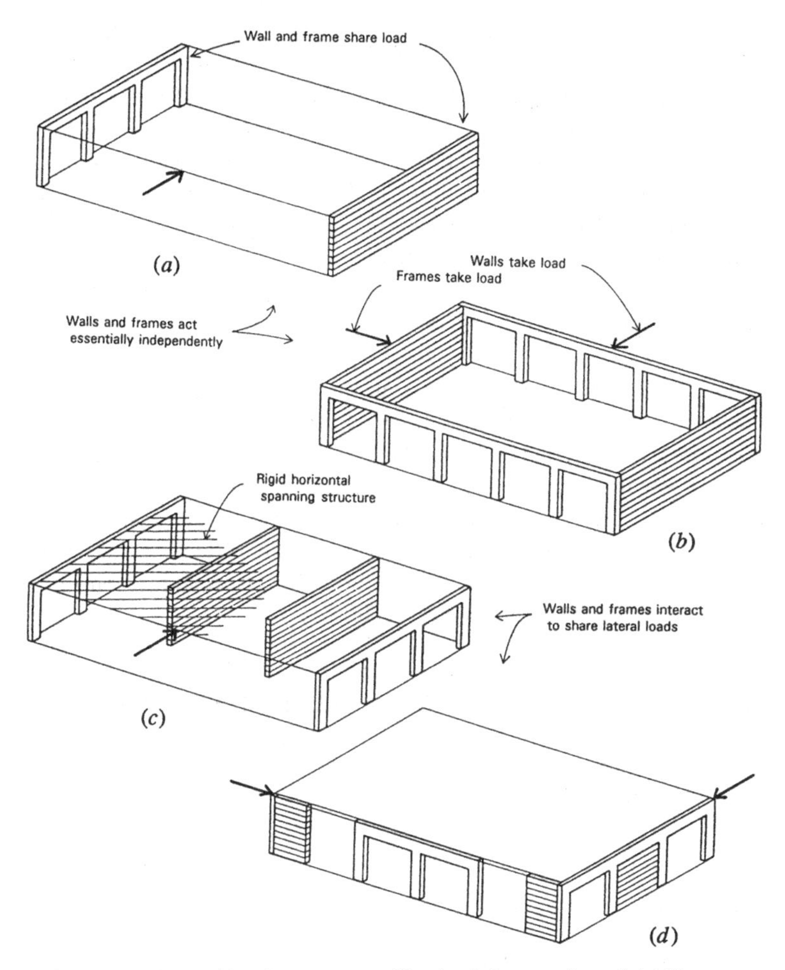

Figure 5.6 Lateral bracing systems with mixed shear walls and rigid frames.

of the building and a parallel frame at the other end for the wind from one direction. These two elements will essentially share equally in the load distribution, regardless of their relative stiffness.

In Figure 5.6*b* walls are used for the lateral loads from one direction and frames for the load in the perpendicular direction. Although some distribution must be made among the walls in one direction and among the frames in the other direction, there is essentially no interaction between the frames and walls, unless there is some significant torsion on the building as a whole.

Figures 5.6*c* and *d* show situations in which walls and frames do interact to share loads. In this case, the walls and frames share the total load from a single direction. If the horizontal structure is reasonably stiff in its own plane, the load sharing will be on the basis of the relative stiffness of the vertical elements. Relative stiffness in this case refers essentially to resistance to deflection under lateral force.

Dual Systems

A dual system for lateral bracing is one in which a shear wall system is made to deliberately share loads with a frame system. In Figure 5.6 the systems shown at *a* and *b* are not dual systems, whereas those shown at *c* and *d* potentially are. The dual system has many advantages for structural performance, but the construction must be carefully designed and detailed to ensure that interactions and deformations do not result in excessive damage to the general construction. Some special problems that can occur are discussed in the next section.

5.5 SPECIAL PROBLEMS OF STEEL RIGID FRAME BENTS

The following are some special problems that may occur in the development of rigid frame bents within a general three-dimensional framing system.

Lack of Symmetry of Columns

A common shape used for steel columns is the W shape. (More truly H-shaped, as used for columns.) This shape has a strong axis (the *x*-axis) and a weak axis (the *y*-axis), and a decision to be made while planning the layout of the framework is that regarding the orientation of the columns in plan. For rigid-frame bent actions, it is desirable to have the beams of the bent connect directly to the column flanges, rather than to the column web. However, other considerations may be involved, including the necessity to make double use of a column as a member of two bents at right angles to each other. General planning of the bents must take note of this issue. See the discussion for the rigid frame alternative for Building Three in Section 10.3.

Pinned versus Rigid Connections

As mentioned, the basic form of the steel framed structure is one with connections between beams and columns that border on being pinned, rather than on developing significant moment resistance. Producing rigid frame bents within a general framework therefore involves the design of special connections. However, an advantage of this situation is the ability to control the extent of the bents—that is, which members of the frame participate in the bent action. Leaving the rest of the frame with essentially pinned connections eliminates their involvement in the bent actions. Most buildings that use rigid bents are developed on this basis, having only selected members functioning in bent actions with the rest of the structure merely tagging along for the ride.

Shop versus Field Work

Economy and general ease of erection of steel frames occurs when most of the fabrication is done in factory conditions (called the *shop*) and connections at the job site (called the *field*) are simple and few in number. Production of large rigid bents usually involves considerable making of connections at the site. This delays the erection in some cases and is usually not able to be done with the quality of workmanship possible in the factory. It adds considerably to the cost of the steel structure and must be carefully considered in making the decision to use rigid frame bents.

General Cost

Rigid frame bents are popular for various reasons. Nevertheless, bear in mind that extra expense is involved. This starts with the extra design effort for investigation of complex behaviors and development of special connections. The major extra cost, however, is that for producing the necessary rigid joints, usually by extensive field welding. Another possible factor is the considerable increase in the size (steel weight) of columns that must be designed for combined compression and bending.

Proportionate Stiffness of Individual Bent Members

Within a single bent, the behavior of the bent and the forces in individual members will be strongly affected by the proportionate stiffness of bent members. If story heights vary and beam spans vary, some very complex and unusual behaviors may be involved. Variations of column and beam stiffnesses may also be a significant factor.

A particular concern is the relative stiffness of all the columns in a single story of the bent. In many cases, the portion of lateral shear in the columns will be distributed on this basis. Thus, the stiffer columns may carry a major part of the lateral force.

Another concern has to do with the relative stiffness of columns in comparison to beams. Most bent analyses assume the column stiffness to be more-or-less equal to the beam stiffness, producing the classic form of lateral deformation shown in Figure 5.3*a*. Individual bent members are assumed to take an S-shaped, inflected form. However, if the columns are exceptionally stiff in relation to the beams, the form of bent deformation may be more like that shown in Figure 5.7*a*, with virtually no inflection in the columns and an excessive deformation in upper beams. In tall frames, this is often the case in lower stories where gravity loads produce large columns.

Conversely, if the beams are exceptionally stiff in comparison to the columns, the form of bent deformation may be more like that shown in Figure 5.7*b*, with columns behaving as if fully fixed at their ends. Deep spandrel beams with relatively small columns commonly produce this situation.

Both of the cases shown in Figure 5.7 can be dealt with for design, although it is important to understand which form of deformation is most likely.

The Captive Frame

In the preceding section, the problem of interaction of parallel bents and walls was discussed. A special problem is the partially restrained column

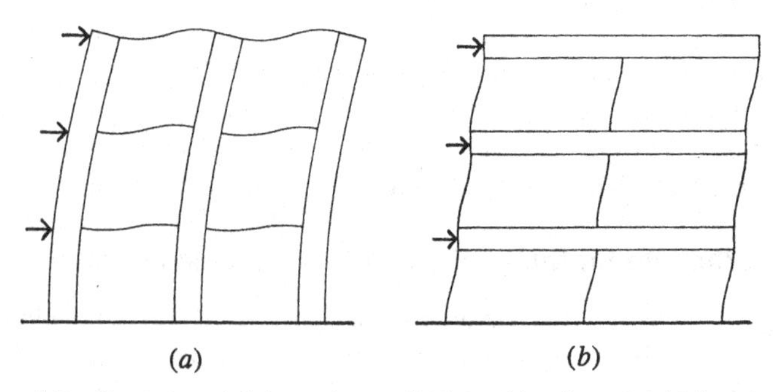

Figure 5.7 Character of deformation under lateral loading of rigid frames with members of disproportionate stiffness: (a) stiff columns and flexible beams; (b) stiff beams and flexible columns.

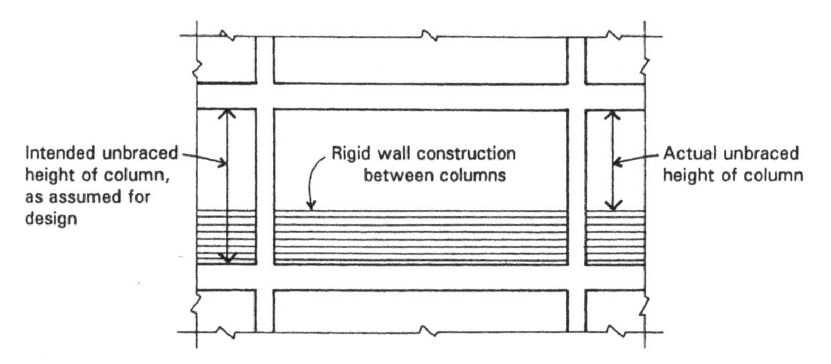

Figure 5.8 Captive columns.

or beam, with inserted construction that alters the form of deformation of bent members. An example of this, as shown in Figure 5.8, is the *captive column*. In the example, a partial height wall is placed between columns. If this wall has sufficient stiffness and strength and is tightly wedged between columns, the laterally unbraced height of the column is drastically altered. As a result, the shear and bending in the column will be considerably different from that of the free column. In addition, the distribution of forces in the bent containing the captive columns may also be affected. Finally, the bent may thus be significantly stiffened, and its share of the load in relation to other parallel bents may be much higher.

This is an issue for the structural designer of the bents, but it must also be considered in cooperation with whomever does the construction detailing for the wall construction. This has been a major source of problems for concrete frames affected by seismic forces.

5.6 TRUSSED FRAMES

The term *braced frame* is used to describe a frame that uses diagonal members to produce some truss action. As used for lateral bracing, this usually means adding some diagonals to the typical rectilinear layouts of vertical planes of beams and columns to produce trussed bents that function as vertically cantilevered trusses. Figure 5.9 shows two forms for such an arrangement of members. In Figure 5.9*a* single diagonals are used to produce a simple truss that is statically determinate. A disadvantage of this form for lateral bracing is that the diagonals must function for lateral force from both directions. Thus, they are sometimes in compression and sometimes in tension. When functioning in compression, the

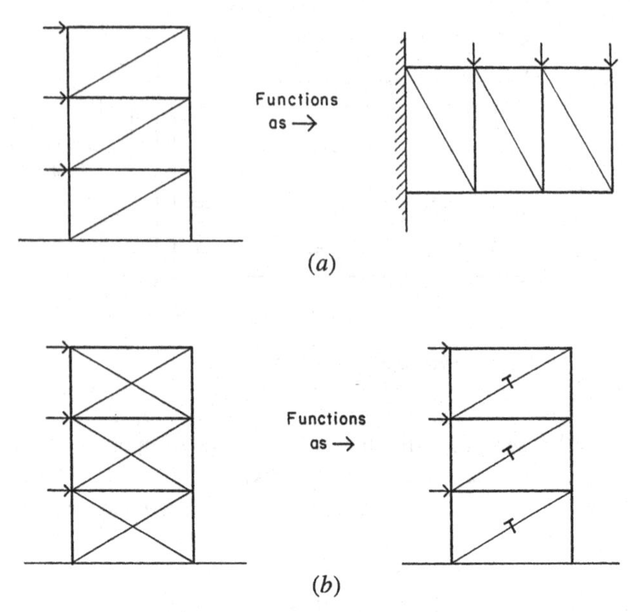

Figure 5.9 Forms and behavior of frames with concentric bracing: (a) basic form of the cantilevered frame under lateral loading; (b) assumed limit condition for the frame with slender X-bracing.

very long diagonals must be quite heavy due to buckling on their unbraced lengths.

A popular layout for trussed bracing is shown in Figure 5.9*b,* called *X-bracing.* One application of this system involves the use of very slender tension members (sometimes even long, round rods), with a single rod functioning in tension for each direction of loading, assuming that the other rod will buckle slightly in compression and become negligible in resistance. If the foregoing action is assumed, the X-braced frame may be analyzed by simple statics, even though it is theoretically indeterminate with all members working. Light X-bracing has been used with some success for wind bracing for many years. However, recent experiences with seismic forces indicate that this is not a desirable form unless the compression diagonals are actually quite stiff.

A problem with both the single and double diagonals is that they use up the space of the braced bay, making placement of windows or doors difficult. This brought about the use of knee bracing and K-bracing, both of which leave more open space in the bay (see Figure 5.10). They also,

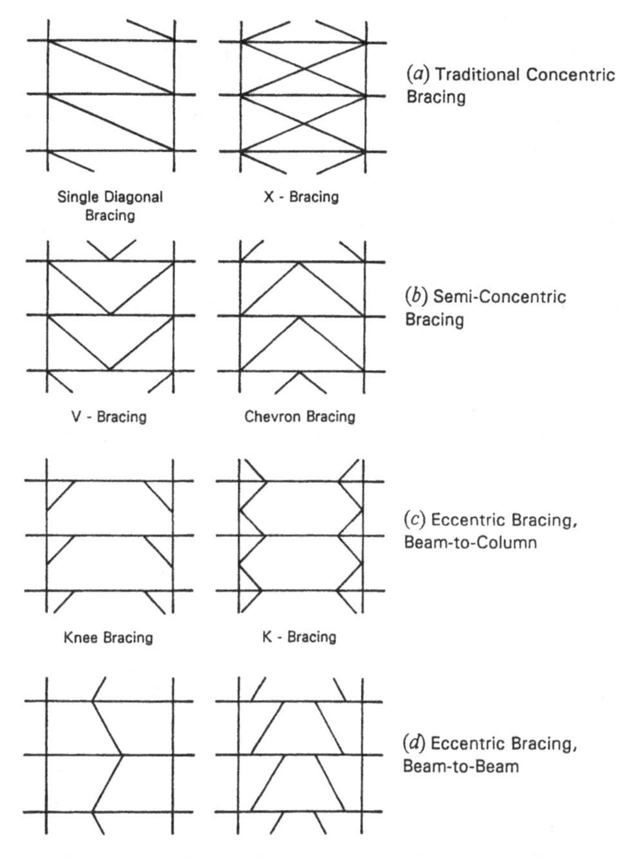

Figure 5.10 Options for development of the braced frame.

however, involve bending in the columns and beams and a form of combined truss and rigid frame action. Except for the lower columns in very tall structures, bending in the columns is mostly objectionable. The V-brace and inverted V, or chevron, brace develop bending only in the beams, which are already designed for bending.

All of the forms of bracing mentioned so far have been successfully used for wind bracing. For seismic actions, the dynamic jerking and the rapid reversals of the direction of forces make all but the V-bracing and chevron bracing less desirable. A possible exception is the X-brace with very stiff members.

A more recent development is the use of *eccentric bracing,* with sloped bracing members connected only at points within the beam spans. With the eccentric braces designed as the weak links, failing in inelastic buckling with some degree of yielding, this system is highly suited to dynamic loading conditions.

Planning of Trussed Bracing

Utilization of trussed bracing involves various considerations for determination of the form of the bracing system and elements. Following are some of these basic considerations:

1. Diagonal members must be placed so as not to interfere with the action of the gravity-resistive structure or with general use of the building. If bracing members are designed essentially as axial stress members, they must be placed in the frame and attached so as to avoid loadings other than those required for their bracing functions. They must also be located so as not to interfere with placement of doors, windows, corridors, roof openings, or with elements of the building services, such as piping, ducts, light fixtures, and so on.
2. As mentioned previously, the reversibility of lateral loads must be considered. As shown in Figure 5.9, this requires either dual-functioning single diagonals or X-bracing.
3. Although diagonals may function only for lateral loads, columns and beams must function for gravity loads as well and must be designed for critical load combinations.
4. Diagonals placed in rectilinear frames are usually quite long. If designed for tension only (as X-bracing), they may be very slender and can sag under their own dead weight, which may indicate need for some support.
5. The trussed structure should be tight; that is, it should be able to sustain reversible loads with little give in the connections. Connections should also not be liable to loosen with repeated loading.
6. To avoid gravity loading on the diagonals, the connections of the diagonals are sometimes not completed until after the rest of the frame is fully assembled and at least the major weight of construction is in place.
7. Lateral deflection of the truss must be considered. This may relate to load distributions or to relative stiffness in a mixed system of

vertical bracing elements. For tall, slender trusses, it may also relate to deformations that are critical for other parts of the building construction.

8. In most cases, the development of trussed bracing involves placing diagonals in only a limited number of bays of the rectilinear framing. Choice for location of bracing must be coordinated with architectural planning and must relate to development of logical placement of vertical bracing elements for the building.

The braced frame may be used for an entire building's vertical bracing elements, but it is also mixed with other systems in a single building. Figure 5.11 shows the use of braced frames for the vertical resistive structure in one direction and shear walls for the other direction. In this example, the two systems act essentially independently, except for torsion effects, and there is no need for a deflection analysis to determine load sharing.

Figure 5.12 shows a structure in which the end bays of the roof framing are truss-braced. For loading in the direction shown, these trussed bays take the highest magnitude of shear in the horizontal structure, allowing the deck diaphragm to be designed for a lower shear stress.

Although individual building elements and some subsystems are often composed of a single material, most buildings employ construction with many different materials. Mixtures of construction are also common with lateral bracing systems. The discussions here often focus on single component systems and single materials for the sake of simplicity, but whole buildings are often complex and their construction quite diverse.

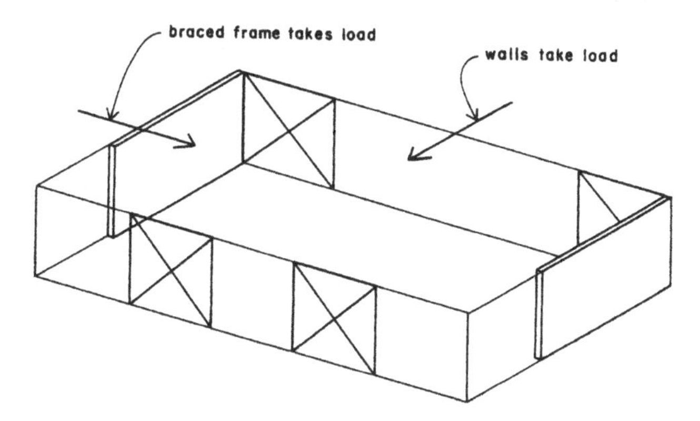

Figure 5.11 Mixed elements for lateral bracing.

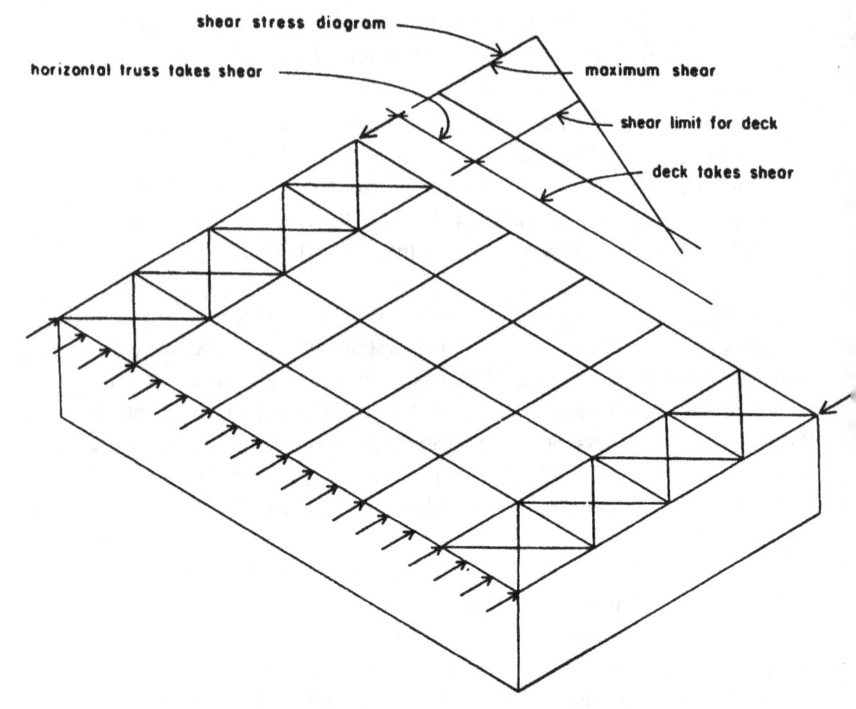

Figure 5.12 Horizontal bracing in a flat roof structure.

Although buildings and structures are often planned and constructed in two-dimensional components (horizontal floor and roof planes and vertical walls or framing bents), buildings are truly three-dimensional. Generation of lateral forces and development of bracing is essentially a three-dimensional problem. Although an individual horizontal or vertical plane of the structure may be stable and adequately load-resistive, the whole system must interact appropriately. Although the single triangle is the basic stabilizing unit for a planar truss, the three-dimensional structure may not be truly stable just because its component planes are braced.

6

MISCELLANEOUS STEEL COMPONENTS AND SYSTEMS

The work in this book deals principally with structures that use rolled structural products and formed sheet steel units for common structural systems. Although the vast majority of steel structures for buildings are formed from this limited inventory of components, there are many other uses of steel for structural components and systems—large and small, simple and complex. This chapter treats a few of these other uses of steel.

6.1 MANUFACTURED SYSTEMS

Many manufacturers produce components that may be used to fashion complete structures or, in some cases, complete buildings. The "package" system may come totally assembled (as with a mobile home) or in a kit for user assembly. Major use of such construction is made for utilitarian buildings for industrial and agricultural applications. Speed of erection and assurance of final cost are major reasons for selection of such construction. A certain stigma of architectural tastelessness is often associated with this construction. Nevertheless, a significant amount of building construction is achieved with these systems.

Various factors favor the use of steel for the structures of these systems. The ability to achieve light, strong structures with dependable performance is a major asset. The noncombustible nature of the material also permits wider uses.

Many of the components of these systems are items patented and produced by a single manufacturer, and the end product systems are industrially produced products. In a sense, however, all of the elements used for steel structures are industrial products produced in controlled and standardized forms. Designing and assembling any steel structure is therefore largely a matter of bringing together many predetermined parts to create end products that are highly predictable in form.

6.2 COMPOSITE STRUCTURAL ELEMENTS

Composite Steel-Concrete Beams

A common form of construction for floors with steel framing consists of a site-cast concrete slab on top of steel beams. Figure 6.1 shows the usual form of such construction. The concrete slab is made to interact with the

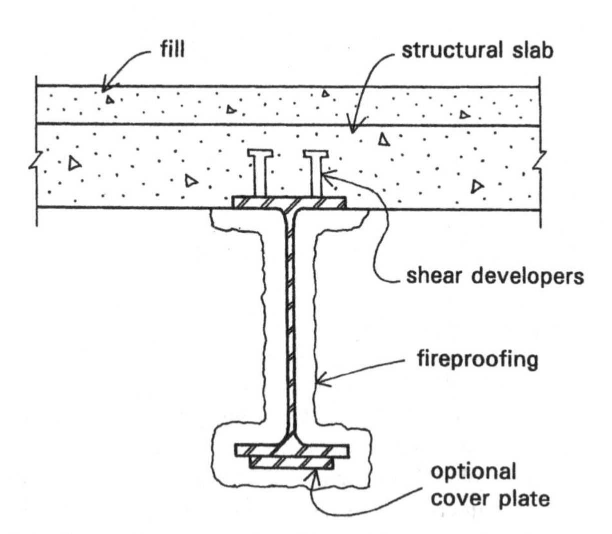

Figure 6.1 Composite construction with steel beams and a site-cast concrete deck.

steel beam to produce a *composite* element—that is, a single structural element with more than one interacting structural material. (Technically, ordinary reinforced concrete is such an element with the interacting concrete and steel reinforcing bars.) As shown in Figure 6.1, the device used to ensure the interaction of the steel beam and concrete slab consists of a series of shear developers, called *studs,* welded to the top of the steel beam.

The composite action in this case produces a stronger element than that represented by the steel beam alone. Thus, there is a higher resistance to bending moment and load-carrying capacity. Although the strength increase may be significant, a major reason for using such construction often relates to the considerable increase in deflection resistance.

Specifications for the design of composite steel–concrete beams, as well as some design aids, are provided in the AISC Manual (Ref. 3). Although the concrete slab can be cast with plywood forming, as shown in Figure 6.1, it is often produced as fill on top of a formed sheet steel deck. The slab forming does not affect the composite action of the steel beam and the slab.

Flitched Beams

In construction using solid sawn-timber beams, there are some occasions when the deflection of a beam is of critical concern. Because heavy timber is generally available only in green-wood condition, some long-term sag is inevitable, especially when considerable dead load is supported. One means commonly used in this situation is to combine a steel plate with wood members in a *flitched beam.* Two forms for flitched beams are shown in Figure 6.2. The most common form uses a single plate sandwiched between two timber members, with the steel and wood bolted together to ensure interaction, as shown in Figure 6.2*a*. For an even

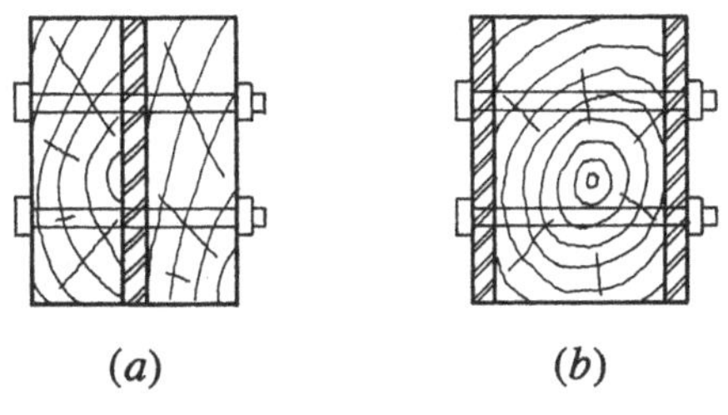

Figure 6.2 Flitched beams.

stronger section, two plates may be used with a single timber section as shown in Figure 6.2*b*.

In most cases, due to the much higher stiffness of the steel, the plates will carry the majority of the load on a flitched beam. A simple design thus consists of considering the plates alone to carry the load and resist deflection. The wood in this case serves primarily to brace the slender plate for lateral and torsional buckling. With the beam used in a general timber structure, the wood is also probably used to achieve necessary connections to other structural elements.

For a true consideration of the composite action, it is usually assumed that the wood and steel interact in proportion to their individual stiffnesses under service load conditions. This means that stress levels are assumed to be well below yield stress in the steel or ultimate failure of the wood. Such a procedure may be developed with the stress method, as described in the following example.

If the two materials truly interact, the assumption is made that they deflect (or deform) the same amount. Thus:

Δ_1 and Δ_2 = deformation per unit length (strain) of the outermost fibers of the two materials

f_1 and f_2 = unit bending stresses in the outermost fibers of the two materials

E_1 and E_2 = moduli of elasticity of the two materials

Because by definition the modulus of elasticity of a material is equal to the unit stress divided by the accompanying unit strain, then

$$E_1 = \frac{f_1}{\Delta_1} \quad \text{and} \quad E_2 = \frac{f_2}{\Delta_2}$$

and transposing:

$$\Delta_1 = \frac{f_1}{E_1} \quad \text{and} \quad \Delta_2 = \frac{f_2}{E_2}$$

Because the two deformations must be equal for interaction:

$$\Delta_1 = \Delta_2 \quad \text{or} \quad \frac{f_1}{E_1} = \frac{f_2}{E_2}$$

and a relationship between the two stresses may thus be expressed as

$$f_2 = f_1 \times \frac{E_2}{E_1}$$

This relationship may be used for investigation or design of a flitched beam by the stress method. The process is demonstrated in the following example.

Example 1. A flitched beam is formed as shown in Figure 6.2*a*, consisting of two 2 × 12 planks of Douglas fir, No. 1 grade, and a steel plate with dimensions of 0.5 by 11.25 in. [13 by 285 mm] and F_y = 36 ksi [248 Mpa]. Determine the allowable uniformly distributed load that this beam can carry on a simple span of 14 ft [4.2 m]. For the steel, use E = 29,000 ksi [200 Gpa] and a maximum allowable bending stress of 22 ksi [150 Mpa]. For the wood, use E = 1800 ksi [12.4 Gpa] and a maximum allowable bending stress of 1500 psi [10.3 Mpa]. For the 2 × 12, S = 31.6 in.^{3} [518 × 10^3 mm^3].

Solution: A procedure that can be used is to assume the maximum allowable stress in the steel and to find the corresponding stress in the wood. Thus:

$$f_w = f_s \times \frac{E_w}{E_s} = (22)\left(\frac{1800}{29{,}000}\right) = 1.366 \text{ ksi} \quad \text{or} \quad 1366 \text{ psi}$$

Because this produces a stress lower than the limit for the wood, the stress in the steel is the limiting value. That is, if a stress higher than 1366 psi is permitted in the wood, the limiting stress for the steel will be exceeded.

Using the stress limit just determined for the wood, one can now find the share of the load that is carried by the wood members. Calling this total load W_w, the moment on the simple beam due to this load is

$$M_w = \frac{W_w L}{8} = \frac{W_w \times 14}{8} = 1.75W_w \text{ (in foot units)}$$

$$\text{or} \quad 1.75 \times 12 = 21W_w \text{ (in inch units)}$$

The limit for this moment may be expressed as:

$$M = f \times S = 1366 \times (2 \times 31.6) = 86{,}311 \text{ in.-lb}$$

and for determination of W_w:

$$M_w = 21W_w = 86{,}311 \text{ in.-lb}$$

$$W_w = \frac{86{,}311}{21} = 4110 \text{ lb } [18.35 \text{ kN}]$$

For the steel plate:

$$S = \frac{bd^2}{6} = \frac{(0.5)(11.25)^2}{6} = 10.55 \text{ in.}^3 \; [176 \times 10^3 \text{ mm}^3]$$

Then:

$$M_s = 21W_s = f_s \times S = (22{,}000)(10.55) = 232{,}100 \text{ in.-lb}$$

$$W_s = \frac{232{,}100}{21} = 11{,}052 \text{ lb}$$

and the total capacity of the composite beam is

$$W = W_w + W_s = 4111 + 11{,}052 = 15{,}163 \text{ lb } [68.6 \text{ kN}]$$

Although the load-carrying capacity of the wood elements is usually reduced in the flitched beam, the resulting total capacity is substantially greater than with the wood elements alone. This major increase in strength is achieved with only a small increase in the size of the beam. In addition, there is a significant reduction in the deflection in most applications—most significantly in the reduction of sag over time.

Sag over time is a natural phenomenon with timber, which is normally sold in an assumed green-wood condition. One effect of this sag in the flitched beam is the steady shift of load from the wood to the steel plates. Thus, in due time, the wood will in fact carry something less than the load just computed, and it is possible that the result can produce an overload on the steel plate. For this reason, the actual permitted load should be something less than the total capacity previously computed.

Design of a flitched beam by the stress method is usually a hit-and-miss situation, requiring some assumptions of properties of the members, a subsequent investigation, and—if necessary—some readjustment of dimensions. It is also necessary to assume some specific amount of reduction of the wood capacity over time. A much simplified design procedure, although somewhat conservative, is to design the steel plate to carry the entire load, relying on the wood members only for lateral and torsional bracing of the plate. This process is demonstrated in the following example, using the strength method for the plate design.

Example 2. Design a flitched beam of the form shown in Figure 6.2*a* with the same wood members as in the preceding example. Assume the steel plate to carry the entire load, and determine the minimum thickness required for the plate in this case. For the design load, assume the service load determined as the total capacity of the beam in Example 1 with the loading being 50 percent dead load and 50 percent live load.

Solution: The required ultimate load is determined with an average load factor of 1.4, thus:

$$W_u = 1.4(15{,}163) = 21{,}228 \text{ lb}$$

and the required ultimate bending moment is thus:

$$M_u = \frac{W_u L}{8} = \frac{(21{,}228)(14)}{8} = 37{,}149 \text{ ft-lb} \quad \text{or} \quad 445{,}792 \text{ in.-lb}$$

Using a resistance factor of 0.9, the required resisting moment of the plate is thus:

$$M_r = \frac{445{,}792}{0.9} = 495{,}325 \text{ in.-lb}$$

For the plate, the resisting moment is determined as:

$$M_r = F_y \times Z$$

For the rectangular form of the plate, Z is determined as follows (see Figure A.11 in the appendix):

$$Z = \frac{bd^2}{4} = \frac{b(11.25)^2}{4} = 31.64b \text{ in.}^3$$

Thus, the thickness of the plate, b, is determined as:

$$M_r = 495{,}325 = (36{,}000)(31.64b) = 1{,}139{,}040b$$

$$b = \frac{495{,}325}{1{,}139{,}040} = 0.435 \text{ in.}$$

This indicates that the plate dimensions given in Example 1 are actually sufficient for the load capacity found by the stress method relying completely on the plate.

Problem 6.2.A. A flitched beam consists of a single 10 × 14 (actual dimensions 9.5 by 13.5 in. [241 by 343 mm]) of Douglas fir, select structural grade, and two A36 steel plates, each 0.5 by 13.5 in. [13 by 343 mm]. (See Figure 6.2*b*.) Using stress methods, find the magnitude of the single concentrated load this beam can safely support at the center of a 16-ft [4.8-m] simple span. Neglect the weight of the beam. Use a value of 22 ksi [152 MPa] for the limiting bending stress in the plates. For the wood, E = 1600 ksi [11.03 Gpa], allowable bending stress is 1600 psi [11.03 Mpa], and S is 228.6 in.^{3} [3.75×10^6 mm^3].

Problem 6.2.B. Using the strength method, find the thickness required for the plates in Problem 6.2.A if the entire load is carried by the plates. Assume the load determined in Problem 6.2.A with an average load factor of 1.4.

6.3 TENSION ELEMENTS AND SYSTEMS

Steel tension elements are used in a number of ways in building structures. Behavior may be simple, as in the case of a single hanger or tie rod, or extremely complex, as in the case of cable networks or restraining cables for tents and pneumatic structures. This section treats the actions of some relatively simple steel tension elements.

Axially Loaded Tension Elements

The simplest case of tension occurs when a linear element is subjected to a tension force that is aligned on an axis corresponding with the centroid of the element's cross section. For such an element, the tension stress is assumed to be distributed evenly on the cross section and the factored resistance is expressed as:

$$\mathrm{T} = \phi F_y A$$

where: ϕ = resistance factor—0.9 for the gross cross section; 0.75 for reduced cross sections
A = cross section area

Elongation as a total dimension is expressed as:

$$e = \frac{f}{E}L = \frac{TL}{AE}$$

The tension force permitted when a limiting elongation is required may be expressed as:

$$T = \frac{AEe}{L}$$

When a tension member is short, as in the case of a short hanger or a truss member, the usable tension capacity is usually based on the limiting stress. However, for very long members, elongation may be critical and may limit the tension capacity to a value below that of the safe stress limit.

Example 3. An arch spans 100 ft [30 m] and is tied at its spring points by a 1-in. [25-mm] diameter round steel rod. Find the limit for the tension force in the rod if F_y = 36 ksi and the total elongation is limited to 1.0 in. [25 mm].

Solution: The cross-sectional area of the rod is

$$A = \pi R^2 = 3.1416(0.5)^2 = 0.785 \text{ in.}^2 \text{ [491 mm}^2\text{]}$$

The maximum tension force based on stress is

$$T = \phi F_y A = (0.9)(36)(0.785) = 25.43 \text{ kips [113 kN]}$$

The maximum tension force based on elongation is

$$T = \frac{AEe}{L} = \frac{(0.785)(29{,}000)(1.0)}{100 \times 12} = 18.47 \text{ kips [81.8 kN]}$$

In this case, the elongation is critical and the limiting tension force is less than the safe stress limit.

Problem 6.3.A. An arch spans 150 ft [45 m] and is tied at its spring points by a 1.5-in. [38-mm] diameter steel rod. Find the limit for the tension force in the rod if F_y = 50 ksi [345 MPa] and total elongation is limited to 2 in. [50 mm].

Problem 6.3.B. A three-hinged steel rigid bent spans 75 ft [23 m] and is tied at its supports by a 1.75-in. [45-mm] diameter steel rod. Find the limit for the tension force in the rod if F_y =75 ksi [517 MPa] and total elongation is limited to 1.5 in. [38 mm].

Net Section and Effective Area

The development of tension in a structural member involves connecting it to something. Achieving tension-resistive connections often involves situations that reduce the net effectiveness of the tension member. Two examples of this are the threaded connection and the bolted connection. Development of force at bolt holes is discussed in Section 8.2. For the bolted member, tension resistance must be considered at two locations: at the section through the bolt holes, called the *net section,* and at the unreduced section of the structural member. This investigation is described in Section 8.2.

For tension members consisting of round steel rods, connections are often achieved by cutting spiral threads into the rod ends. This also produces an effective net section for which properties are given in the AISC Manual (Ref. 3). This situation also occurs with bolts that are loaded in tension.

It sometimes happens that the practical considerations of achieving connections make it difficult to fully develop the potential tension resistance of the connected members. Figure 6.3 shows a typical situation involving a steel angle in which one leg of the angle is attached by welding to a supporting element. At the connection, the tension force is developed only in the connected leg of the angle. For this situation, a conservative design would consider the tension member to consist essentially only of a plate equal to the connected leg. Further out from the connection, the member may be fully developed, but not at the connection. This situation will also be true if the attachment is achieved with bolts.

Flexible Elements

Tension elements are somewhat unique in that there is little basis for establishment of limits on slenderness. By comparison, columns with height-to-width ratios of 20 or more tend to approach a condition of slenderness

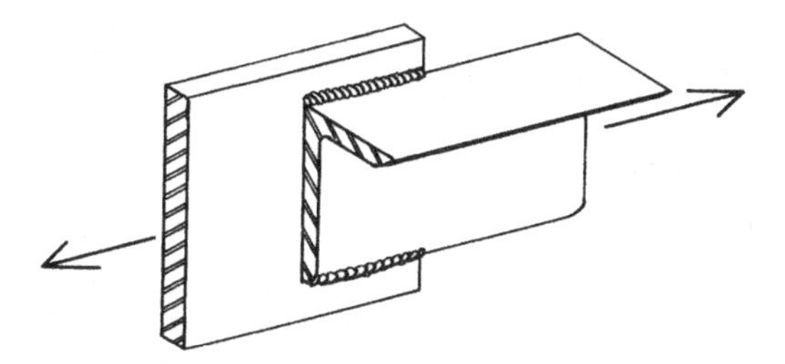

Figure 6.3 Typical connection for a single steel angle with tension force developed primarily in the connected angle leg only.

that has the potential to significantly reduce effectiveness in compression resistance. In a similar manner, beams with a span-to-depth ratio of 20 or more tend to approach a condition where deflection is a critical concern. For members in pure tension, lateral stiffness may be virtually zero, as it is for a fiber rope or a chain. It may also be quite negligible for very long tension rods, wires, or cables.

The slender and flexible tension member will assume a profile that follows very precisely the development of the tension force; it has no other way of functioning. This severely limits the manner in which such elements can be used, and the designer must take care to determine the logical pattern of the loads that act on such an element.

The hanger rod, tie rod, or truss member that functions as a simple two-force member is no problem; it will automatically assume a straight form to resolve the tension load, even if it has to straighten itself to do so. The only problem in these cases is to ensure that the connection details do not result in something other than the pure axial transfer of tension force to the member. For such members, length is not related to force resistance and is limited only by considerations for elongation (stretch), sag (if horizontal), vibration or flutter in a low-stress situation, and available lengths from suppliers (which may require splicing for very long members).

Tension members that have significant stiffness may also function in pure axially loaded situations. However, they also have some potential for other actions, such as bending or torsion, and may thus develop some combined force interaction.

When the super-flexible tension element is used for spanning, it cannot assume the rigid, minutely deflected form of a beam. It must, instead, assume a profile that permits it to act in pure tension, and the form for this pure tension element must be determined from the loading and support conditions. It cannot be an arbitrary form concocted by the designer; it must be a true form determined by the structure itself.

Spanning Cables

Super-flexible tension elements may be visualized as a rope or chain. The steel rope is composed of bundled steel wires and is known most commonly as a cable, although it may be technically qualified as a rope, cable, or strand, depending on its form. For this discussion, the familiar name of *cable* will be used.

Consider the single-span cable shown in Figure 6.4*a,* spanning horizontally and supporting only its own dead weight. The draped shape assumed by the cable is a catenary curve whose profile is described as

$$y = \frac{a}{2}(e^{x/a} + e^{-x/a})$$

Except for cables that actually do carry only their own weight (such as electrical power transmission lines) or that carry loads proportionally small with respect to their weight, this equation is not particularly useful.

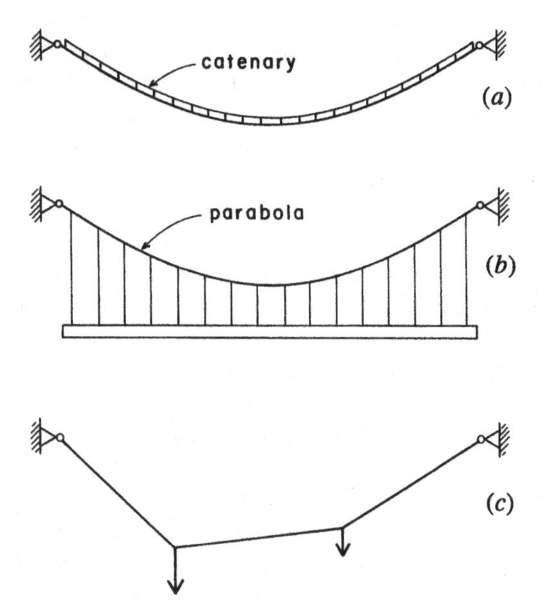

Figure 6.4 Response of the super-flexible structure to various loadings.

For the problems dealt with in this section, the weight of the cable can be ignored without significant error. When this is assumed, the cable profile becomes a pure response to the static resolution of the loads and support forces. The cable will thus assume a simple parabolic form (Figure 6.4*b*) when loaded with a uniformly distributed load, and a form consisting of straight segments (Figure 6.4*c*) when loaded with individual concentrated loads.

Consider the cable shown in Figure 6.5*a*, supporting a single concentrated load W and having four exterior reaction components: H_1, V_1, H_2, and V_2. Without consideration of the internal nature of the structure, the analysis is indeterminate with regard to the use of static equilibrium conditions alone. A special condition that can be used is that the element has

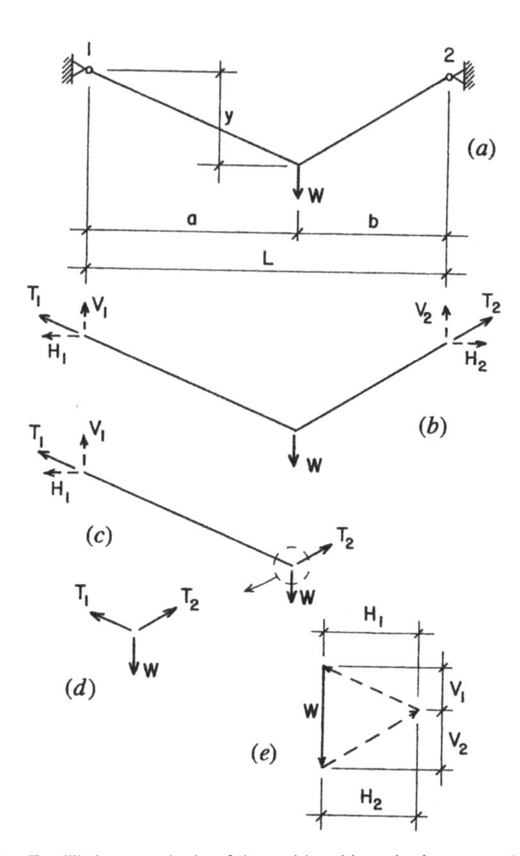

Figure 6.5 Equilibrium analysis of the cable with a single concentrated load.

no flexural resistance; therefore, the sum of the moments of forces at any point along the cable must be zero. Another special condition is that each segment of the cable operates as a two-force member; thus, the direction of the tension force must correspond with the form of the cable.

Referring to the free-body diagram of the whole cable in Figure 6.5*b*, consider a summation of moments about support 2, with clockwise moment as positive:

$$\Sigma M_2 = -(W \times b) + (V_1 \times L) = 0$$

from which

$$V_1 = \frac{b}{L} W$$

Similarly, using moments about support 1:

$$V_2 = \frac{a}{L} W$$

Now, considering the free-body diagram of the left portion of the cable, as shown in Figure 6.5*c*, take moments about the point of the load and consider clockwise moment as positive:

$$\Sigma M = 0 = (V_1 \times a) - (H_1 \times y)$$

from which:

$$H_1 = \frac{a}{y} V_1$$

Referring to the free-body diagram for the entire cable (Figure 6.5*b*), note that the two horizontal reaction components are the only horizontal forces on the structure. Therefore, $H_1 = H_2$. There are now sufficient relationships established to determine the four reaction components and the values of T_1 and T_2.

Observe also that the single load and the two cable tension forces form a concentric force system at the point of the load (Figure 6.5*d*). This system can be analyzed graphically by construction of the force triangle

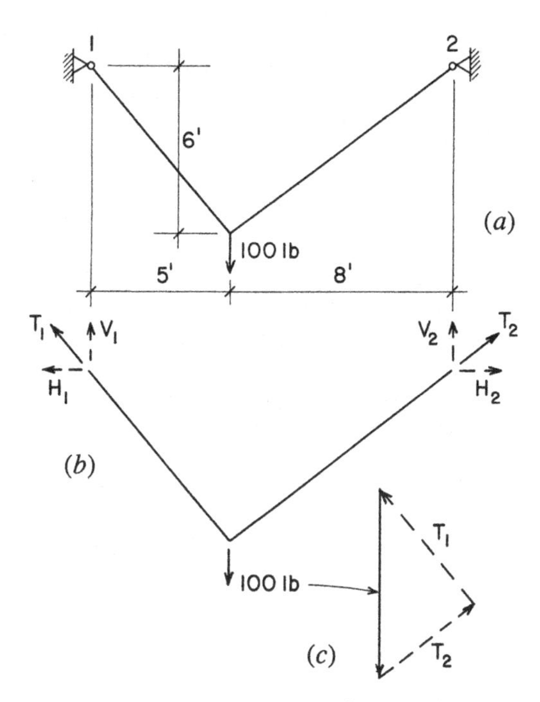

Figure 6.6 Reference figure for Example 4.

shown in Figure 6.5*e*. If desired, the values for the reaction components can also be determined from the graphical construction.

Example 4. Find the horizontal and vertical components of the reactions and the tension forces in the cable for the system shown in Figure 6.6*a*.

Solution: Using the relationships derived previously for the structure in Figure 6.5:

$$V_1 = \frac{b}{L}W = \frac{8}{13}(100) = 61.54 \text{ lb}$$

$$V_2 = \frac{a}{L}W = \frac{5}{13}(100) = 38.46 \text{ lb}$$

$$H_1 = H_2 = \frac{a}{y} V_1 = \frac{5}{6}(61.54) = 51.28 \text{ lb}$$

$$T_1 = \sqrt{(V_1)^2 + (H_1)^2} = \sqrt{(61.54)^2 + (51.28)^2} = 80.1 \text{ lb}$$

$$T_2 = \sqrt{(V_2)^2 + (H_2)^2} = \sqrt{(38.46)^2 + (51.28)^2} = 64.1 \text{ lb}$$

When the two supports are not at the same elevation, the preceding problem becomes somewhat more complex. The solution is still determinate, however, and may be accomplished as follows.

Example 5. Find the horizontal and vertical components of the reactions and the forces in the cables for the system shown in Figure 6.7*a*.

Solution: Referring to the free-body diagram in Figure 6.7*b*, note that

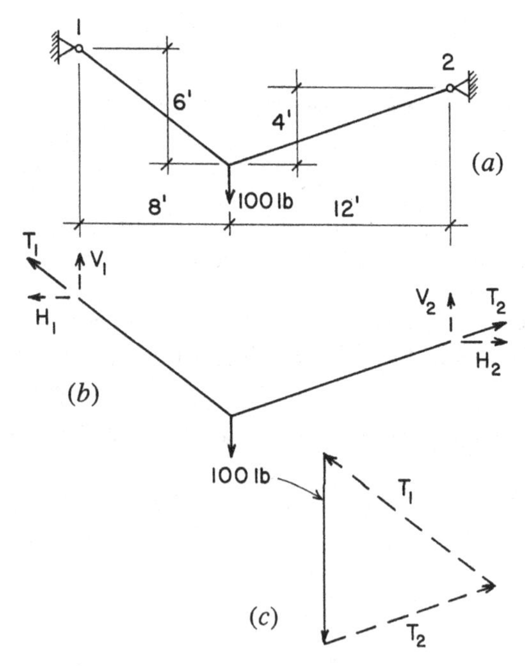

Figure 6.7 Reference figure for Example 5.

$$\Sigma F_v = 0 = V_1 + V_2 - 100; \quad \text{thus,} \quad V_1 + V_2 = 100$$

$$\Sigma F_h = 0 = H_1 + H_2; \quad \text{thus,} \quad H_1 = H_2$$

and, considering clockwise moments about support 2 as positive:

$$\Sigma M_2 = 0 = +(V_1 \times 20) - (H_1 \times 2) - (100 \times 12)$$

$$20V_1 - 2H_1 = 1200$$

From the geometry of T_1, observe that

$$H_1 = \frac{8}{6} V_1$$

Substituting this into the moment equation:

$$20V_1 - 2\left(\frac{8}{6} V_1\right) = 1200$$

$$\frac{104}{6} V_1 = 1200$$

$$V_1 = \frac{6}{104}(1200) = 69.23 \text{ lb}$$

Then:

$$H_1 = \frac{8}{6}(V_1) = \frac{8}{6}(69.23) = 92.31 \text{ lb} = H_2$$

$$V_2 = 100 - V_1 = 100 - 69.23 = 30.77 \text{ lb}$$

$$T_1 = \sqrt{(69.23)^2 + (92.31)^2} = 115.4 \text{ lb}$$

$$T_2 = \sqrt{(30.77)^2 + (92.31)^2} = 97.3 \text{ lb}$$

A cable loaded with a uniformly distributed load along a horizontal span (not along the cable itself), as shown in Figure 6.8*a,* assumes a simple parabolic (second-degree) curve profile. Referring to the free-body diagram in Figure 6.8*b,* observe that the horizontal component of the internal tension is the same for all points along the cable due to the equilibrium of horizontal forces. The vertical component of the internal tension varies as the slope of the cable changes, becoming a maximum value at the support and zero at the center of the span. Thus, the maximum internal tension occurs at the support and the minimum internal tension occurs at the center of the span.

Referring to Figure 6.8*c,* which is a free-body diagram of the left half of the cable, a summation of moments about the left support yields

$$\Sigma M = 0 = (H \times y) + \left(\frac{wL}{2} \times \frac{L}{4}\right)$$

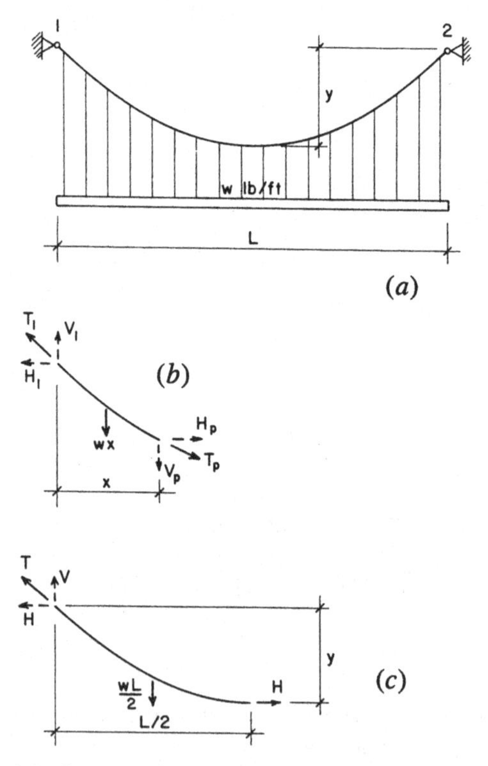

Figure 6.8 Behavior of the cable with a uniformly distributed load.

$$H = \frac{wL^2}{8y}$$

which is the general expression for horizontal force at all points along the cable.

Using the solutions for the vertical reaction components and the horizontal force in the cable, the maximum tension at the supports can be determined.

Problem 6.3.C, D. Find the maximum tension force in the cable in Figure 6.9 if T = 10 kips. (c) $x = y = 10$ ft, $s = t = 4$ ft; (d) $x = 12$ ft, $y = 16$ ft, $s = 8$ ft, $t = 12$ ft.

Combined Action: Tension Plus Bending

Although it is not a frequent occurrence, some situations in steel structures involve members subjected to a combination of tension and bending. For investigation and design, this condition is treated as one of interaction of the two distinct phenomena. This is essentially the same concept as that used for members subjected to combined compression and bending—that is, mostly columns with bending as discussed in Section 4.5.

A major difference with tension members, as compared to columns, is the lack of concern for slenderness and buckling. Otherwise, however, the AISC stipulates the use of the same basic formulas for interaction. As given in Section 4.5, two formulas can be used, depending on the ratio of the required axial force to the axial load resistance of the member. The two formulas are as follows:

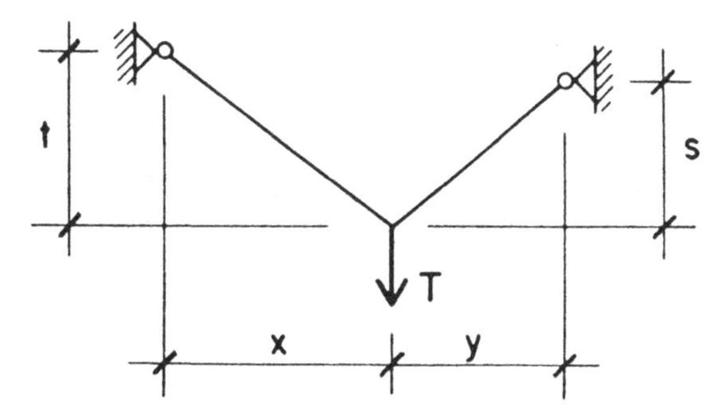

Figure 6.9 Reference figure for Problems 6.3.C and D.

If $P_u/\phi_t P_n < 0.2$:

$$\frac{P_u}{\phi_t P_n} + \frac{8}{9}\left(\frac{M_{ux}}{\phi_b M_{nx}} + \frac{M_{uy}}{\phi_b M_{ny}}\right) \le 1.0$$

If $P_u/\phi_t P_n < 0.2$:

$$\frac{P_u}{2\phi_t P_n} + \left(\frac{M_{ux}}{\phi_b M_{nx}} + \frac{M_{uy}}{\phi_b M_{ny}}\right) \le 1.0$$

In these equations, P_u and M_u are the required ultimate tension and bending strengths and P_n and M_n are the nominal tension and bending strengths of the member.

The following example demonstrates the interaction investigation for a simple steel hanger.

Example 6. Figure 6.10 shows a hanger consisting of a square steel bar, 2 in. on a side. Welded to one face of the bar is a steel plate with provision for attachment of the hanging factored load of 10 kips. The form of the hanger results in an eccentricity of 6 in. between the load and the axis of the 2-in. bar. Investigate the bar for combined tension and bending. Yield strength of the bar is 36 ksi.

Solution: For the interaction investigation, first determine the ratio of $P_u/\phi_t P_n$:

$$P_u = 10 \text{ kips}$$

$$\phi_t P_n = 0.9(36)(2)^2 = 129.6 \text{ kips}$$

$$\frac{P_u}{\phi_t P_n} = \frac{10}{129.6} = 0.0774$$

Because this ratio is less than 0.2, the second of the two interaction formulas is chosen. For the investigation, proceed as follows:

$$M_u = 10 \times 6 = 60 \text{ kips-in.}$$

$$Z = \frac{bd^2}{4} = \frac{2(2)^2}{4} = 2.0 \text{ in.}^3$$

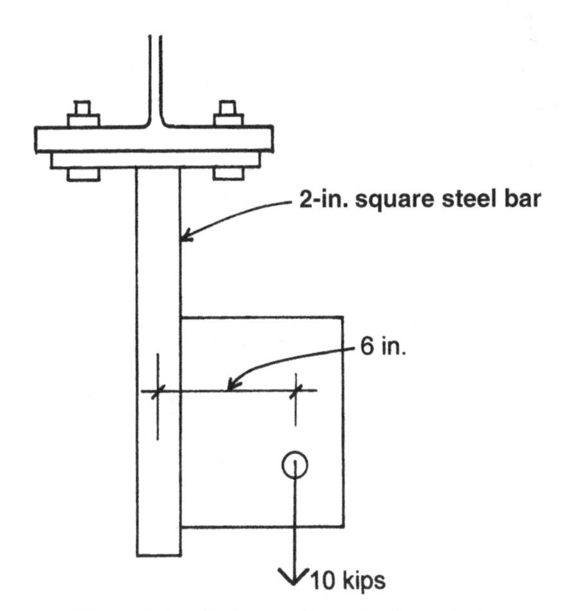

Figure 6.10 Reference figure for Example 6.

$$\phi_b M_n = 0.9F_y Z = 0.9(36)(2.0) = 64.8 \text{ kip-in.}$$

$$\frac{10}{2(129.6)} + \frac{60}{64.8} = 0.0387 + 0.9260 = 0.9647$$

Because the result is less than 1.0, the 2-in. bar is adequate.

Problem 6.3.E, F. Investigate the combined tension and bending in a hanger similar in form to that shown in Figure 6.10 for the following data: (e) bar is 2.5 in. square, load eccentricity is 8 in., P_u is 12 kips; (f) bar is 3 in. square, load eccentricity is 8 in., P_u is 20 kips.

7

HORIZONTAL-SPAN STEEL TRUSSES

Trussing, or triangulated framing, is a means for developing stability with a light frame. It is also a means for producing very light two-dimensional or three-dimensional structural elements for towers, spanning systems, or structures in general. In this chapter, we deal with some uses of simple, planar trusses for building structures, with a concentration on roof trusses—an application that generally makes fullest use of the potential lightness and freedom of form of the truss.

7.1 GENERAL CONSIDERATIONS

A historically common use of the truss is to achieve the simple, double-slope, gabled roof form. This is typically done by use of sloping members and a horizontal bottom member, as shown in Figure 7.1. Depending on the size of the span, the interior of the simple triangle formed by these three members may be filled by various arrangements of triangulated members. Some of the terminology used for the components of such a truss, as indicated in Figure 7.1 are as follows:

Chord members. These are the top and bottom boundary members of the truss, analogous to the top and bottom flanges of a steel beam. For trusses of modest size, these members are often made of a single element that is continuous through several joints, with a total length limited only by the maximum piece ordinarily obtainable from suppliers for the element selected.

Web members. The interior members of the truss are called web members. Unless there are interior joints, these members are of a single piece between chord joints.

Panels. Most trusses have a pattern that consists of some repetitive modular unit. This unit ordinarily is referred to as the panel of the truss; joints are sometimes referred to as panel points.

A critical dimension of a truss is its overall height, which is sometimes referred to as its *rise* or its *depth.* For the truss illustrated, this dimension relates to the establishment of the roof pitch and also determines the length of the web members. A critical concern with regard to the efficiency of the truss as a spanning structure is the ratio of the span of the truss to its height. Although beams and joists may be functional with span/height ratios as high as 20 to 30, trusses generally require much lower ratios.

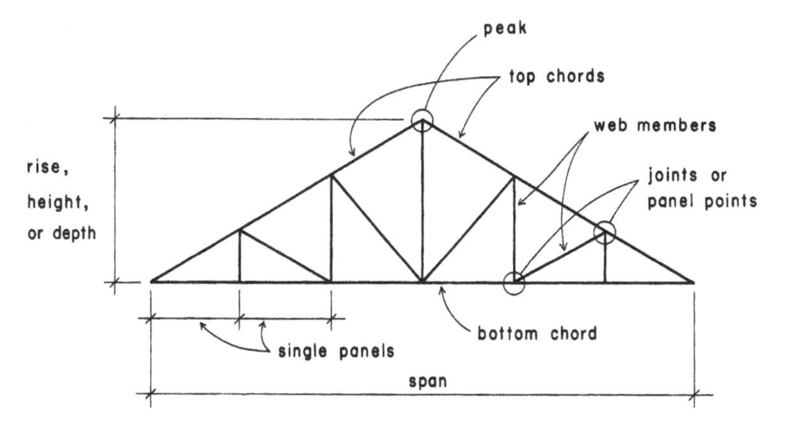

Figure 7.1 Truss elements.

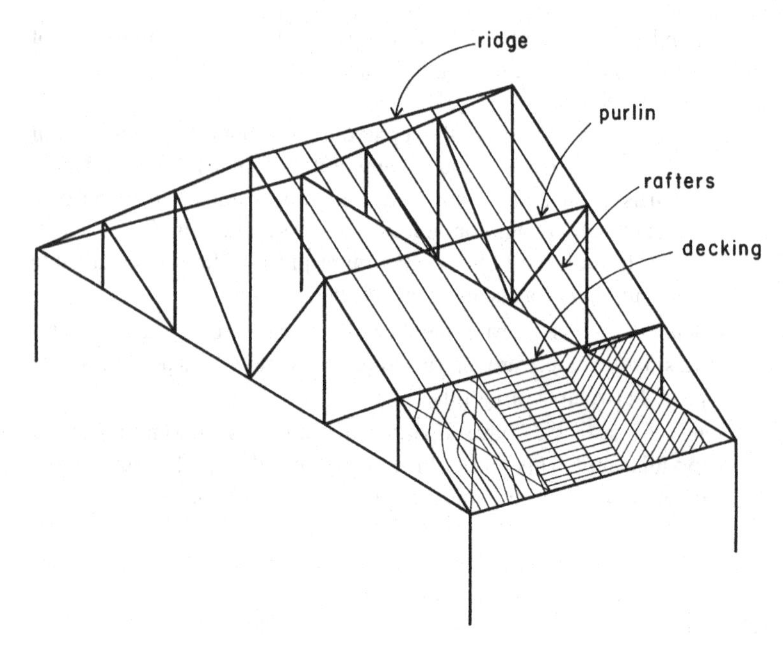

Figure 7.2 A roof structure with trusses.

Trusses may be used in a number of ways as part of the total structural system for a building. Figure 7.2 shows a series of single-span, planar trusses in the form shown in Figure 7.1 with the other elements of the building structure that develop the roof system and provide support for the trusses. In this example, the trusses are spaced a considerable distance apart. In this situation, it is common to use purlins to span between the trusses, usually supported at the top chord joints of the trusses to avoid bending in the chords. The purlins, in turn, support a series of closely spaced rafters that are parallel to the trusses. The roof deck is then attached to the rafters so that the roof surface actually floats above the level of the top of the trusses.

Figure 7.3 shows a similar structural system of trusses with parallel chords. This system may be used for a floor or a flat roof.

When the trusses are slightly closer together, it may be more practical to eliminate the purlins and to increase the size of the top chords to accommodate the additional bending due to the rafters. As an extension of this idea, if the trusses are really close, it may be possible to eliminate the

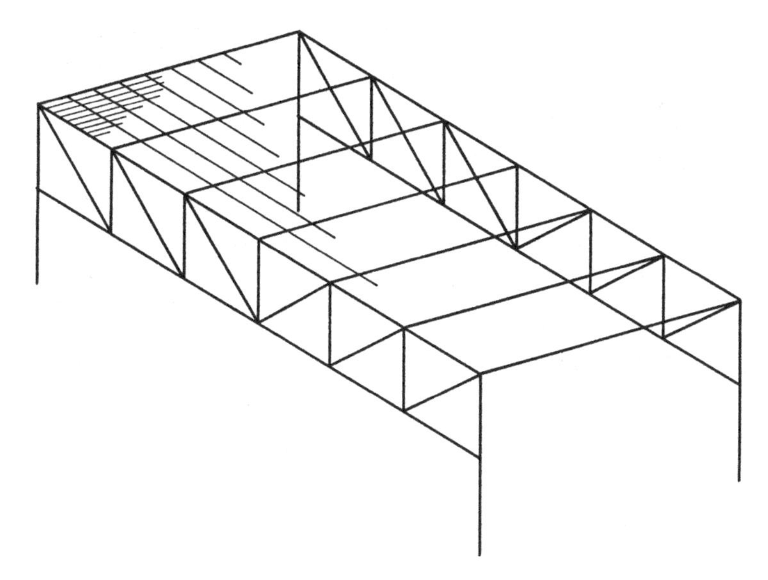

Figure 7.3 Flat-spanning, parallel-chorded trusses.

rafters as well and to place the deck directly on the top chords of the trusses.

For various situations, additional elements may be required for the complete structural system. If a ceiling is required, another framing system is used at the level of the bottom chords or suspended some distance below it. If the supported deck and framing and the ceiling framing do not provide it adequately, it may be necessary to use some bracing system perpendicular to the trusses in order to brace the laterally unstable trusses.

Truss patterns are derived from a number of considerations, starting with the basic profile of the truss. For various reasons, a number of classic truss patterns have evolved and have become standard parts of our structural vocabulary. Some of these carry the names of the designers who first developed them. Several common truss forms are shown in Figure 7.4.

The two most common forms of steel trusses of small to medium size are those shown in Figure 8.13. In both cases, the members may be connected by rivets, bolts, or welds. The most common practice is to use welding for connections that are assembled in the fabricating shop, high-strength bolts (torque tensioned) for permanent field connections, and unfinished bolts for temporary field connections.

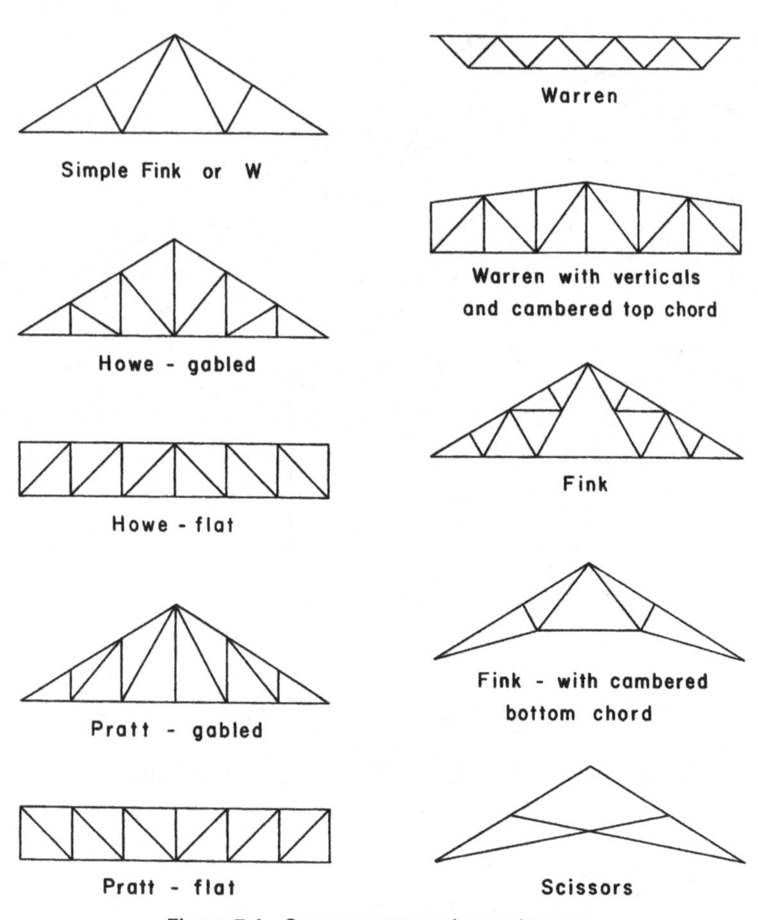

Figure 7.4 Common patterns for steel trusses.

7.2 BRACING FOR TRUSSES

Single planar trusses are very thin structures that require some form of lateral bracing. The compression chord of the truss must be designed for its laterally unbraced length. In the plane of the truss, the chord is braced by other truss members at each joint. However, if there is no lateral bracing, the unbraced length of the chord in a direction perpendicular to the plane of the truss becomes the full length of the truss. Obviously, it is not feasible to design a slender compression member for this unbraced length.

In most buildings, other elements of the construction ordinarily provide some or all of the necessary bracing for the trusses. In the structural system shown in Figure 7.5*a,* the top chord of the truss is braced at each truss joint by the purlins. If the roof deck is a reasonably rigid planar structural element and is adequately attached to the purlins, this constitutes a very adequate bracing of the compression chord—which is the main problem for the truss. However, it is also necessary to brace the truss generally for out-of-plane movement throughout its height. In Figure 7.5*a* this is done by providing a vertical plane of X-bracing at every other panel point of the truss. The purlin does an additional service by serving as part of this vertical plane of trussed bracing. One panel of this bracing is actually capable of bracing a pair of trusses, so that it would be possible to place it only in alternate bays between the trusses. However, the bracing may be part of the general bracing system for the building, as well as providing for the bracing of the individual trusses. In the latter case, it would probably be continuous.

Light trusses that directly support a deck, as shown in Figure 7.5*b,* are usually adequately braced by the deck. This constitutes continuous bracing, so that the unbraced length of the chord in this case may be virtually zero (depending on the form of attachment of the deck). Additional bracing in this situation often is limited to a series of continuous steel rods or single small angles that are attached to the bottom chords as shown in the illustration.

Another form of bracing is shown in Figure 7.5*c.* In this case, a horizontal plane of X-bracing is placed between two trusses at the level of the bottom chords. This single braced bay may be used to brace several other bays of trusses by connecting them to the X-braced trusses with horizontal struts. As in the previous example, with vertical planes of bracing, the top chord is braced by the roof construction. It is likely that bracing of this form is also part of the general lateral bracing system for the building so that its use, location, and details are not developed strictly for the bracing of the trusses.

7.3 LOADS ON TRUSSES

The first step in the design of a roof truss consists of computing the loads the truss will be required to support. These are dead and live loads. The former includes the weight of all construction materials supported by the truss; the latter includes loads resulting from snow and wind, and, on flat

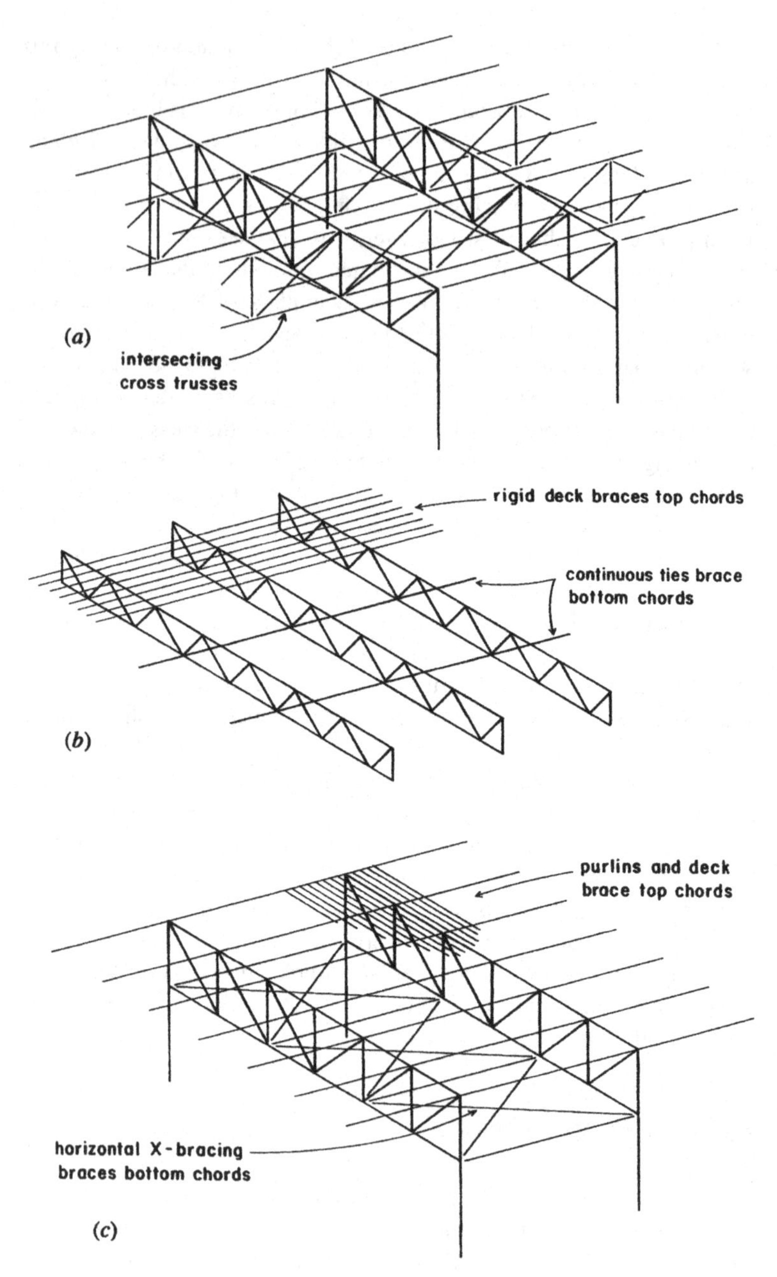

Figure 7.5 Forms of lateral bracing for steel trusses.

roofs, occupancy loads and an allowance for the possible ponding of water due to impaired drainage.

The following items constitute the materials to be considered in computing the dead loads: roof covering and roof deck, purlins and sway bracing, ceiling and any suspended loads, and the weight of the truss itself. Obviously, all cannot be determined exactly before the truss is designed, but all may be checked later to see whether a sufficient allowance has been made. The dead loads are downward vertical forces; hence, the end reactions of the truss are also vertical with respect to these loads. Table 9.1 gives the weights of certain roofing materials, and Table 7.1 provides estimated weights of steel trusses for various spans and pitches. With respect to the latter, one procedure is to establish an estimate in pounds per square foot of roof surface and consider this load as acting at the panel points of the upper chord. A more exact method would be to apportion a part of such loads to the panel points of the lower chord, but this is customary only in trusses with exceptionally long spans. After the truss has been designed, its actual weight may be computed and compared with the estimated weight.

The weight allowance for snow load depends primarily on the geographical location of the structure and the roof slope. Freshly fallen snow may weigh as much as 10 lb per cu ft [0.13 kg/m^3], and accumulations of wet or packed snow may exceed this value. The amount of snow retained on a roof over a given period depends on the type of roofing as well as the slope; for example, snow slides off a metal or slate roof more readily than from a wood shingle surface; also, the amount of insulation in the roof construction will influence the period of retention. Local building codes are the usual source for minimum snow loads.

Required general design live loads for roofs are also specified by local building codes. When snow is a potential problem, the load is usually based on anticipated snow accumulation. Otherwise, the specified live

TABLE 7.1 Approximate Weight of Steel Trusses in Pounds per Square Foot of Supported Roof Surface

Span		Slope of Roof			
ft	m	45°	30°	20°	Flat
Up to 40	Up to 12	5	6	7	8
40–50	12–15	6	7	7	8
50–65	15–20	7	8	9	10
65–80	20–25	9	9	10	11

load is intended essentially to provide some capacity for sustaining loads experienced during construction and maintenance of the roof. The basic required load can usually be modified when the roof slope is of some significant angle and on the basis of the total roof surface area supported by the structure.

Magnitudes of design roof loads and wind pressures and various other requirements for design are specified by local building codes. The code in force for a specific building location should be used for any design work. For a general explanation of the analysis and design of the effects of wind and earthquake forces on buildings, refer to *Simplified Building Design for Wind and Earthquake Forces* by Ambrose and Vergun (Ref. 6).

7.4 INVESTIGATION FOR INTERNAL FORCES IN PLANAR TRUSSES

Planar trusses, composed of linear elements assembled in triangulated frameworks, have been used for spanning structures in buildings for many centuries. Investigation for internal forces in trusses is typically performed by the basic methods illustrated in the preceding sections. In this section, these procedures are demonstrated using both algebraic and graphical methods of solution.

Graphical Analysis of Planar Trusses

When the so-called *method of joints* is used, finding the internal forces in the members of a planar truss consists of solving a series of concurrent force systems. Figure 7.6 shows a truss with the truss form, the loads, and the reactions displayed in a space diagram. Below the space diagram is a figure consisting of the free-body diagrams of the individual joints of the truss. These are arranged in the same manner as they are in the truss in order to show their interrelationships. However, each joint constitutes a complete concurrent planar force system that must have its independent equilibrium. Solving the problem consists of determining the equilibrium conditions for all of the joints. The procedures used for this solution will now be illustrated.

Figure 7.7 shows a single span, planar truss subjected to vertical gravity loads. This example will be used to illustrate the procedures for determining the internal forces in the truss—that is, the tension and compression forces in the individual members of the truss. The space diagram in the figure shows the truss form and dimensions, the support condi-

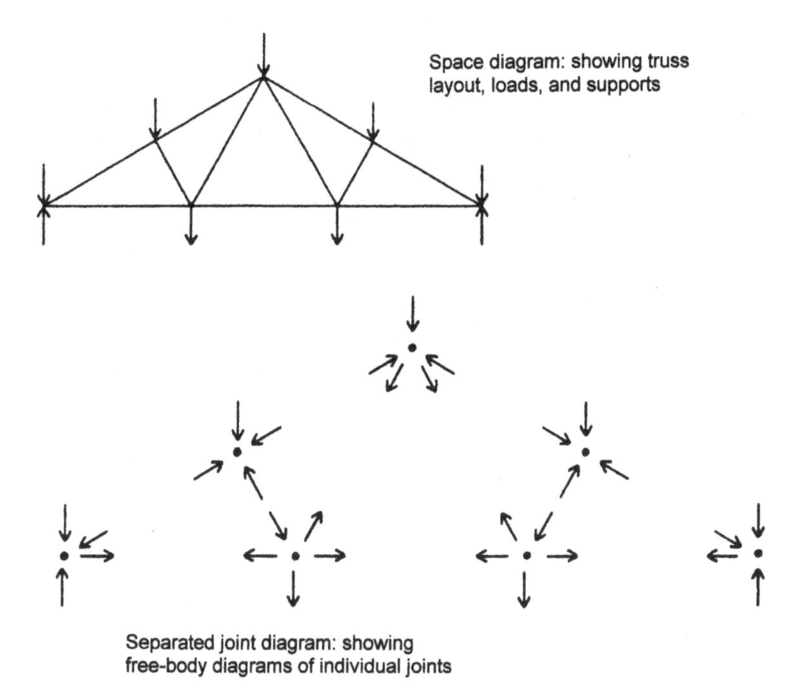

Figure 7.6 Examples of diagrams used to represent trusses and their actions.

tions, and the loads. The letters on the space diagram identify individual forces at the truss joints. The sequence of placement of the letters is arbitrary; the only necessary consideration is to place a letter in each space between the loads and the individual truss members so that each force at a joint can be identified by a two-letter symbol.

The separated joint diagram in the figure provides a useful means for visualization of the complete force system at each joint as well as the interrelation of the joints through the truss members. The individual forces at each joint are designated by two-letter symbols that are obtained by simply reading around the joint in the space diagram in a clockwise direction. Note that the two-letter symbols are reversed at the opposite ends of each of the truss members. Thus, the top chord member at the left end of the truss is designated as *BI* when shown in the joint at the left support (joint 1) and is designated as *IB* when shown in the first interior upper chord joint (joint 2). The purpose of this procedure will be demonstrated in the following explanation of the graphical analysis.

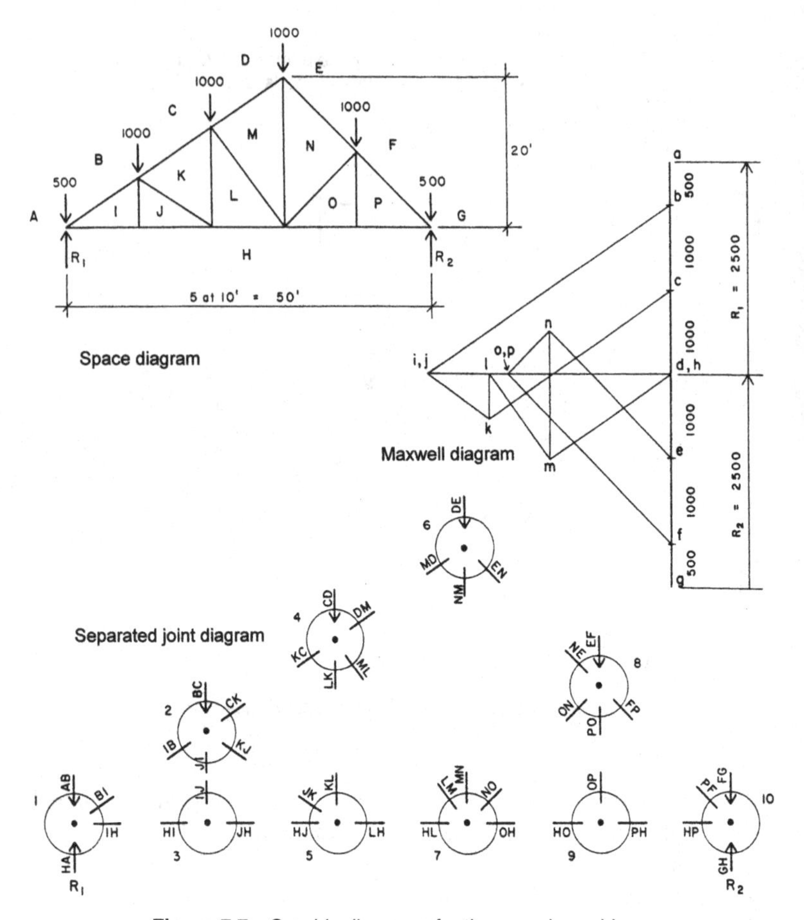

Figure 7.7 Graphic diagrams for the sample problem.

The third diagram in Figure 7.7 is a composite force polygon for the external and internal forces in the truss. It is called a *Maxwell diagram* after one of its early promoters, James Maxwell, a British engineer. The construction of this diagram constitutes a complete solution for the magnitudes and senses of the internal forces in the truss. The procedure for this construction is as follows:

Construct the force polygon for the external forces. Before this can be done, the values for the reactions must be found. There are graphic techniques for finding the reactions, but it is usually much simpler and faster to find them with an algebraic solution. In this example, although the truss is not symmetrical, the loading is, and it may simply be observed that the reactions are each equal to one-half of the total load on the truss, or 5000/2 = 2500 lb. Because the external forces in this case are all in a single direction, the force polygon for the external forces is actually a straight line. Using the two-letter symbols for the forces and starting with the letter *A* at the left end, we read the force sequence by moving in a clockwise direction around the outside of the truss. The loads are thus read as *AB, BC, CD, DE, EF,* and *FG,* and the two reactions are read as *GH* and *HA.* Beginning at *A* on the Maxwell diagram, the force vector sequence for the external forces is read from *A* to *B, B* to *C, C* to *D,* and so on, ending back at *A,* which shows that the force polygon closes and the external forces are in the necessary state of static equilibrium. Note that we have pulled the vectors for the reactions off to the side in the diagram to indicate them more clearly. Note also that we have used lowercase letters for the vector ends in the Maxwell diagram, whereas uppercase letters are used on the space diagram. The alphabetic correlation is thus retained (*A* to *a*), while any possible confusion between the two diagrams is prevented. The letters on the space diagram designate open spaces, whereas the letters on the Maxwell diagram designate points of intersection of lines.

Construct the force polygons for the individual joints. The graphic procedure for this consists of locating the points on the Maxwell diagram that correspond to the remaining letters, *I* through *P,* on the space diagram in Figure 7.7. When all the lettered points on the diagram are located, the complete force polygon for each joint may be read on the diagram. To locate these points, we use two relationships. The first is that the truss members can resist only forces that are parallel to the members' positioned directions. Thus, we know the directions of all the internal forces. The second relationship is a simple one from plane geometry: A point may be located at the intersection of two lines. Consider the forces at joint 1, as shown in the separated joint diagram in Figure 7.7. Note that there are four forces and that two of them are known (the load and the reaction) and two are unknown (the internal forces in the truss members). The force polygon for this joint, as shown on the Maxwell diagram, is read as *ABIHA. AB* represents the load, *BI* the force in the upper chord member, *IH* the force in the lower chord member, and *HA* the reaction. Thus, the location of point *i* on the Maxwell diagram is determined by noting that *i* must be in a horizontal direction from *h* (corresponding to the horizontal position of the lower chord) and in a direction from *b* that is parallel to the position of the upper chord.

The remaining points on the Maxwell diagram are found by the same process, using two known points on the diagram to project lines of known direction whose intersection will determine the location of an unknown point. Once all the points are located, the diagram is complete and can be used to find the magnitude and sense of each internal force. The process for construction of the Maxwell diagram typically consists of moving from joint to joint along the truss. Once one of the letters for an internal space is determined on the Maxwell diagram, it may be used as a known point for finding the letter for an adjacent space on the space diagram. The only limitation of the process is that it is not possible to find more than one unknown point on the Maxwell diagram for any single joint. Consider joint 7 on the separated joint diagram in Figure 7.7. To solve this joint first, knowing only the locations of letters *a* through *h* on the Maxwell diagram, it is necessary to locate four unknown points: *l, m, n,* and *o.* This is three more unknowns than can be determined in a single step, so three of the unknowns must be found by using other joints.

Solving for a single unknown point on the Maxwell diagram corresponds to finding two unknown forces at a joint, because each letter on the space diagram is used twice in the force identification for the internal forces. Thus, for joint 1 in the previous example, the letter *I* is part of the identity of forces *BI* and *IH,* as shown on the separated joint diagram. The graphic determination of single points on the Maxwell diagram, therefore, is analogous to finding two unknown quantities in an algebraic solution. Two unknowns are the maximum that can be solved for the equilibrium of a coplanar, concurrent force system, which is the condition of the individual joints in the truss.

When the Maxwell diagram is completed, the internal forces can be read from the diagram as follows:

The magnitude is determined by measuring the length of the line in the diagram, using the scale that was used to plot the vectors for the external forces.

The sense of individual forces is determined by reading the forces in clockwise sequence around a single joint in the space diagram and tracing the same letter sequences on the Maxwell diagram.

Figure 7.8*a* shows the force system at joint 1 and the force polygon for these forces as taken from the Maxwell diagram. The forces known initially are shown as solid lines on the force polygon, and the unknown forces are shown as dashed lines. Starting with letter *A* on the force system, we read the forces in a clockwise sequence as *AB, BI, IH,* and *HA.* Note that on the Maxwell diagram moving from *a* to *b* is moving in the order of the sense of the force—that is, from tail to end of the force vector that represents the external load on the joint. Using this sequence on the Maxwell diagram, this force sense flow will be a continuous one. Thus, reading from *b* to *i* on the Maxwell diagram is reading from tail to head of the force vector, which indicates that force *BI* has its head at the left end. Transferring this sense indication from the Maxwell diagram to the joint diagram indicates that force *BI* is in compression; that is, it is pushing, rather than pulling, on the joint. Reading from *i* to *h* on the Maxwell diagram shows that the arrowhead for this vector is on the right, which translates to a tension effect on the joint diagram.

Having solved for the forces at joint 1 as described, the fact that the forces in truss members *BI* and *IH* are known can be used to consider the adjacent joints, 2 and 3. However, it should be noted that the sense reverses at the opposite ends of the members in the joint diagrams. Referring to the separated joint diagram in Figure 7.7, if the upper chord member shown as force *BI* in joint 1 is in compression, its arrowhead is at the lower left end in the diagram for joint 1, as shown in Figure 7.8*a.* However, when the same force is shown as *IB* at joint 2, its pushing effect on the joint will be indicated by having the arrowhead at the upper right end in the diagram for joint 2. Similarly, the tension effect of the lower chord is shown in joint 1 by placing the arrowhead on the right end of the force *IH,* but the same tension force will be indicated in joint 3 by placing the arrowhead on the left end of the vector for force *HI.*

If the solution sequence of solving joint 1 and then joint 2 is chosen, it is now possible to transfer the known force in the upper chord to joint 2. Thus, the solution for the five forces at joint 2 is reduced to finding three unknowns, because the load *BC* and the chord force *IB* are now known. However, it is still not possible to solve joint 2, because there are two unknown points on the Maxwell diagram (*k* and *j*) corresponding to the three unknown forces. An option, therefore, is to proceed from joint 1 to joint 3, at which there are presently only two unknown forces. On the Maxwell diagram, the single unknown point *j* can be found by projecting vector *IJ* vertically from *i* and projecting vector *JH* horizontally from point *h.* Because point *i* is also located horizontally from point *h,* this shows that the vector *IJ* has zero magnitude, since both *i* and *j* must be

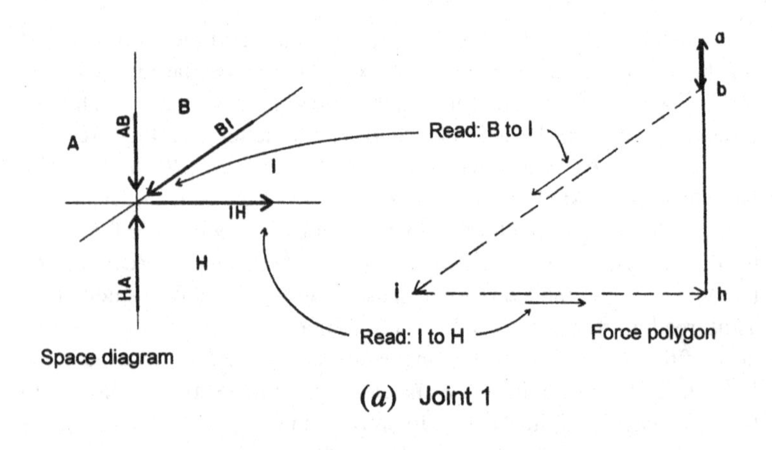

(*a*) Joint 1

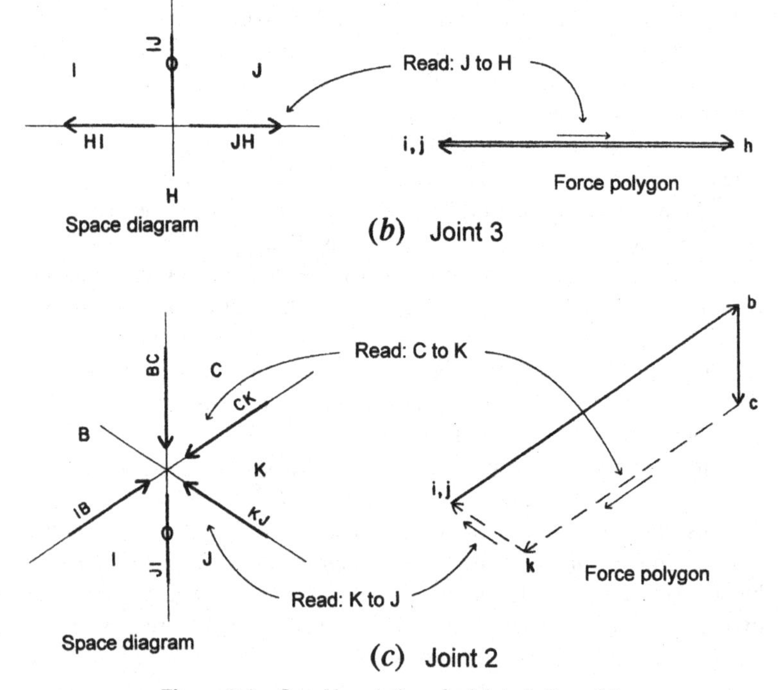

(*b*) Joint 3

(*c*) Joint 2

Figure 7.8 Graphic solutions for joints 1, 2, and 3.

on a horizontal line from *h* in the Maxwell diagram. This indicates that there is actually no stress in this truss member for this loading condition and that points *i* and *j* are coincident on the Maxwell diagram. The joint force diagram and the force polygon for joint 3 are as shown in Figure 7.8*b*. In the joint force diagram, place a zero, rather than an arrowhead, on the vector line for *IJ* to indicate the zero stress condition. In the force polygon in Figure 7.8*b*, the two force vectors are slightly separated for clarity, although they are actually coincident on the same line.

Having solved for the forces at joint 3, proceed to joint 2, because there remain only two unknown forces at this joint. The forces at the joint and the force polygon for joint 2 are shown in Figure 7.8*c*. As for joint 1, read the force polygon in a sequence determined by reading clockwise around the joint: *BCKJIB*. Following the continuous direction of the force arrows on the force polygon in this sequence, it is possible to establish the sense for the two forces *CK* and *KJ*.

It is possible to proceed from one end and to work continuously across the truss from joint to joint to construct the Maxwell diagram in this example. The sequence in terms of locating points on the Maxwell diagram would be *i-j-k-l-m-n-o-p*, which would be accomplished by solving the joints in the following sequence: 1,3,2,5,4,6,7,9,8. However, it is advisable to minimize the error in graphic construction by working from both ends of the truss. Thus, a better procedure would be to find points *i-j-k-l-m*, working from the left end of the truss, and then to find points *p-o-n-m*, working from the right end. This would result in finding two locations for *m*, whose separation constitutes the error in drafting accuracy.

Problems 7.4.A, B. Using a Maxwell diagram, find the internal forces in the truss in Figure 7.9.

Algebraic Analysis of Planar Trusses

Graphical solution for the internal forces in a truss using the Maxwell diagram corresponds essentially to an algebraic solution by the *method of joints*. This method consists of solving the concentric force systems at the individual joints using simple force equilibrium equations. The process will be illustrated using the previous example.

As with the graphic solution, first determine the external forces, consisting of the loads and the reactions. Then proceed to consider the equilibrium of the individual joints, following a sequence as in the graphic solution. The limitation of this sequence, corresponding to the limit of

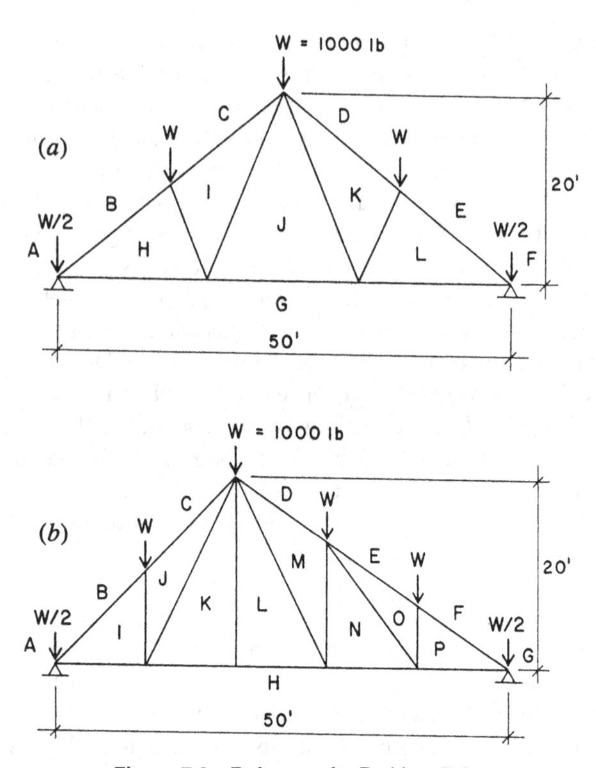

Figure 7.9 Reference for Problem 7.4.

finding only one unknown point in the Maxwell diagram, is that only two unknown forces at any single joint can be found in a single step. (Two conditions of equilibrium produce two equations.) With reference to Figure 7.10, the solution for joint 1 is as follows.

The force system for the joint is drawn with the sense and magnitude of the known forces shown, but with the unknown internal forces represented by lines without arrowheads, because their senses and magnitudes initially are unknown. For forces that are not vertical or horizontal, replace the forces with their horizontal and vertical components. Then consider the two conditions necessary for the equilibrium of the system: The sum of the vertical forces is zero and the sum of the horizontal forces is zero.

If the algebraic solution is performed carefully, the sense of the forces will be determined automatically. However, it is recommended that whenever possible the sense be predetermined by simple observations of the joint conditions, as will be illustrated in the solutions.

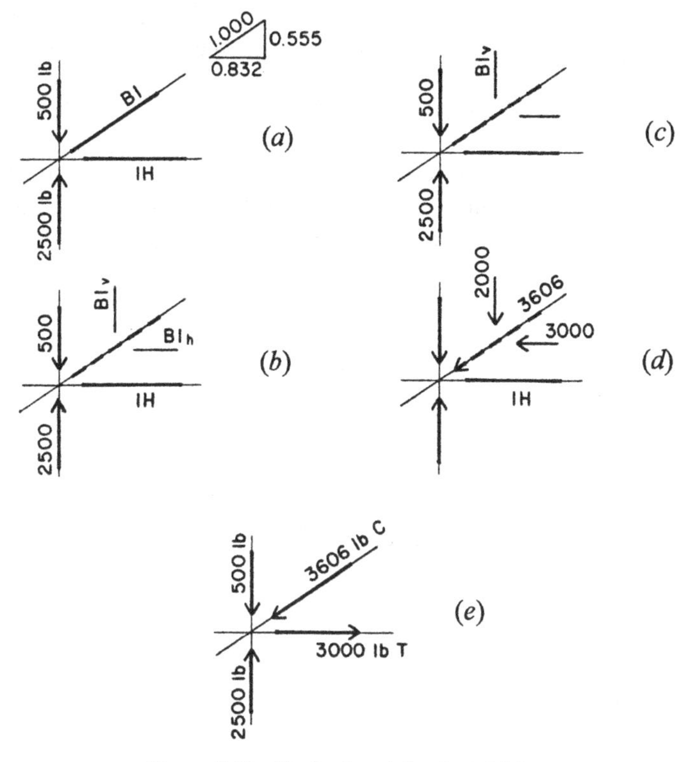

Figure 7.10 Algebraic solution for joint 1.

The problem to be solved at joint 1 is as shown in Figure 7.10*a*. In Figure 7.10*b* the system is shown with all forces expressed as vertical and horizontal components. Note that although this now increases the number of unknowns to three (IH, BI_v, and BI_h), there is a numeric relationship between the two components of BI. When this condition is added to the two algebraic conditions for equilibrium, the number of usable relationships totals three, so that the necessary conditions to solve for the three unknowns are present.

The condition for vertical equilibrium is shown at (*c*) in Figure 7.10. Because the horizontal forces do not affect the vertical equilibrium, the balance is between the load, the reaction, and the vertical component of the force in the upper chord. Simple observation of the forces and the known magnitudes makes it obvious that force BI_v must act downward, indicating that BI is a compression force. Thus, the sense of BI is established by simple visual inspection of the joint, and the algebraic equation for vertical equilibrium (with upward force considered positive) is

$$\Sigma F_v = 0 = +2500 - 500 - BI_v$$

From this equation BI_v is determined to have a magnitude of 2000 lb. Using the known relationships between BI, BI_v, and BI_h, the values of these three quantities can be determined if any one of them is known. Thus:

$$\frac{BI}{1.000} = \frac{BI_v}{0.555} = \frac{BI_h}{0.832}$$

from which:

$$BI_h = \left(\frac{0.832}{0.555}\right)(2000) = 3000 \text{ lb}$$

and

$$BI = \left(\frac{1.000}{0.555}\right)(2000) = 3606 \text{ lb}$$

The results of the analysis to this point are shown at (*d*) in Figure 7.10, from which it may be observed that the conditions for equilibrium of the horizontal forces can be expressed. Stated algebraically (with force sense toward the right considered positive), the condition is

$$\Sigma F_h = 0 = IH - 3000$$

from which it is established that the force in *IH* is 3000 lb.

The final solution for the joint is then as shown at (*e*) in the figure. On this diagram, the internal forces are identified as to sense by using *C* to indicate compression and *T* to indicate tension.

As with the graphic solution, proceed to consider the forces at joint 3. The initial condition at this joint is as shown at (*a*) in Figure 7.11, with the single known force in member *HI* and the two unknown forces in *IJ* and *JH*. Because the forces at this joint are all vertical and horizontal, there is no need to use components. Consideration of vertical equilibrium

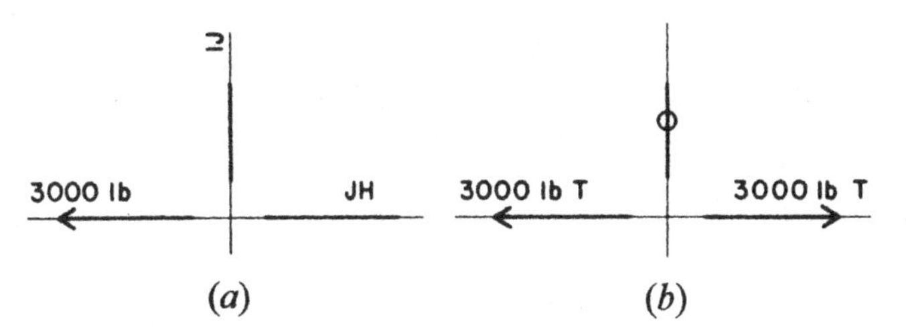

Figure 7.11 Algebraic solution for joint 3.

makes it obvious that it is not possible to have a force in member *IJ*. Stated algebraically, the condition for vertical equilibrium is

$$\Sigma F_v = 0 = IJ \text{ (because } IJ \text{ is the only force)}$$

It is equally obvious that the force in *JH* must be equal and opposite to that in *HI,* because they are the only two horizontal forces. That is, stated algebraically:

$$\Sigma F_v = 0 = JH - 3000$$

The final answer for the forces at joint 3 is as shown at (*b*) in Figure 7.11. Note the convention for indicating a truss member with no internal force.

Now proceed to consider joint 2; the initial condition is as shown at (*a*) in Figure 7.12. Of the five forces at the joint only two remain unknown. Following the procedure for joint 1, first resolve the forces into their vertical and horizontal components, as shown at (*b*) in Figure 7.12.

Because the sense of forces *CK* and *KJ* is unknown, use the procedure of considering them to be positive until proven otherwise. That is, if they are entered into the algebraic equations with an assumed sense, and the solution produces a negative answer, then the assumption was wrong. However, be careful to be consistent with the sense of the force vectors, as the following solution will illustrate.

Arbitrarily assume that force *CK* is in compression and force *KJ* is in tension. If this is so, the forces and their components will be as shown at (*c*) in Figure 7.12. Then consider the conditions for vertical equilibrium; the forces involved will be those shown at (*d*) in Figure 7.12, and the equation for vertical equilibrium will be

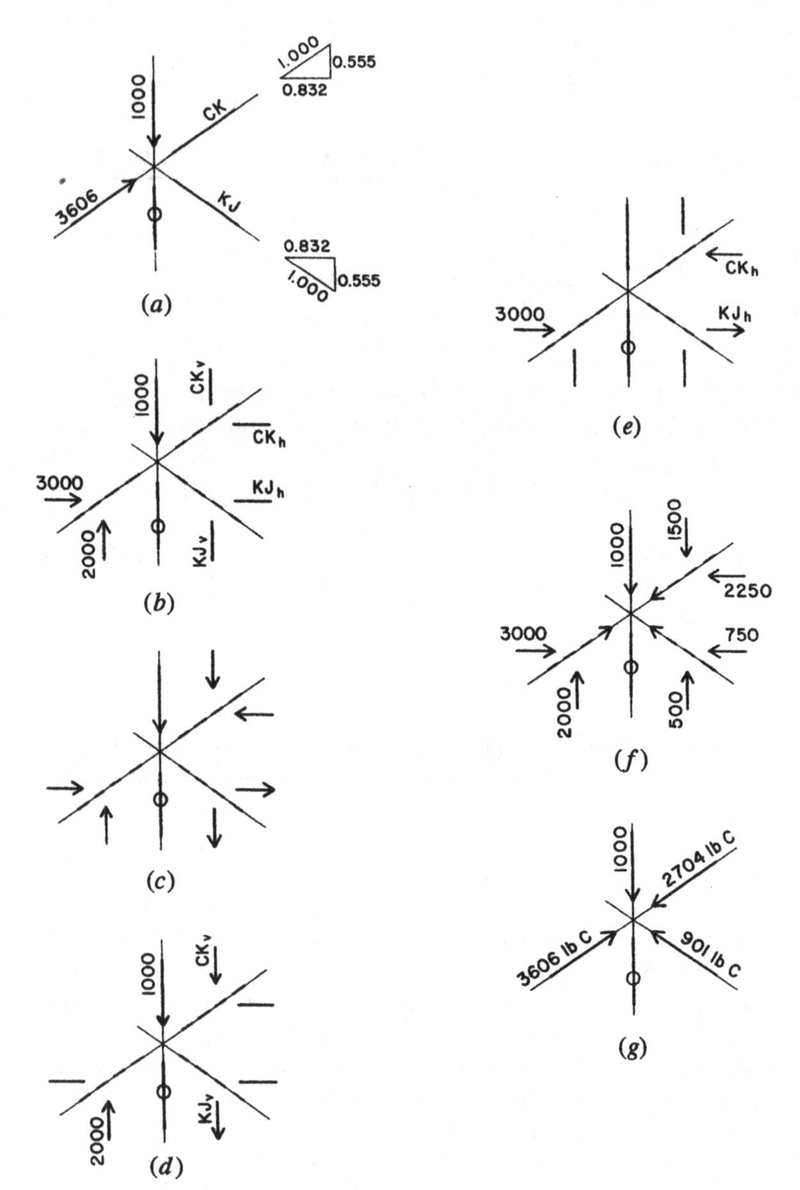

Figure 7.12 Algebraic solution for joint 2.

$$\Sigma F_v = 0 = -1000 + 2000 - CK_v - KJ_v$$

or:

$$0 = +\ 1000 - 0.555CK - 0.555KJ \qquad (7.4.1)$$

Now consider the conditions for horizontal equilibrium; the forces will be as shown at (*e*) in Figure 7.12, and the equation will be

$$\Sigma F_h = 0 = +\ 3000 - CK_h + KJ_h$$

or

$$0 = +\ 3000 - 0.832CK + 0.832KJ \qquad (7.4.2)$$

Note the consistency of the algebraic signs and the sense of the force vectors, with positive forces considered as upward and toward the right. Now solve these two equations simultaneously for the two unknown forces as follows:

1. Multiply equation (7.4.1) by 0.832/0.555:

$$0 = \left(\frac{0.832}{0.555}\right)(+1000) + \left(\frac{0.832}{0.555}\right)(-0.555CK) + \left(\frac{0.832}{0.555}\right)(-0.555KJ)$$

or:

$$0 = +\ 1500 - 0.832CK - 0.832KJ$$

2. Add this equation to equation (7.4.2) and solve for *CK*:

$$0 = +\ 4500 - 1.664CK,\ \mathrm{CK} = \frac{4500}{1.664} = 2704 \text{ lb}$$

Note that the assumed sense of compression in *CK* is correct, because the algebraic solution produces a positive answer. Substituting this value for *CK* in equation (7.4.1):

$$0 = +\ 1000 - 0.555(2704) - 0.555(KJ)$$

and:

$$KJ = \frac{-500}{0.555} = -901 \text{ lb}$$

Because the algebraic solution produces a negative quantity for *KJ,* the assumed sense for *KJ* is wrong and the member is actually in compression.

The final answers for the forces at joint 2 are as shown at (g) in Figure 7.12. To verify that equilibrium exists, however, the forces are shown in the form of their vertical and horizontal components at (f) in the illustration.

When all of the internal forces have been determined for the truss, the results may be recorded or displayed in a number of ways. The most direct way is to display them on a scaled diagram of the truss, as shown in Figure 7.13*a*. The force magnitudes are recorded next to each member

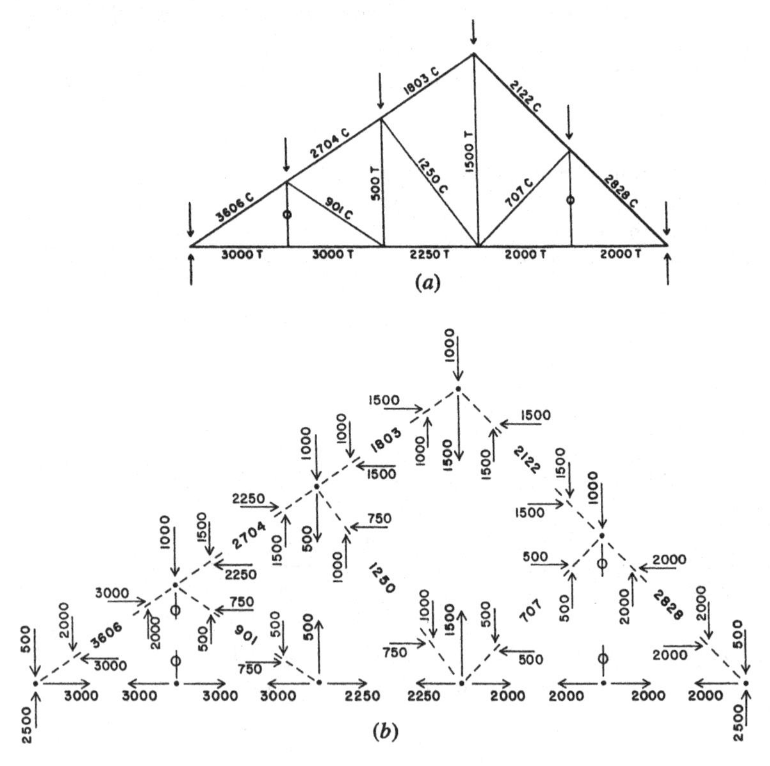

Figure 7.13 Presentation of the internal forces in the truss.

with the sense shown as *T* for tension or *C* for compression. Zero stress members are indicated by the conventional symbol consisting of a zero placed directly on the member.

When solving by the algebraic method of joints, the results may be recorded on a separated joint diagram as shown in Figure 7.13*b*. If the values for the vertical and horizontal components of force in sloping members are shown, it is a simple matter to verify the equilibrium of the individual joints.

Problem 7.4.C, D. Using the algebraic method of joints, find the internal forces in the truss in Figure 7.9.

Internal Forces Found by Coefficients

Figure 7.14 shows a number of simple trusses of both parallel-chorded and gable form. Table 7.2 lists coefficients that may be used to find the values for the internal forces in these trusses. For the gable-form trusses, coefficients are given for three different slopes of the top chord: 4 in 12, 6 in 12, and 8 in 12. For the parallel-chorded trusses, coefficients are given for two different ratios of the truss depth to the truss panel length: 1 to 1 and 3 to 4. Loading results from gravity loads and is assumed to be applied symmetrically to the truss.

Values in Table 7.2 are based on a unit load of 1.0 for *W;* thus, true forces may be simply proportioned for specific values of *W* once actual loading is determined. Because all the trusses are symmetrical, values are given for only half of each truss.

7.5 DESIGN FORCES FOR TRUSS MEMBERS

The primary concern in analysis of trusses is determining the critical forces for which each member of the truss must be designed. The first step in this process is deciding which combinations of loading must be considered. In some cases, the potential combinations may be quite numerous. When both wind and seismic actions are potentially critical and more than one type of live loading occurs (e.g., roof loads plus hanging loads), the theoretically possible combinations of loadings can be overwhelming. However, designers are usually able to exercise judgment in reducing the sensible combinations to a reasonable number; for example, it is statistically improbable that a violent windstorm will occur simultaneously with a major earthquake shock.

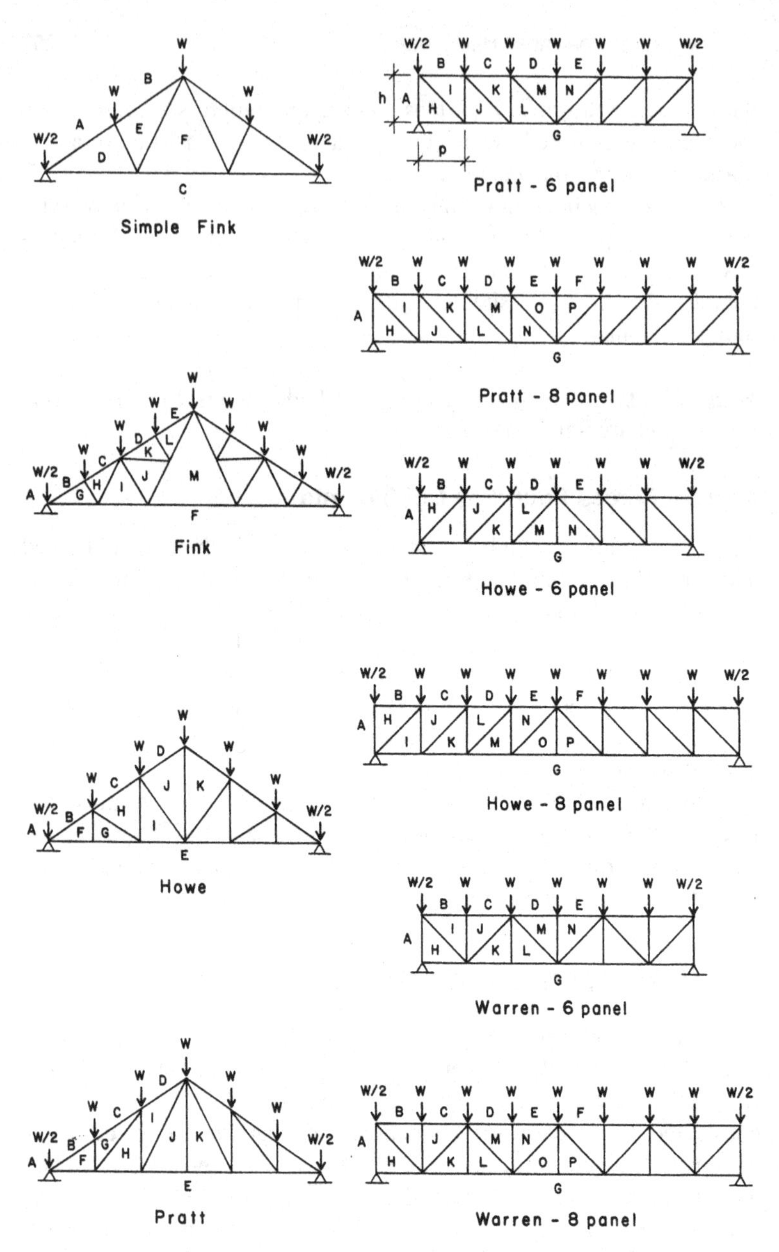

Figure 7.14 Simple planar trusses of gabled and parallel-chorded form. Reference figures for Table 7.2.

TABLE 7.2 Coefficients for Internal Forces in Simple Trusses[a]

Force in members = (table coefficient) X (panel load, W)

T indicates tension, C indicates compression

Gable Form Trusses

Truss Member	Type of Force	Roof Slope 4/12	Roof Slope 6/12	Roof Slope 8/12
Truss 1 - Simple Fink				
AD	C	4.74	3.35	2.70
BE	C	3.95	2.80	2.26
DC	T	4.50	3.00	2.25
FC	T	3.00	2.00	1.50
DE	C	1.06	0.90	0.84
EF	T	1.06	0.90	0.84
Truss 2 - Fink				
BG	C	11.08	7.83	6.31
CH	C	10.76	7.38	5.76
DK	C	10.44	6.93	5.20
EL	C	10.12	6.48	4.65
FG	T	10.50	7.00	5.25
FI	T	9.00	6.00	4.50
FM	T	6.00	4.00	3.00
GH	C	0.95	0.89	0.83
HI	T	1.50	1.00	0.75
IJ	C	1.90	1.79	1.66
JK	T	1.50	1.00	0.75
KL	C	0.95	0.89	0.83
JM	T	3.00	2.00	1.50
LM	T	4.50	3.00	2.25
Truss 3 - Howe				
BF	C	7.90	5.59	4.51
CH	C	6.32	4.50	3.61
DJ	C	4.75	3.35	2.70
EF	T	7.50	5.00	3.75
EI	T	6.00	4.00	3.00
GH	C	1.58	1.12	0.90
HI	T	0.50	0.50	0.50
IJ	C	1.81	1.41	1.25
JK	T	2.00	2.00	2.00
Truss 4 - Pratt				
BF	C	7.90	5.59	4.51
CG	C	7.90	5.59	4.51
DI	C	6.32	4.50	3.61
EF	T	7.50	5.00	3.75
EH	T	6.00	4.00	3.00
EJ	T	4.50	3.00	2.25
FG	C	1.00	1.00	1.00
GH	T	1.81	1.41	1.25
HI	C	1.50	1.50	1.50
IJ	T	2.12	1.80	1.68

Flat - Chorded Trusses

Truss Member	Type of Force	6 Panel Truss $\frac{h}{p}=1$	6 Panel Truss $\frac{h}{p}=\frac{3}{4}$	8 Panel Truss $\frac{h}{p}=1$	8 Panel Truss $\frac{h}{p}=\frac{3}{4}$
Truss 5 - Pratt					
BI	C	2.50	3.33	3.50	4.67
CK	C	4.00	5.33	6.00	8.00
DM	C	4.50	6.00	7.50	10.00
EO	C	–	–	8.00	10.67
GH	O	0	0	0	0
GJ	T	2.50	3.33	3.50	4.67
GL	T	4.00	5.33	6.00	8.00
GN	T	–	–	7.50	10.00
AH	C	3.00	3.00	4.00	4.00
IJ	C	2.50	2.50	3.50	3.50
KL	C	1.50	1.50	2.50	2.50
MN	C	1.00	1.00	1.50	1.50
OP	C	–	–	1.00	1.00
HI	T	3.53	4.17	4.95	5.83
JK	T	2.12	2.50	3.54	4.17
LM	T	0.71	0.83	2.12	2.50
NO	T	–	–	0.71	0.83
Truss 6 - Howe					
BH	O	0	0	0	0
CJ	C	2.50	3.33	3.50	4.67
DL	C	4.00	5.33	6.00	8.00
EN	C	–	–	7.50	10.00
GI	T	2.50	3.33	3.50	4.67
GK	T	4.00	5.33	6.00	8.00
GM	T	4.50	6.00	7.50	10.00
GO	T	–	–	8.00	10.67
AH	C	0.50	0.50	0.50	0.50
IJ	T	1.50	1.50	2.50	2.50
KL	T	0.50	0.50	1.50	1.50
MN	T	0	0	0.50	0.50
OP	O	–	–	0	0
HI	C	3.53	4.17	4.95	5.83
JK	C	2.12	2.50	3.54	4.17
LM	C	0.71	0.83	2.12	2.50
NO	C	–	–	0.71	0.83
Truss 7 - Warren					
BI	C	2.50	3.33	3.50	4.67
DM	C	4.50	6.00	7.50	10.00
GH	O	0	0	0	0
GK	T	4.00	5.33	6.00	8.00
GO	T	–	–	8.00	10.67
AH	C	3.00	3.00	4.00	4.00
IJ	C	1.00	1.00	1.00	1.00
KL	O	0	0	0	0
MN	C	1.00	1.00	1.00	1.00
OP	O	–	–	0	0
HI	T	3.53	4.17	4.95	5.83
JK	C	2.12	2.50	3.54	4.17
LM	T	0.71	0.83	2.12	2.50
NO	C	–	–	0.71	0.83

[a]See Figure 7.14 for truss forms and member identifications.

Once the required design loading conditions are established, the usual procedure is to perform separate analyses for each of the loadings. The values obtained can then be combined at will for each member to ascertain the particular combination that establishes the critical result for the member. This means that in some cases certain members will be designed for one combination and others for different combinations.

In most cases, design codes permit an increase in allowable stress for design of members when the critical loading includes forces due to wind or seismic loads.

7.6 COMBINED ACTIONS IN TRUSS MEMBERS

In analyzing trusses, the usual procedure is to assume that the loads will be applied to the truss joints. This results in the members themselves being loaded only through the joints and thus having only direct tension or compression forces. In some cases, however, truss members may be directly loaded—for example, when the top chord of a truss supports a roof deck without benefit of purlins or rafters. Thus, the chord member is directly loaded with a linear uniform load and functions as a beam between its end joints.

The usual procedure in these situations is to accumulate the loads at the truss joints and analyze the truss as a whole for the typical joint loading arrangement. The truss members that sustain the direct loading are then designed for the combined effects of the axial force caused by the truss action and the bending caused by the direct loading.

A typical situation for a roof truss is one in which the actual loading consists of the roof load distributed continuously along the top chords and a ceiling loading distributed continuously along the bottom chords for a combination of axial tension plus bending. This will, of course, result in somewhat larger members being required for both chords, and any estimate of the truss weight should account for this anticipated additional requirement.

7.7 DESIGN CONSIDERATIONS FOR STEEL TRUSSES

Design of a steel truss begins with the decision that establishes the use of the truss; that is, with the selection of a truss—in particular, a steel truss—over other possible structures for a building design project. Some of the conditions that favor such a selection may include the following:

Need for a relatively efficient structure (for example, in lieu of a solid web beam).

Need to achieve a long span, particularly with a light load (as for most roofs).

Need to achieve a roof profile geometry that other structures may not work for as well. (The truss is the most geometrically adaptable structure.)

Need to have an open structure. Permitting air circulation, vision, or light through the structure or passage of items such as ducts, piping, wiring, stairs, and so on.

Once the need for the steel truss is determined, design proceeds to essential concerns for any truss, including the following:

Truss profile and member layout

Selection of member form

Selection of form of connections

Development of necessary bracing

All of these issues are discussed in this chapter or in others in this book.

A particularly critical decision is that of establishing the form of connections. Because of the large number of connections, this is a major consideration for achieving economy in cost of the construction. The type of connection used and its range of both form variations and load capacity must work right for the truss design. Considerations for connections relate to just about all the issues in the truss design, for example:

Size of the structure. What works for small trusses may not be able to be expanded to larger sizes and greater connection loads.

Truss member form. Different connecting methods may be necessary for angles, W shapes, round pipes, tubes, and so on.

Layout of the truss. Frequency of joints may make an expensive connection method unfeasible; many members meeting at single joints may be difficult to accommodate.

Supported elements. Roof, floor, or ceiling structures—or other suspended items—must usually be attached at truss joints; this adds to the functions of the truss joint.

Because of the wide-ranging concerns for truss design, it is best to see the design problem in as broad a context as possible. This is the primary purpose for the building case study examples in Chapter 10, and several examples of use of trusses are presented there. These include the following:

Use of roof trusses for Buildings One and Four

Use of open web steel joist construction for Buildings One, Two, and Three

Use of a two-way spanning truss system for Building Six

7.8 TWO-WAY TRUSSES

An ordinary truss system consisting of a set of parallel, planar trusses may be turned into a two-way spanning system by connecting the parallel trusses with cross-trussing. However, this system lacks stability in the horizontal plane. If viewed in plan, the two sets of vertical, planar trusses form rectangles, which do not have the inherent triangulation needed for stability in the horizontal plane. Of course, you can rectify this situation by adding horizontal trussing.

Whereas the planar triangle is the basic unit of planar trusses, the tetrahedron is the basic unit of the spatial truss (see Figure 7.15). If a three-dimensional system is developed with three orthogonal planes (*x*, *y*, *z* coordinate system), the system basically defines rectangles in each plane and cubical forms in three dimensions. Triangulation of each or-

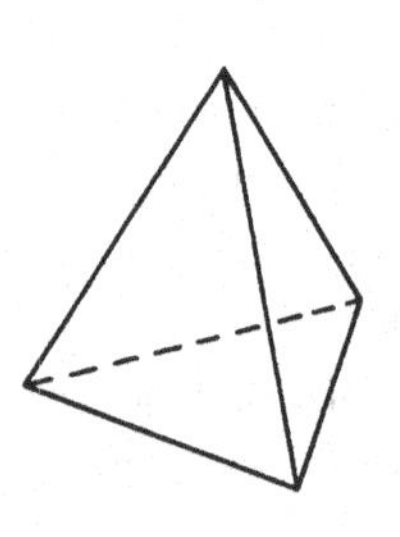

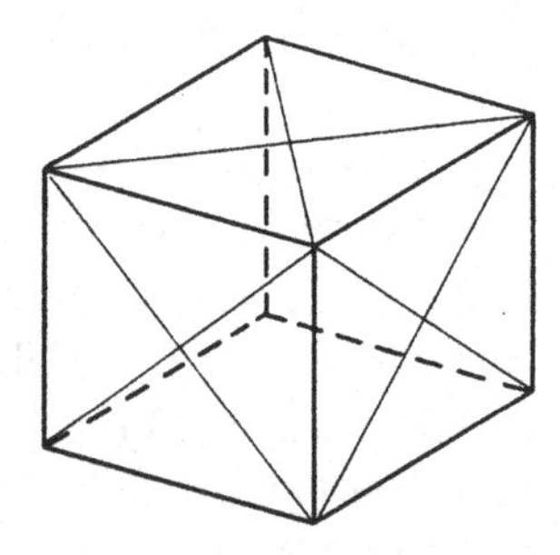

Figure 7.15 Basic units for three-dimensional trusses: the tetrahedron and the hexahedron (cube).

thogonal plane actually produces sets of tetrahedra in space, as shown in Figure 7.15.

When you use the mutually perpendicular vertical, planar truss system, you can easily form squared corners at the edge of the system and in relation to plan layouts beneath it. The roof edge is cleanly formed by the basic truss system, and the exterior walls meet the bottom of the truss at natural locations of the truss chords.

A purer form for the spatial truss derives from the basic spatial triangulation of the tetrahedron. If all the edges of the tetrahedron (or truss members) are equal in length, the solid form described is not orthogonal; that is, it does not describe the usual *x, y, z* system of mutually perpendicular planes (see Figure 7.16). If used for a two-way spanning truss system, a flat structure in the horizontal plane may be developed, but it will not inherently develop rectangles in plan: Its natural plan form will be in multiples of triangles, diamonds, and hexagons.

For small trusses, the fully triangulated system can save you money because it can be fully formed with all members of the same length

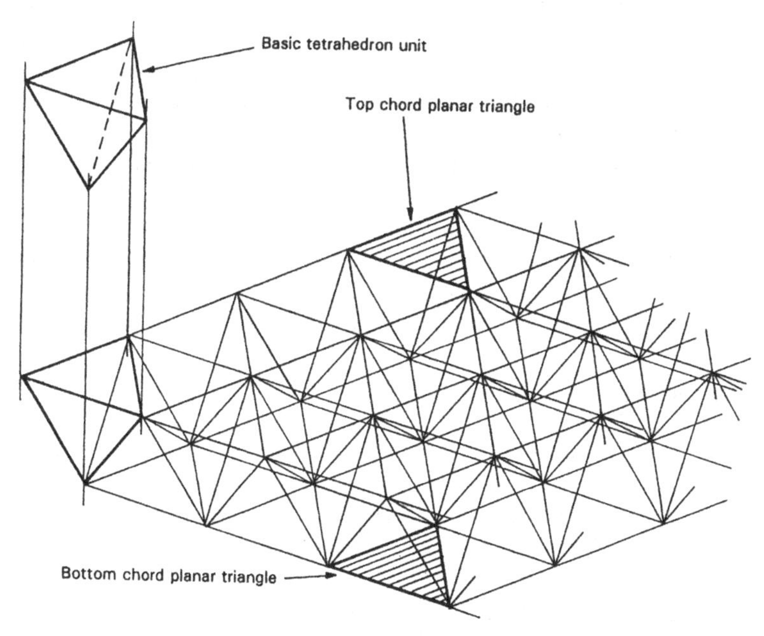

Figure 7.16 A two-way spanning truss system with a tetrahedron as the basic spatial unit and with triangulation in the top and bottom planes.

and a single, simple joint can be used for most of the truss system. The pure triangulated space system, however, allows you to form truly three-dimensional structures and not just flat, two-way spanning systems.

A compromise form—somewhere between the pure triangular and the orthogonal systems—is the offset grid. This system consists of two horizontal planes, each constituted as square, rectangular grids. However, the upper plane grid (top chord plan) has its grid intersections located over the centers of the grids in the lower plane (bottom chord plan). The truss web members are arranged to connect the top grid intersections with the bottom grid intersections. The typical joint therefore consists of the meeting of four chord members and four web diagonal members. The web members describe vertical planes in diagonal plan directions but do not form vertical planes with the chords.

The offset grid permits development of square plan layouts (usually preferred by designers) while retaining some features of the triangulated system (members the same length, joints the same form). Some details of offset grid systems are discussed for Building Six in Section 10.6.

Truss geometry must relate reasonably to the purpose of the structure—often the forming of a building's roof. For less constricting design situations, such as the forming of a canopy, a structural sculpture, or a theme structure, the space frame may be developed more dramatically.

Spatial trussing can be used to produce just about any form of structural unit, including columns, trussed mullions, freestanding towers, beams, surfaces of large multiplaned structures, and even arched surfaces formed as vaults (cylinders) or domes.

In many situations, trussing offers a practical, economically efficient structural solution. It also offers extreme lightness and a visual openness, which are often major design considerations. Light transmission, for example, may be a major factor, not just aesthetically, but functionally.

Span and Support Considerations

When planning structures that have spatial trussing, attention must be paid to the nature of the supports if you want to realize optimal use of the two-way action. The locations of supports define not only the size of spans to be achieved but also many aspects of the structural behavior of the two-way truss system.

Figure 7.17 shows possible support systems for a single square panel of a two-way spanning system. In Figure 7.17*a* support is provided by four columns placed at the outside corners, resulting in a maximum span

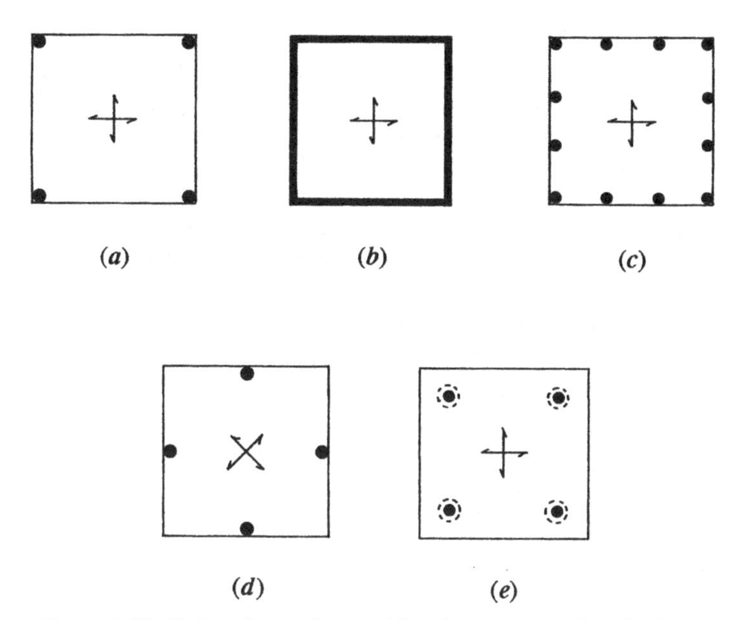

Figure 7.17 Various forms of support for a two-way spanning structure.

condition for the system's interior portion. This form of support also results in a very high shear condition in the single quadrant of the corner and requires that the edges act as one-way spanning supports for the two-way system. As a result, the edge chords will be very heavy and the corner web members will be heavily loaded for transfer of the vertical force to the columns.

Figures 7.17*b* and *c* show supports that eliminate the edge spanning and corner shear by providing either bearing walls or closely spaced perimeter columns. The trade-off is a lower truss cost for higher costs for the support system and its foundations. Such support is more restrictive architecturally.

Figure 7.17*d* shows an interesting possibility. The placement of four columns in the centers of the sides requires a considerable edge structure to achieve the corner cantilevers, but it actually reduces the span of the interior system and further reduces its maximum moment by the overhang effect of the cantilever corners. The clear space of the building interior remains the same, and the high shear at the four columns is the same as in Figure 7.17*a*.

For an ideal column shear condition, the solution shown in Figure 7.17*e* places the columns inside the edges. This provides a wider edge spanning strip as well as a total perimeter of the truss system at each column with shear divided between more truss members. It reduces the clear interior span to something less than the full width of the truss. If the exterior wall is at the roof edge, the interior columns may be intrusive in the plan. However, if a roof overhang is desired, the wall may be placed at the column lines and the structure will relate well to the architectural plan.

Single units of two-way spanning system should describe a square as closely as possible. Even though the assemblage has a potential for two-way spanning, if the support arrangement provides oblong units, the structural action may be essentially one-way in function.

Figure 7.18 shows three forms for a span unit, with sides in ratios of 1:1, 1:1.5, and 1:2. If the system is otherwise fully symmetrical, the square unit will share the spanning effort equally in each direction. If the ratio of sides gets as high as 1:1.5, the shorter span becomes much stiffer for deflection and will attract as much as 75 percent of the total load. This ratio is usually the maximum one for any practical consideration of two-way action.

If the ratio of the sides gets as high as 1:2, scarcely any load will be carried in the long direction, except for that near the ends of the plan unit, adjacent to the short sides. Two-way action relates generally to the development of two-way curvature (inverted domed or dishlike form), and it should be clear that the long, narrow unit will bend mostly in single, inverted arch-like form.

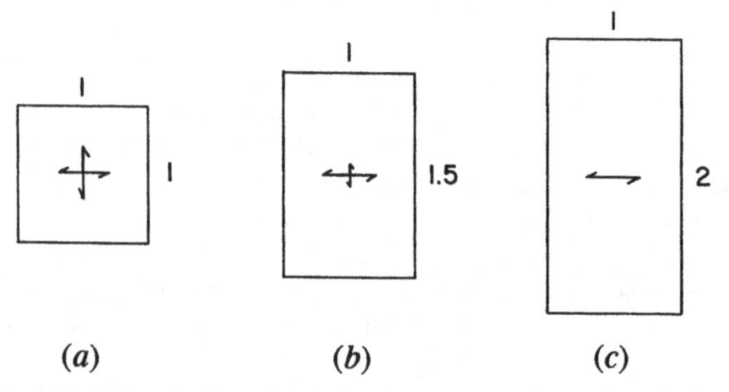

Figure 7.18 Different ratios of spans in a two-way spanning system. As the ratio approaches 2:1, the structure essentially becomes a one-way spanning structure, in spite of support conditions.

Two-way systems—especially those with multiple spans—are often supported on columns. This generally results in a high shear condition at the supporting columns, and various means are used to relieve the interior force concentrations in the spanning structure. A common solution for the two-spanning concrete flat slab has the effective perimeter size of the column extended by an enlarged top (called a *column capital*) and the strength of the slab increased locally by a thickened portion at the column (called a *drop panel*). These elements have their analogous counterparts in systems using two-way trussing.

In some truss systems, it is possible to develop something analogous to the drop panel in the concrete slab. Thus, an additional layer of trussing, or a single dropped-down unit, may be provided to achieve the local strengthening within the truss itself.

Joints and System Assembly

Many of the issues to consider in developing joints for spatial truss systems are essentially the same as those considered for planar truss systems: selection of materials, shape of truss member, arrangement of members at a joint, and magnitude of loads. These primary decisions preface selection of jointing methods (welding, bolting, nailing, and so on) and the use of intermediate devices (gusset plates, nodal units, and so on).

For the spatial system, two additional concerns are significant. The first is that most joints are of a three-dimensional character, relating to the specific geometry of the system. This generally calls for a somewhat more complex joint development than the simple alternatives often possible for planar trusses. A second concern is that the usual large number of joints requires a relatively simple, economically feasible joint construction.

Proprietary truss systems are often characterized by a special jointing system that accommodates the variety of joints for the system, is inexpensive when mass-produced, and can be quickly and easily assembled. Although a particular joint system may relate specifically to a single truss member shape, the member selection is usually not as critical a design problem as the joint development.

Most spatial trusses are of steel, and the primary jointing methods are as follows:

By welding of members directly to each other or of all members to a primary joint element (gusset, node, and so on)

By bolting, most likely to a joint element of some form

By direct connection with threaded, snap-in, or other attachments

As with other truss systems, it is often possible and desirable to prepare large units of the total structure in a fabricating shop and transport them to the building site, where they are bolted together and to their supports. Careful design of the field joints and of the shapes and sizes of shop-fabricated units can minimize erection problems. This procedure is more often the case with relatively large-scale trusses of individual design than with proprietary systems, which generally consist of individual members whose joints are individually field-assembled.

As with all truss systems, joints may also be developed to facilitate other functions besides that of the truss assemblage. Support of roofs, floors, ceilings, or other items is usually involved, because these loads must be attached at the truss joints.

Joints may also need to accommodate thermal expansion, seismic separation, or some specific controlled structural action (for example, pinned joint response to avoid transfer of bending moments). Support of suspended elements is often a requirement as well. Elements supported may relate to the module of the truss as defined by the joints or the member spacing.

A special joint is one that occurs at a support for the truss system. Beyond the usual requirement for direct compression bearing, there may be need for tension uplift resistance or transfer of lateral force for wind or seismic actions. It may also be necessary to achieve some pin joint response to avoid transfer of bending to column tops as the truss deflects. Varying chord length due to thermal expansion or live load stress changes may also present some problems. See the discussion of control joints in Section 8.9.

8

STEEL CONNECTIONS

Making a steel structure for a building typically involves the connecting of many parts. The technology available for achieving connections is subject to considerable variety, depending on the form and size of the connected parts, the structural forces transmitted between parts, and the nature of the connecting materials. At the scale of building structures, the primary connecting methods used presently involve electric arc welding and high-strength steel bolts; these are the methods principally treated in this chapter.

8.1 BASIC CONSIDERATIONS

Choices for basic connecting methods and for the specific forms and sizes of connecting materials for individual joints are subject to several considerations. These must all be addressed in the design of any single connection, although many situations result in a highly limited set of choices for practical reasons.

Types of Connections

Basic forms of connection for steel parts include the following:

Direct. Parts may be directly joined in some cases, without recourse to intermediate elements. The threaded connection is of such a form, using the basic male and female coupling elements with cut spiral grooves. Other means of this type include interlocking, folding, and direct bearing. These means are indeed used in various cases for building construction in general, but not much for connecting of structural elements.

Fusion. Separate metal pieces may be fused at their interface by welding; this is a major means for achieving structural connections.

Adhesion. Various adhesive materials may be used to form attachments, by simple contact adhesion or by some form of fusion of the connected materials. Development of composite units of construction for walls, roofs, and floor panels is achieved this way, but the method is not widely used for structural connections.

Intermediate connector. Mechanical devices of various form are frequently used to achieve connection, including nails, staples, screws, rivets, and bolts. In some cases, a separate element may also be used, such as the angle pieces commonly used to develop beam connections. This is a major means for achieving structural connections; especially for joints that are assembled at the job site.

Selection of connection means has a great deal to do with the size of the connected parts and the magnitude of transmitted forces. There are also a great number of practical matters relating to the achieving of the work and the geometric details of the connections themselves. The following concerns address various issues that influence the choice of basic connecting means.

Structural Functions

For structural connections, a major concern is the nature of the forces required to be transmitted between connected parts. An individual connection may be designed for a single loading condition and a singular force action, or it may be required to develop several different actions for different loading conditions. Individual forms of actions are the following:

Direct force. Direct forces include tension, compression, and shear.

Bending. Bending is a rotational effect occurring in a plane that includes the axis of the linear connected parts. Single connections usually transmit a single bending moment for any given loading condition but may also be subjected to other bending moments for different loadings.

Torsion. Torsion is also a rotational effect; it occurs in a plane at right angles to the linear axis of a connected member. Thus, both bending and torsion may occur together, each producing different effects.

Reversible forces. Connections may need to resist opposite force effects at different times: compression versus tension (push/pull); back-and-forth shear effects; clockwise rotation versus counterclockwise rotation in either bending or torsion. This often calls for some dual-resistance capability by a connection.

Dynamic effects. Ordinary effects of gravity are usually static in nature, but wind, earthquakes, mechanical vibrations, or intense sounds may cause dynamic force actions on connections. What works for static resistance may loosen, develop fatigue in the metal, or otherwise fail under dynamic loading conditions. For example, nuts may work off of bolts because of repeated dynamic effects.

Every structural connection must be investigated for the specific types and magnitudes of forces that the connection is expected to transmit. Optimized for singular actions (tension only, for example), a connection may take a unique form. However, the multiple loading condition is a common situation, resulting in some popularity of connecting means that have a range of resistances, rather than a singular one.

Special Concerns

There are many possible special concerns for connections. Common considerations are for economy, ease of achieving (often preferably with low skill required), speed of achieving, and ready availability of necessary materials. Some others are the following:

Shop versus field. There are often major differences in the form of connections that relate to whether they are performed under factory conditions (in the *shop*) or at the job site (the *field*). This may affect the basic means (bolting versus welding) or specific details of the process. Automation and high quality control are generally possible in the shop; ease, speed, and low skill required are generally desirable in the field.

Permanence of connections. Building structures are usually designed to stay in place indefinitely once they are assembled. Choices for means of connection are usually made without consideration for ease of undoing the connections. However, some structures are deliberately created to be demountable to allow for replacement of parts or to facilitate future recycling of assembled components. Bolts, screws, and interlocking connections may permit some ease of disassembly; welds, adhesives, rivets, and staples are less accommodating. However, disassembly may be only partial, and thus affect only selected joints.

Movement-selective connections. Most connections are developed to resist movements in the directions of applied forces, thus preventing separation, slipping, rotation, and so on. However, for purposes of accommodating thermal expansion, achieving seismic separation, or controlling selected structural responses, joints may be designed to selectively resist some movements and facilitate others.

Some aspects of connection design may relate to the general type of structure being assembled—for example, trusses, beam-and-column frameworks, and rigid frames. For extensive structures with many repetitive parts, similar connections may be repeated many times—with their economy or high cost rapidly multiplying. Similar tasks for connections occur repeatedly, and "standard" forms are widely used, having been refined over many years of use. The custom-designed, once-only connection may be a design challenge taking much effort in its refinement. However, mostly, steel frameworks employ very common connecting means with very familiar devices—usually selected from handbook tabulations.

8.2 BOLTED CONNECTIONS

Elements of steel are often connected by mating flat parts with common holes and inserting a pin-type device to hold them together. In times past, the device was a rivet; today it is usually a bolt. Many types and sizes of bolt are available, as are many connections in which they are used.

Structural Actions of Bolted Connections

Figures 8.1*a* and *b* show the plan and section of a simple connection between two steel bars that functions to transfer a tension force from one bar to the other. Although this is a tension-transfer connection, it is also referred to as a *shear connection* because of the manner in which the connecting device (the bolt) works in the connection (see Figure 8.1*c*). For structural connections, this type of joint is now achieved mostly with so-called *high-strength bolts,* which are special bolts that are tightened in a controlled manner that induces development of yield stress in the bolt shaft. For a connection using such bolts, many possible forms of failure must be considered, including the following:

Bolt shear. In the connection shown in Figure 8.1*a* and *b,* the failure of the bolt involves a slicing (shear) failure that is developed as a shear stress on the bolt cross section. The resistance factor (ϕ_v) is taken as 0.75. The factored design shear strength ($\phi_v R_n$) of the bolt can be expressed as a nominal shear stress (F_v) times the nominal cross-sectional area of the bolt, or

$$\phi_v R_n = \phi F_v A_b$$

With the size of the bolt and the grade of steel known, it is a simple matter to establish this limit. In some types of connections, it may

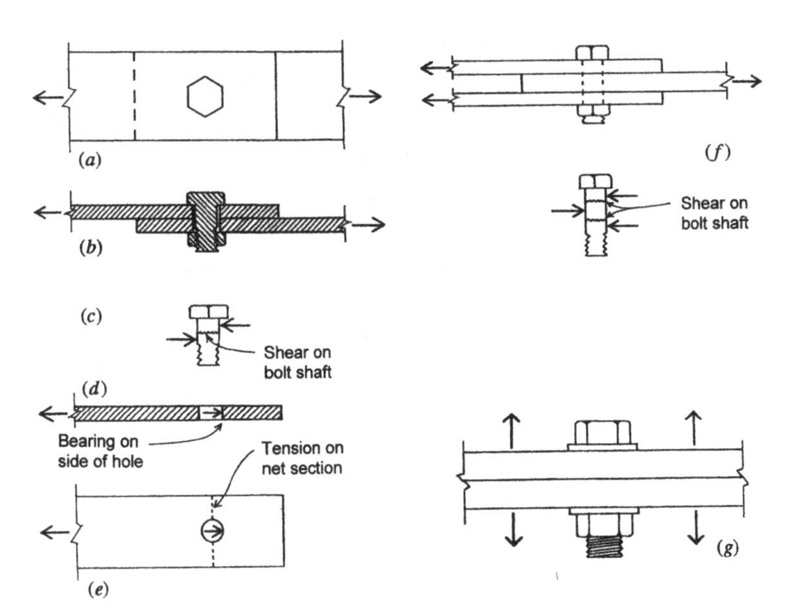

Figure 8.1 Actions of bolted joints.

be necessary to slice the same bolt more than once to separate the connected parts. This is the case in the connection shown in Figure 8.1*f*, which shows that the bolt must be sliced twice to make the joint fail. When the bolt develops shear on only one section (Figure 8.1*c*), it is said to be in *single shear;* when it develops shear on two sections (Figure 8.1*f*), it is said to be in *double shear.*

Bearing. If the bolt tension (due to tightening of the nut) is relatively low, the bolt serves primarily as a pin in the matched holes, bearing against the sides of the holes, as shown in Figure 8.1*d*. When the bolt diameter is larger or the bolt is made of very strong steel, the connected parts must be sufficiently thick if they are to develop the full capacity of the bolts. The design bearing strength ($\phi_v R_n$) permitted for this situation by the AISC Specification is the smaller of

$$\phi_v R_n = L_c \times t \times F_u$$

and,

$$\leq 3.0 \times d \times t \times F_u$$

Where ϕ_v = 0.75
L_c = distance between edge of hole and edge of next hole
t = thickness of connected material (in.)
F_u = ultimate tensile strength of connected material (ksi)
D = diameter of bolt (in.)

Tension on net section of connected parts. For the connected bars in Figure 8.1*b*, the tension stress in the bars will be a maximum at a section across the bar at the location of the hole. This reduced section is called the *net section* for tension resistance. Although this is indeed a location of critical stress, it is possible to achieve yield here without serious deformation of the connected parts. For this reason, design strength ($\phi_t P_n$) at the net section is based on the ultimate—rather than the yield—strength of the connected parts. The value used for the design tensile strength is

$$\phi_t P_n = \phi_t \times F_u \times A_e$$

Where ϕ_t = 0.75

F_u = ultimate tensile strength of the connected material (ksi)

A_e = reduced (or net) cross-sectional area (in.^{2})

Bolt tension. Although the shear (slip-resisting) connection shown in Figure 8.1*a* and *b* is common, some joints employ bolts for their resistance in tension, as shown in Figure 8.1*g*. For the threaded bolt, the maximum tension stress is developed at the net section through the cut threads. However, it is also possible for the bolt to have extensive elongation if yield stress develops in the bolt shaft (at an unreduced section). However stress is computed, bolt tension resistance is established on the basis of data from destructive tests.

Bending in the connection. Whenever possible, bolted connections are designed to have a bolt layout that is symmetrical with regard to the directly applied forces. This is not always possible, because in addition to the direct force actions, the connection may be subjected to twisting due to a bending moment or torsion induced by the loads. Figure 8.2 shows some examples of this situation.

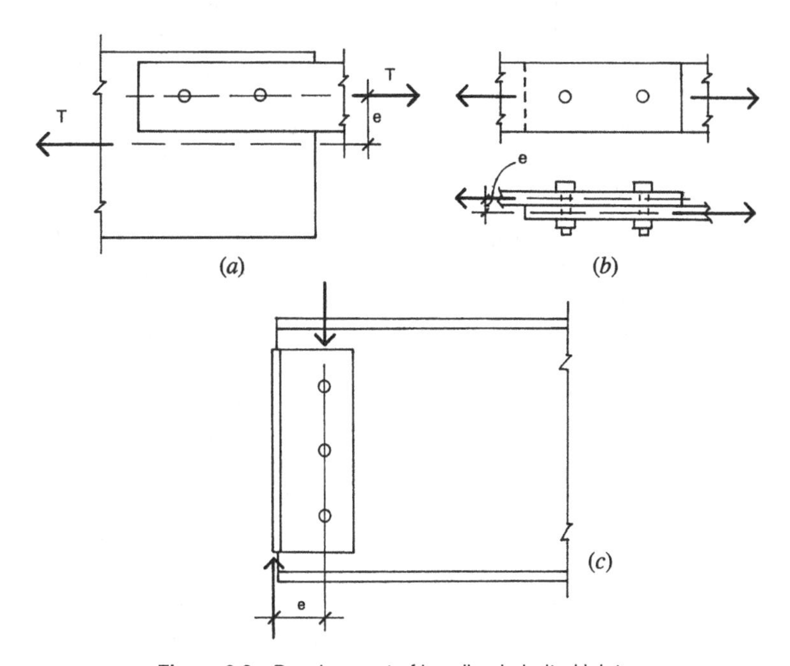

Figure 8.2 Development of bending in bolted joints.

In Figure 8.2*a* two bars are connected by bolts, but the bars are not aligned in a way to transmit tension directly between the bars. This may induce a rotational effect on the bolts, with a torsional twist equal to the product of the tension force and the eccentricity due to misalignment of the bars. Shearing forces on individual bolts will be increased by this twisting action. And, of course, the ends of the bars will also be twisted.

Figure 8.2*b* shows the single-shear joint, as shown in Figures 8.1*a* and *b*. When viewed from the top, such a joint may appear to have the bars aligned; however, the side view shows that the basic nature of the single-shear joint is such that a twisting action is inherent in the joint. This twisting increases with thicker bars. It is usually not highly critical for steel structures, where connected elements are usually relatively thin. For connecting wood elements, however, it is not a favored form of joint.

Figure 8.2*c* shows a side view of a beam end with a typical form of connection that employs a pair of angles. As shown, the angles grasp the beam web between their legs and turn the other legs out to fit flat against a column or the web of another beam. Vertical load from the beam, vested in the shear in the beam web, is transferred to the angles by the connection of the angles to the beam web—with bolts as shown here. This load is then transferred from the angles at their outward-turned face, resulting in a separated set of forces caused by the eccentricity shown. This action must be considered with others in design of these connections.

Slipping of connected parts. Highly tensioned, high-strength bolts develop a very strong clamping action on the mated flat parts being connected, analogous to the situation shown in Figure 8.3*a*. As a result, there is a strong development of friction at the slip face, which is the initial form of resistance in the shear-type joint. Development of bolt shear, bearing, and even tension on the net section will not occur until this slipping is allowed. For service-level loads, therefore, this is the *usual* form of resistance, and the bolted joint with high-strength bolts is considered to be a very rigid form of joint.

Block shear. One possible form of failure in a bolted connection is that of tearing out the edge of one of the attached members. This is called a *block shear* failure. The diagrams in Figure 8.3*b* show this potentiality in a connection between two plates. The failure in this

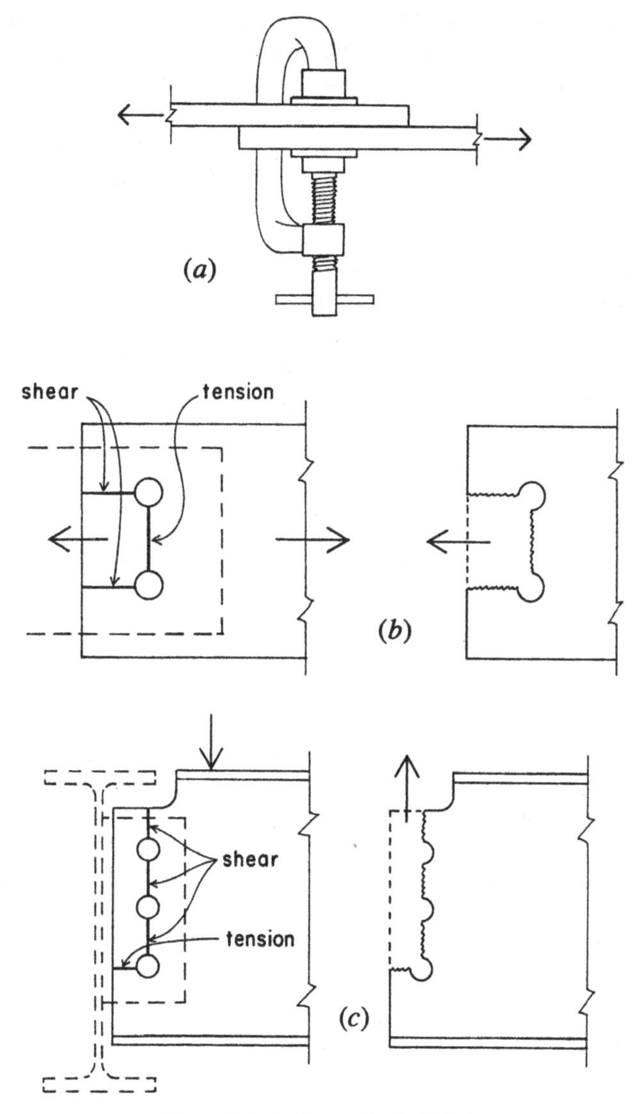

Figure 8.3 Actions of bolted joints.

case involves a combination of shear and tension to produce the torn-out form shown. The total tearing force is computed as the sum required to cause both forms of failure. The design strength ($\phi_t P_n$) of the net tension area is computed as described before for net cross sections. The design strength ($\phi_v R_n$) of the shear areas is specified as $0.75F_v A_c$, where A_c is the cross-sectional area experiencing shear stress.

With the edge distance, hole spacing, and diameter of the holes known, the net widths for tension and shear are determined and multiplied by the thickness of the part in which the tearing occurs. These areas are then multiplied by the appropriate stress to find the total tearing force that can be resisted. If this force is greater than the connection design load, the tearing problem is not critical.

Another case of potential tearing is shown in Figure 8.3*c*. This is the common situation for the end framing of a beam in which support is provided by another beam, whose top is aligned with that of the supported beam. The end portion of the top flange of the supported beam must be cut back to allow the beam web to extend to the side of the supporting beam. With the use of a bolted connection, the tearing condition shown is developed.

Types of Steel Bolts

Bolts used for the connection of structural steel members come in two basic types. Bolts designated A307 and called *unfinished* have the lowest load capacity of the structural bolts. The nuts for these bolts are tightened just enough to secure a snug fit of the attached parts; because of this low resistance to slipping, plus the oversizing of the holes to achieve practical assemblage, there is some movement in the development of full resistance. These bolts are generally not used for major connections, especially when joint movement or loosening under vibration or repeated loading may be a problem. They are, however, used extensively for temporary connections during erection of frames.

Bolts designated A325, F1852, or A490 are called *high-strength bolts*. The nuts of these bolts are tightened to produce a considerable tension force, which results in a high degree of friction resistance between the attached parts. Different specifications for installation of these bolts results in different classifications of their strength, relating generally to the critical mode of failure.

When loaded in shear-type connections, bolt capacities are based on the development of shearing action in the connection. The shear capacity of a single bolt is further designated as *S* for single shear (Figure 8.1*c*) or *D* for double shear (Figure 8.1*f*). In high-strength bolts, the shear capacity is affected by the bolt threads. If the threads are present in the shear plane being considered, the cross-sectional area is reduced and therefore the capacity of the bolt is also reduced. The capacities of structural bolts in both tension and shear are given in Table 8.1.

These bolts range in size from ½ to 1½ in. in diameter, and capacities for these sizes are given in tables in the AISC Manual. The most commonly used sizes for light structural steel framing are ¾ and ⅞ in. However, for larger connections and large frameworks, sizes of 1 to 1¼ are also used. This is the size range for which data is given in Table 8.1: ¾ to 1¼.

Bolts are ordinarily installed with a washer under both head and nut. Some manufactured high-strength bolts have specially formed heads or nuts that in effect have self-forming washers, eliminating the need for a separate, loose washer. When a washer is used, it is sometimes the limiting dimensional factor in detailing for bolt placement in tight locations, such as close to the fillet (inside radius) of angles or other rolled shapes.

TABLE 8.1 Design Strength of Structural Bolts (kips)[a]

ASTM Designation	Loading Condition[b]	Thread Condition	Nominal Diameter of Bolts (in.)				
			¾	⅞	1	1⅛	1¼
A307	S		7.95	10.8	14.1	17.9	22.1
	D		15.9	21.6	28.3	35.8	44.2
	T		14.9	20.3	26.5	33.5	41.4
A325	S	Included	15.9	21.6	28.3	35.8	44.2
		Excluded	19.9	27.1	35.3	44.7	55.2
	D	Included	31.8	43.3	56.5	71.6	88.4
		Excluded	39.8	54.1	70.7	89.5	110
	T		29.8	40.6	53	67.1	82.8
A490	S	Included	19.9	27.1	35.3	44.7	55.2
		Excluded	24.9	33.8	44.2	55.9	69
	D	Included	39.8	54.1	70.7	89.5	110
		Excluded	49.7	67.6	88.4	112	138
	T		37.4	51	66.6	84.2	104

[a] Slip-critical connections; assuming there is no bending in the connection and that bearing on connected materials is not critical.

[b] *S* = single shear; *D* = double shear; *T* = tension.

Source: Compiled from data in the *Manual of Steel Construction,* with permission of the publishers, American Institute of Steel Construction.

For a given diameter of bolt, a minimum thickness is required for the bolted parts in order to develop the full shear capacity of the bolt. This thickness is based on the bearing stress between the bolt and the side of the hole. The stress limit for this situation may be established by either the bolt steel or the steel of the bolted parts.

Steel rods are sometimes threaded for use as anchor bolts or tie rods. When they are loaded in tension, their capacities are usually limited by the stress on the reduced section at the threads. Tie rods are sometimes made with *upset ends,* which consist of larger-diameter portions at the ends. When these enlarged ends are threaded, the net section at the thread is the same as the gross section in the remainder of the rods; the result is no loss of capacity for the rod.

Layout of Bolted Connections

Designing bolted connections generally involves a number of considerations in the dimensional layout of the bolt-hole patterns for the attached structural members. This section presents some basic factors that often must be included in the design of bolted connections. In some situations, the ease or difficulty of achieving a connection may affect the choice for the form of the connected members.

Figure 8.4*a* shows the layout of a bolt pattern with bolts placed in two parallel rows. Two basic dimensions for this layout are limited by the size (nominal diameter) of the bolt. The first is the center-to-center spacing of the bolts, usually called the *pitch.* The AISC Specification limits this dimension to an absolute minimum of two and one-half times the bolt diameter. The preferred minimum, however, which is used in this book, is three times the diameter.

The second critical layout dimension is the *edge distance,* which is the distance from the center line of the bolt to the nearest edge of the member containing the bolt hole. There is also a specified limit for this as a function of bolt size and the nature of the edge; the latter refers to whether the edge is formed by rolling or cutting. Edge distance may also be limited by edge tearing in block shear, as previously discussed.

Table 8.2 gives the recommended limits for pitch and edge distance for the bolt sizes used in ordinary steel construction.

In some cases, bolts are staggered in parallel rows (Figure 8.4 *b*). In this case the diagonal distance, labeled *m* in the illustration, must also be considered. For staggered bolts the spacing in the direction of the rows is usually referred to as the *pitch;* the spacing of the rows is called the *gage.*

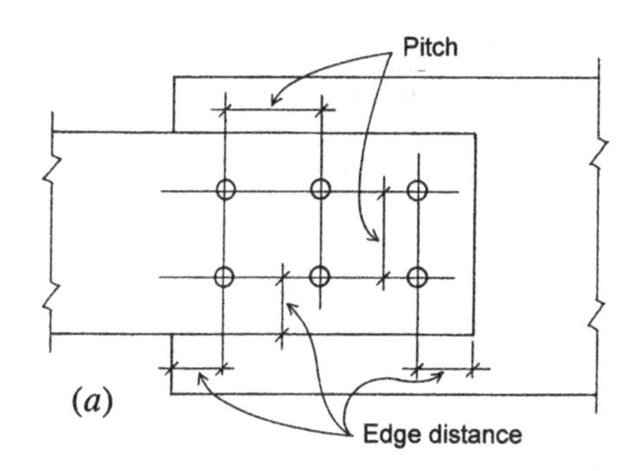

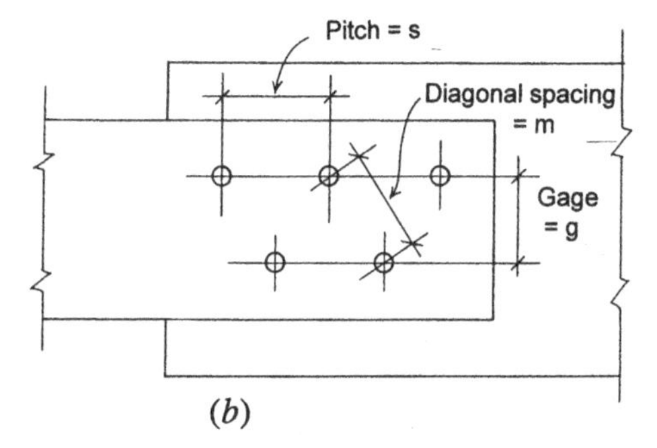

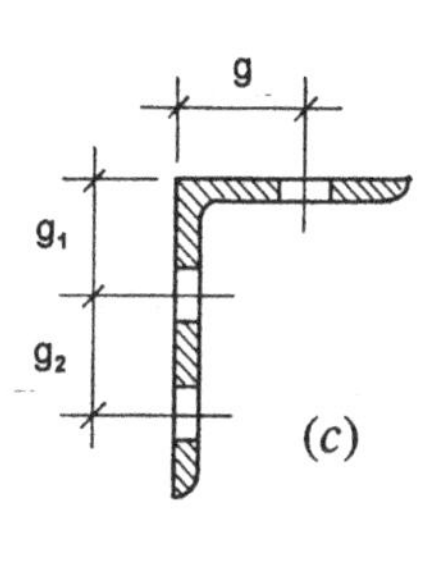

Figure 8.4 Layout considerations for bolted joints.

The usual reason for staggering the bolts is that sometimes the rows must be spaced closer (gage spacing) than the minimum spacing required for the bolts selected. However, staggering the bolt holes also helps to create a slightly less critical net section for tension stress in the steel member with the holes.

Location of bolt lines is often related to the size and type of structural members being attached. This is especially true of bolts placed in the legs of angles or in the flanges of W, M, S, C, and structural tee shapes. Figure 8.4*c* shows the placement of bolts in the legs of angles. When a single row is placed in a leg, its recommended location is at the distance labeled g from the back of the angle. When two rows are used, the first row is placed at the distance g_1, and the second row is spaced a distance g_2 from the first. Table 8.3 gives the recommended values for these distances.

When placed at the recommended locations in rolled shapes, bolts will end up a certain distance from the edge of the part. Based on the recom-

TABLE 8.2 Pitch and Edge Distances for Bolts

	Minimum Edge Distance for Punched, Reamed, or Drilled Holes (in.)		Minimum Recommended Pitch, Center-to-Center (in.)	
Rivet or Bolt Diameter d (in.)	At Sheared Edges	At Rolled Edges of Plates, Shapes, or Bars, or Gas-Cut Edges[a]	$2.667d$	$3d$
0.625	1.125	0.875	1.67	1.875
0.750	1.25	1.0	2.0	2.25
0.875	1.5[b]	1.125	2.33	2.625
1.000	1.75[b]	1.25	2.67	3.0

[a]May be reduced 0.125 in. when the hole is at a point where stress does not exceed 25% of the maximum allowed in the connected element.

[b]May be 1.25 in. at the ends of beam connection angles.

Source: Adapted from data in the *Manual of Steel Construction*, with permission of the publishers, American Institute of Steel Construction.

mended edge distance for rolled edges given in Table 8.2, it is thus possible to determine the maximum size of bolt that can be accommodated. For angles, the maximum fastener may be limited by the edge distance, especially when two rows are used; however, other factors may in some cases be more critical. The distance from the center of the bolts to the inside fillet of the angle may limit the use of a large washer where one is required. Another consideration may be the stress on the net section of the angle, especially if the member load is taken entirely by the attached leg.

Tension Connections

When tension members have reduced cross sections, two stress investigations must be considered. This is the case for members with holes for bolts. For the member with a hole, the design tension strength ($\phi_t P_n$) at the reduced cross section through the hole is

$$\phi_t P_n = \phi_t \times F_u \times A_e$$

where A_e = reduced (or net) cross-sectional area (in.^{2})

ϕ_t = 0.75

F_u = ultimate tensile strength of the connected material (ksi)

TABLE 8.3 Usual Gage Dimensions for Angles (in.)

Gage Dimension	Width of Angle Leg								
	8	7	6	5	4	3.5	3	2.5	2
g	4.5	4.0	3.5	3.0	2.5	2.0	1.75	1.375	1.125
g_1	3.0	2.5	2.25	2.0					
g_2	3.0	3.0	2.5	1.75					

Source: Adapted from data in the *Manual of Steel Construction*, with permission of the publishers, American Institute of Steel Construction.

The resistance at the net section must be compared with the resistance at the unreduced section of the member for which the reduction factor (ϕ_t) is 0.90.

For steel bolts, the design strength is specified as a value based on the type of bolt.

Angles used as tension members are usually connected by only one leg. In a conservative design, the effective net area is only that of the connected leg, less the reduction caused by bolt holes.

Rivet holes and bolt holes are punched larger in diameter than the nominal diameter of the fastener. The punching damages a small amount of the steel around the perimeter of the hole; consequently the diameter of the hole to be deducted in determining the net section is ⅛ in. greater than the nominal diameter of the fastener.

When only one hole is involved, as in Figure 8.2*b*, or in a similar connection with a single row of fasteners along the line of stress, the net area of the cross section of one of the plates is found by multiplying the plate thickness by its net width (width of member minus diameter of hole).

When holes are staggered in two rows along the line of stress (Figure 8.5), the net section is determined somewhat differently. The AISC Specification reads as follows:

> In the case of a chain of holes extending across a part in any diagonal or zigzag line, the net width of the part shall be obtained by deducting from the gross width the sum of the diameters of all the holes in the chain and adding, for each gage space in the chain, the quantity $s^2/4g$, where
>
> s = longitudinal spacing (pitch) in inches for any two successive holes.
>
> g = transverse spacing (gage) in inches for the same two holes.

The critical net section of the part is obtained from the chain that gives the least net width.

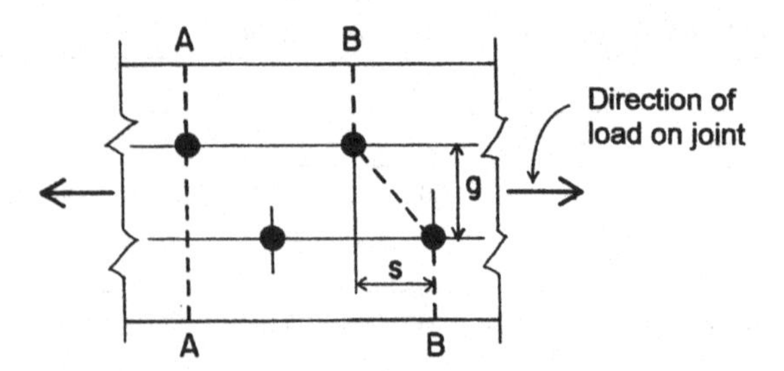

Figure 8.5 Determination of net cross-sectional area for the connected members in a bolted joint.

8.3 DESIGN OF A BOLTED CONNECTION

The issues raised in the preceding sections are illustrated in the following design example.

Example 1. The connection shown in Figure 8.6 consists of a pair of narrow plates that transfer a tension force that is produced by a dead load of 50 kips [222 kN] and a live load of 90 kips [495 kN]. All plates are of A36 steel with F_y = 36 ksi [250 MPa] and F_u = 58 ksi [400 MPa] and are attached with ¾-in. A325 bolts placed in two rows with threads included in the planes of shear. Using data from Table 8.1, determine the number of bolts required, the width and thickness of the narrow plates, the thickness of the wide plate, and the layout for the connection.

Solution. The process begins with the determination of the ultimate load:

$$P_u = 1.2D + 1.6L = 1.2(50) + 1.6(90) = 204 \text{ kips [907 kN]}$$

From Table 8.1, the capacity of a single bolt in double shear is found as 31.8 kips [141 kN]. The required number of bolts for the connection is thus:

$$n = \frac{204}{31.8} = 6.41, \text{ or } 7$$

Although placement of seven bolts in the connection is possible, most designers would choose to have a symmetrical arrangement with eight bolts, four to a row. The average bolt load is thus:

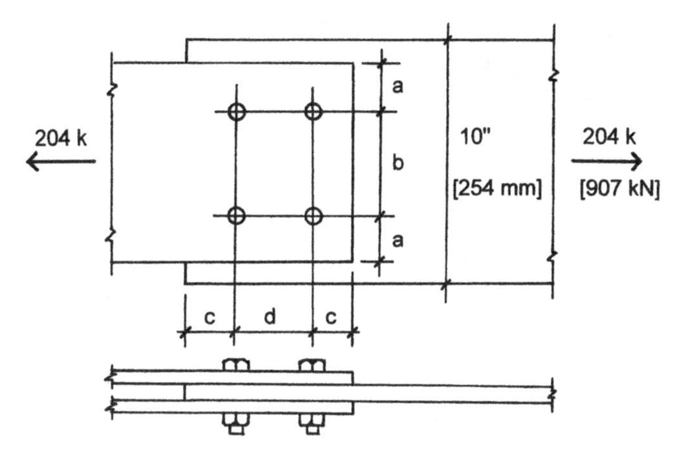

Figure 8.6 Reference figure for Example 1.

$$V_u = \frac{204}{8} = 25.5 \text{ kips [113 kN]}$$

From Table 8.2, for the 3/4-in. bolts, minimum edge distance for a cut edge is 1.25 in. and minimum recommended spacing is 2.25 in. The minimum required width for the plates is thus (see Figure 8.6):

$$w = b + 2(a) = 2.25 + 2(1.25) = 4.75 \text{ in. [121 mm]}$$

If space is tightly constrained, this actual width could be specified for the narrow plates. For this example a width of 6 in. is used. Checking for the requirement of stress on the gross area of the plate cross section:

$$A_g = \frac{P_u}{\phi_t F_y} = \frac{204}{0.9(36)} = 6.30 \text{ in.}^2 \text{ [4070 mm}^2\text{]}$$

and, with the 6-in. width, the required thickness is

$$t = \frac{6.30}{2 \times 6} = 0.525 \text{ in. [13 mm]}$$

This permits the use of a minimum thickness of 9⁄16 in. [14 mm]. The next step is to check the stress on the net section. For the computations, use a bolt hole size at least 1⁄8 in. larger than the bolt diameter. This allows for the true over-size (usually 1⁄16-in.) and some loss due to the roughness of the hole edges. Thus, the hole is assumed to be 7⁄8-in. (0.875) in diameter, and the net width and area are

$$w = 6 - 2(0.875) = 4.25 \text{ in. } [108 \text{ mm}]$$

$$A_e = w \times t = 4.25 \times \frac{9}{16} = 2.39 \text{ in.}^2$$

and the design strength at the net section is

$$\phi_t P_n = 0.75 \times F_u \times A_e = 0.75 \times 58 \times (2 \times 2.39) = 208 \text{ kips } [935 \text{ kN}]$$

Because this is greater than the factored load (P_u), the narrow plates are adequate for tension stress.

The bolt capacities in Table 8.1 are based on a slip-critical condition, which assumes a design failure limit to be that of the friction resistance (slip resistance) of the bolts. However, the backup failure mode is the one in which the plates slip to permit development of the pin action of the bolts against the sides of the holes; this then involves the shear capacity of the bolts and the bearing resistance of the plates. Bolt shear capacities are higher than the slip failures, so the only concern for this is the bearing on the plates.

Assuming a bolt spacing of 2.25 in., bearing design strength ($\phi_v R_n$) is computed for a single bolt as:

$$\phi_v R_n = 2[0.75 \times L_c \times t \times F_u] = 2\left[0.75 \times 1.375 \times \frac{9}{16} \times 58\right]$$

$$= 67.3 \text{ kips } [299 \text{ kN}]$$

or

$$\phi_v R_n = 2\{\phi_v(3.0 \times d \times t \times F_u)\} = 2(0.75)(3.0)\left(\frac{3}{4}\right)\left(\frac{9}{16}\right)(58)$$

$$= 110 \text{ kips}$$

which is clearly not a critical concern, because the factored load on each bolt is 25.5 kips.

For the middle plate, the procedure is essentially the same, except that the width is given and there is a single plate. As before, the stress on the unreduced cross section requires an area of 6.30 in.^{2}, so the required thickness of the 10-in.-wide plate is

$$t = \frac{6.30}{10} = 0.630 \text{ in. [16 mm]}$$

which indicates the use of a ⅝-in. thickness.

For the middle plate the width at the net section is

$$w = 10 - (2 \times 0.875) = 8.25 \text{ in. [210 mm]}$$

and the design strength at the net section is

$$\begin{aligned} \phi_t P_n &= 0.75 \times F_u \times A_e = 0.75 \times 58 \times (0.625 \times 8.25) \\ &= 224 \text{ kips [9296 kN]} \end{aligned}$$

which is greater than the factored load of 204 kips.

The computed bearing design strength on the sides of the holes in the middle plate is

$$\begin{aligned} \phi_v R_n &= 0.75 \times L_c \times t \times F_u = 0.75 \times 1.375 \times \frac{5}{8} \times 58 \\ &= 37.4 \text{ kips [166 kN]} \end{aligned}$$

or

$$\begin{aligned} \phi_v R_n &= \phi_v(3.0 \times d \times t \times F_u) = 0.75(3.0)\left(\frac{3}{4}\right)\left(\frac{5}{8}\right)(58) \\ &= 61.2 \text{ kips} \end{aligned}$$

which is greater than the factored load of 25.5 kips, as determined previously.

A final problem that must be considered is the possibility for tearing out of the two bolts at the end of a plate in a block shear failure (Figure 8.3*a*). Because the combined thicknesses of the outer plates is greater than that of the middle plate, the critical case for this connection is that of the middle plate. Figure 8.7 shows the condition for tearing, which involves a combination of tension on the section labeled "1" and shear on the two sections labeled "2." For the tension section:

$$\text{Net } w = 3 - 0.875 = 2.125 \text{ in. [54 mm]}$$

$$A_e = 2.125 \times \frac{5}{8} = 1.328 \text{ in.}^2 \text{ [857 mm}^2\text{]}$$

and the design strength for tension is

$$\phi_t P_n = \phi_t \times F_u \times A_e = 0.75 \times 58 \times 1.328 = 57.8 \text{ kips [257 kN]}$$

For the two shear sections:

$$\text{Net } w = 2\left(1.25 - \frac{0.875}{2}\right) = 1.625 \text{ in. [41.3 mm]}$$

$$A_e = 1.625 \times \frac{5}{8} = 1.016 \text{ in.}^2 \text{ [655 mm}^2\text{]}$$

and the design strength for shear is

$$\phi_v R_n = \phi_v \times F_u \times A_c = 0.75 \times 58 \times 1.016 = 44.2 \text{ kips [197 kN]}$$

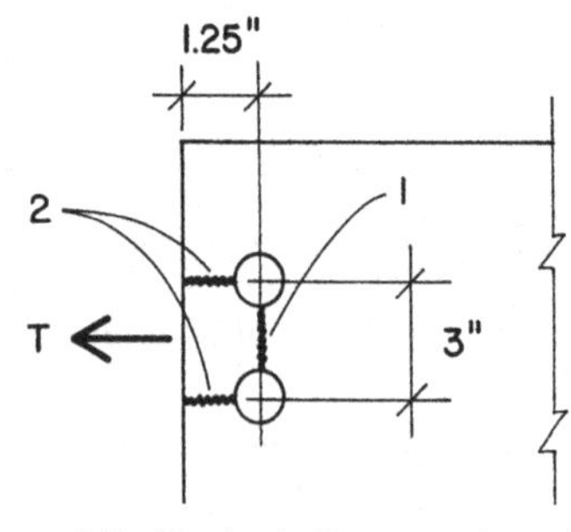

Figure 8.7 Tearing in the example problem.

The total resistance to tearing is thus:

$$T = 57.8 + 44.2 = 102 \text{ kips [454 kN]}$$

Because this is greater than the combined load on the two end bolts (51 kips), the plate is not critical for tearing in block shear.

The solution for the connection is displayed in the top and side views in Figure 8.8.

Connections that transfer compression between the joined parts are essentially the same with regard to the bolt stresses and bearing on the parts. Stress on the net section in the joined parts is not likely to be critical, because the compression members are likely to be designed for a relatively low stress due to column action.

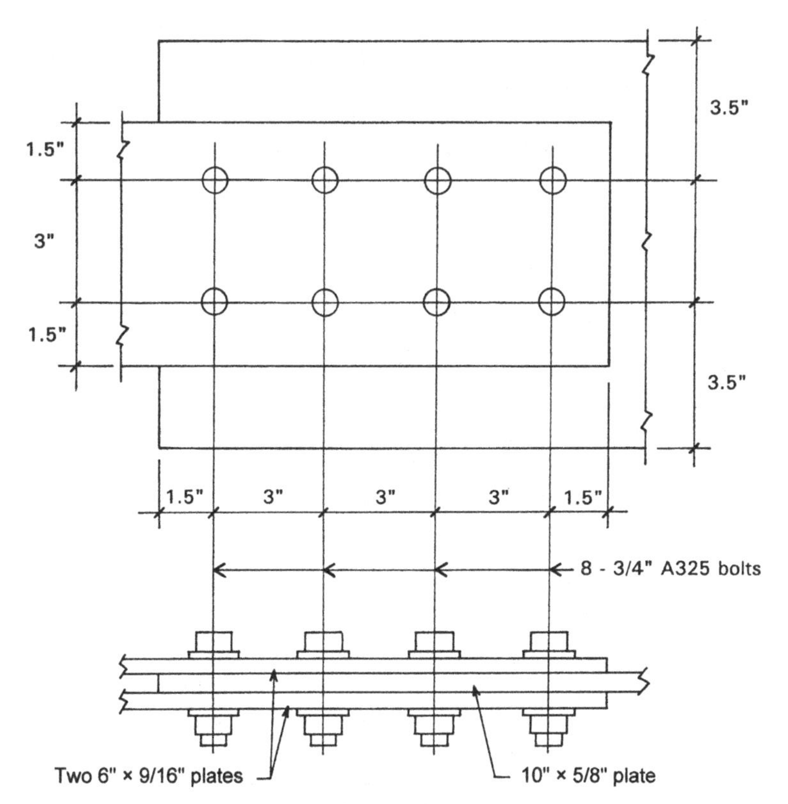

Figure 8.8 Solution for Example 1.

Problem 8.3.A. A bolted connection of the general form shown in Figure 8.6 is to be used to transmit a tension force of 75 kips [334 kN] dead load and 100 kips live load by using 7/8-in. A325 bolts and plates of A36 steel. The outer plates are to be 8 in. wide [200 mm], and the center plate is to be 12 in. wide [300 mm]. Find the required thicknesses of the plates and the number of bolts needed if the bolts are placed in two rows. Sketch the final layout of the connection.

Problem 8.3.B. Design the connection for the data in Problem 8.3.A, except that the outer plates are 9 in. wide and the bolts are placed in three rows.

8.4 BOLTED FRAMING CONNECTIONS

How structural steel members are joined depends on the connected parts, the connecting method, and the forces that are to be transferred between members. Figure 8.9 shows a number of common connections that are used to join steel columns and beams consisting of rolled shapes.

In the joint shown in Figure 8.9*a,* a steel beam is connected to a supporting column by the simple means of resting it on top of the column. The connecting device used is a steel plate that is welded to the top of the column and bolted to the bottom flange of the beam. For vertical loads only, the bolts in this case carry no computed loads; they are used merely to secure the members in the frame. A stress condition that must be considered in this joint is vertical compression in the beam web, as discussed in Section 3.11.

The remaining details in Figure 8.9 illustrate situations in which beam reactions are transferred to the supporting members by attachment of the beam web. This is an appropriate form of connection because the vertical shear in the beam is vested primarily in the web.

The most common form of connection is one that uses a pair of steel angles for a connecting device, as shown in Figure 8.9*b*. The two most frequent situations for this connection are the joining of a steel beam to the face of a supporting column (Figure 8.9*b*) and to the side of a supporting beam (Figure 8.9*d*). A beam may also be connected to the web of a column in this manner if the size of the column allows for the connection.

A variation of this connection is shown in Figure 8.9*c,* where a single angle is used. This lopsided connection results in twisting of the joint, and it should only be used for beams used for bracing or other non-load-carrying functions.

When the two intersecting beams must have their tops at the same

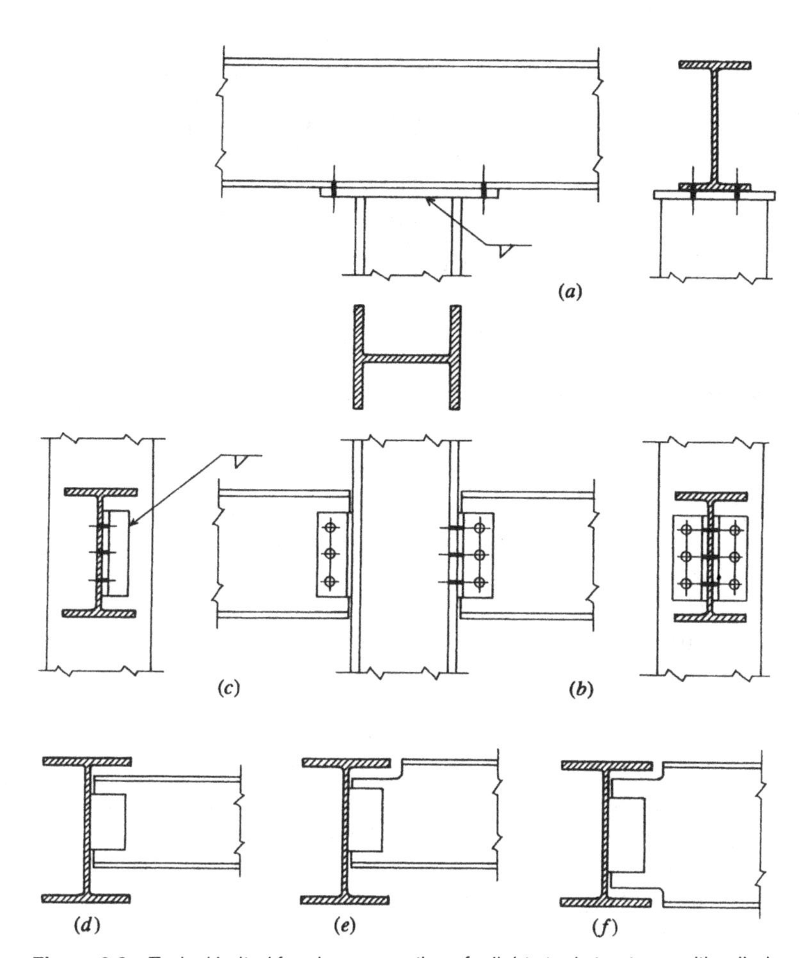

Figure 8.9 Typical bolted framing connections for light steel structures with rolled shapes.

level, the flange of the supported beam must be cut back as shown in Figure 8.9*e*. Further reduction of the supported beam occurs if the beams have the same depth, as shown in Figure 8.9*f*. If the top of the supported beam can be placed lower than the top of the supporting beam, as shown in Figure 8.9*d*, it may not be necessary to cut the flange of the supported beam. This somewhat simplifies the steel fabrication work, but it is not a common occurrence. Figure 8.10 shows a possible use of this arrangement, but it requires some fine work by the installers of the steel deck.

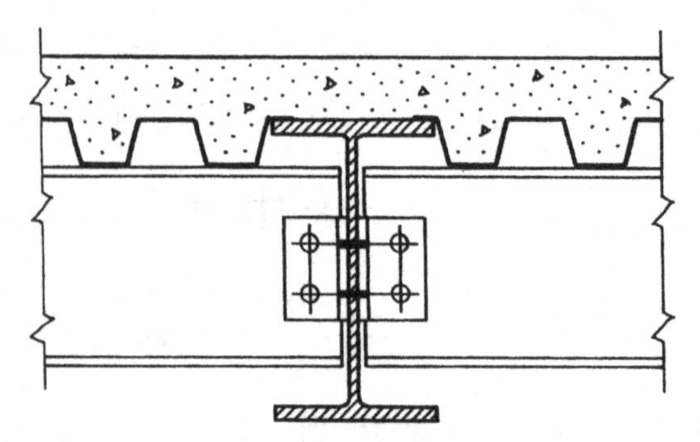

Figure 8.10 Special construction with deck-supporting structure; simplifies beam-to-girder connections.

Figure 8.11 shows additional framing details that are used for special situations. The connection shown in Figure 8.11*a* is sometimes used when the connected beam is shallow. The vertical load in this case is transferred through the seat angle, which may be welded or bolted to the web of the supporting beam. The connection to the web of the supported beam in this situation provides resistance to rollover, or torsional rotation, on the part of the beam. Figure 8.11*b* shows the use of a similar connection for joining a beam to a column.

Figures 8.11*c* and *d* show connections commonly used when the supporting column is a round pipe or rectangular tube. Because the lopsided connection in Figure 8.11*c* produces some twisting of the connection, the detail in Figure 8.11*d* is favored when the beam end reaction is large.

Framing connections commonly involve the use of both welding and bolting in a single connection. In general, welding is favored for connection made in the fabricating shop and bolting is favored for connection achieved at the erection site. If this practice is recognized, the connection must be developed with a view to the overall fabrication and erection process, and decisions must be made regarding where the work is done.

Development of connection details is particularly critical for structures in which a great number of connections occur. The truss is one such structure, and details for trusses are discussed in Sections 8.5 and 8.8.

Framed Beam Connections

The connection shown in Figure 8.12*a* is the type used most frequently in the development of framed structures that consist of I-shaped beams

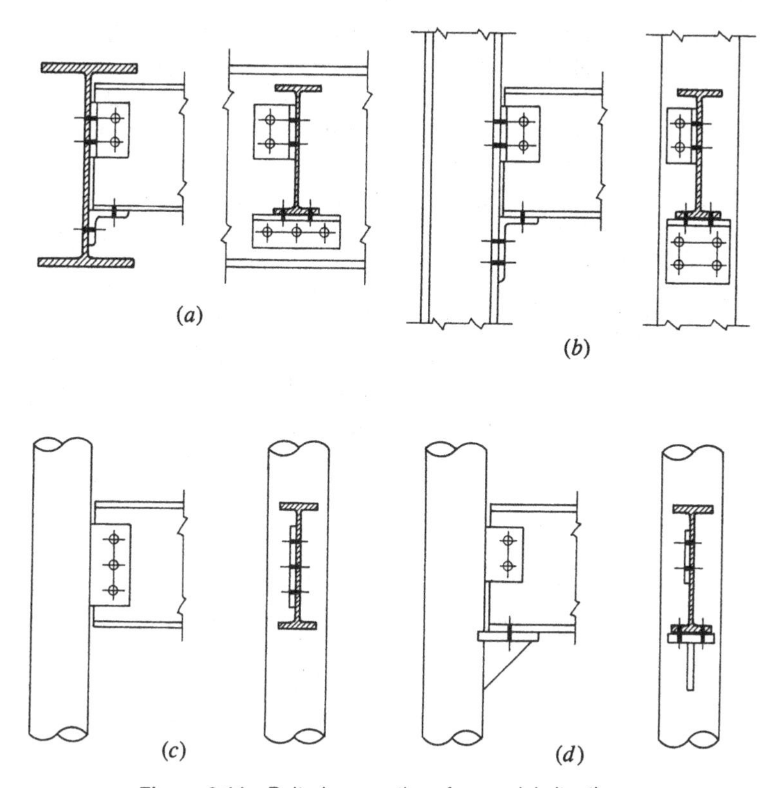

Figure 8.11 Bolted connections for special situations.

and H-shaped columns. This device is referred to as a *framed beam connection,* for which there are several design considerations:

Type of fastening. Fastening of the angles to the supported beam and to the support may be accomplished with welds or with any of several types of structural bolt. The most common practice is to weld the angles to the supported beam's web in the fabricating shop and to bolt the angles to the support (column face or supporting beam's web) in the field (the erection site).

Number of fasteners. If bolts are used, this refers to the number of bolts used on the supported beam's web; twice this number of bolts is used in the outstanding legs of the angles. The capacities are matched, however, because the web bolts are in double shear and the others in single shear. For smaller beams, or for light loads in general, angle leg sizes are typically narrow—just enough to accommodate a single row of bolts, as shown in Figure 8.12*b*. However, for very large beams and for greater loads, a wider leg may be used to accommodate two rows of bolts.

Size of the angles. Leg width and thickness of the angles depend on the size of fasteners and the magnitude of loads. Width of the outstanding legs may also depend on space available, especially if attachment is to the web of a column.

Length of the angles. Length must be that required to accommodate the number of bolts. Standard layout of the bolts is that shown in Figure 8.12, with bolts at 3-in. spacing and end distance of 1.25 in. This will accommodate up to 1-in.-diameter bolts. However, the angle length is also limited to the distance available on the beam web—that is, the total length of the flat portion of the beam web (see Figure 8.12*a*).

The AISC Manual (Ref. 3) provides considerable information to support the design of these frequently used connecting elements. Data is provided for both bolted and welded fastenings. Predesigned connections are tabulated and can be matched to magnitudes of loadings and to sizes (primarily depths) of the beams (mostly W shapes) that can accommodate them.

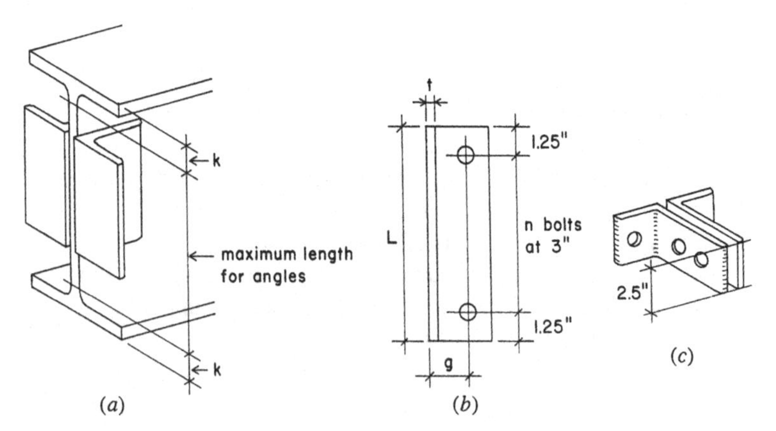

Figure 8.12 Framed beam connections for rolled shapes, using intermediate connecting angles.

Although there is no specified limit for the minimum size of a framed connection to be used with a given beam, a general rule is to use one with the angle length at least one-half of the beam depth. This rule is intended in the most part to ensure some minimum stability against rotational effects at the beam ends. For very shallow beams, the special connector shown in Figure 8.12*c* may be used. Regardless of loading, this requires the angle leg at the beam web to accommodate two rows of bolts (one bolt in each row) simply for the stability of the angles.

There are many structural effects to consider for these connections. Of special concern is the inevitable bending in the connection that occurs as shown in Figure 8.2*c*. The bending moment arm for this twisting action is the gage distance of the angle leg—dimension *g* in Figure 8.12*b*. It is a reason for choosing a relatively narrow angle leg.

If the top flange of the supported beam is cut back, as it commonly is when connection is to another beam, either vertical shear in the net cross section or block shear failure (Figure 8.3*c*) may be critical. Both of these conditions will be aggravated when the supported beam has a very thin web, which is a frequent condition because the most efficient beam shapes are usually the lightest shapes in their nominal size categories.

Another concern for the thin beam web is the possibility of critical bearing stress in the bolted connection. Combine a choice for a large bolt with one for a beam with a thin web, and this is likely to be a problem.

8.5 BOLTED TRUSS CONNECTIONS

A major factor in the design of trusses is the development of the truss joints. Because a single truss typically has several joints, the joints must be relatively easy to produce and economical, especially if there is a large number of trusses of a single type in the building structural system. Considerations involved in the design of connections for the joints include the truss configuration, member shapes and sizes, and the fastening method—usually welding or high-strength bolts.

In most cases, the preferred method of fastening for connections made in the fabricating shop is welding. Trusses are usually shop-fabricated in the largest units possible, which means the whole truss for modest spans or the maximum-sized unit that can be transported for large trusses. Bolting is mostly used for connections made at the building site. For the small truss, bolting is usually done only for the connections to supports and to supported elements or bracing. For the large truss, bolting may also be done at splice points between shop-fabricated units. All of this is subject

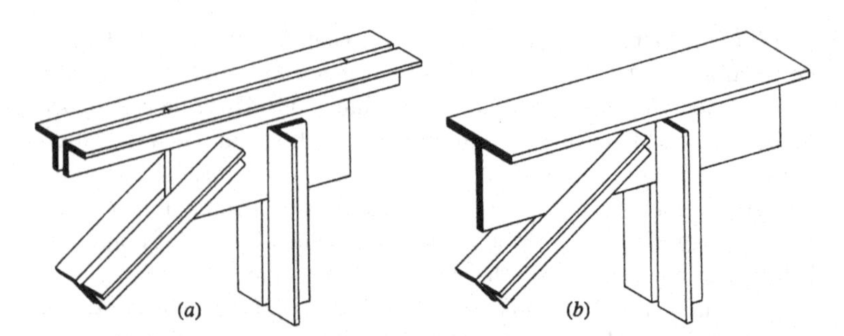

Figure 8.13 Common fabrication details for light steel trusses: (a) with chords consisting of double angles and joints using gusset plates, and (b) with chords consisting of structural tees.

to many considerations relating to the nature of the rest of the building structure, the particular location of the site, and the practices of local fabricators and erectors.

Two common forms for light steel trusses are shown in Figure 8.13. In Figure 8.13*a* the truss members consist of pairs of angles and the joints are achieved by using steel gusset plates to which the members are attached. For top and bottom chords, the angles are often made continuous through the joint, reducing the number of connectors required and the number of separate cut pieces of the angles. For flat-profiled, parallel-chord trusses of modest size, the chords are sometimes made from tees, with interior members fastened directly to the tee web (Figure 8.13*b*).

Figure 8.14 shows a layout for several joints of a light roof truss, employing the system shown in Figure 8.13*a*. This is a form commonly used in the past for roofs with high slopes and with many short-span trusses fabricated in a single piece in the shop, usually with riveted joints. Trusses of this form are now mostly welded or use high-strength bolts as shown in Figure 8.14.

Development of the joint designs for the truss shown in Figure 8.14 would involve many considerations, including:

Truss member size and load magnitude. This determines primarily the size and type of connector (bolt) required, based on individual connector capacity.

Angle leg size. This relates to the maximum diameter of bolt that can be used, based on angle gages and minimum edge distances. (See Table 8.3.)

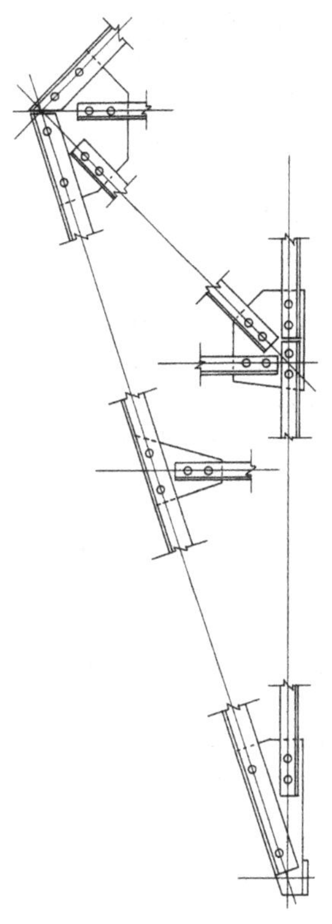

Figure 8.14 Typical form of a light steel truss with double-angle truss members and bolted joints with gusset plates.

Thickness and profile size of gusset plates. The preference is to have the lightest weight added to the structure (primarily for the cost per pound of the steel), which is achieved by reducing the plates to a minimum thickness and general minimum size.

Layout of members at joints. The aim is to have the action lines of the forces (vested in the rows of bolts) all meet at a single point, thus avoiding twisting in the joint.

Many of the points mentioned are determined by data. Minimum edge distances for bolts (Table 8.2) can be matched to usual gage dimensions for angles (Table 8.3). Forces in members can be related to bolt capacities in Table 8.1; the general intent is to keep the number of bolts to a minimum in order to make the required size of the gusset plate smaller.

Other issues involve some judgment or skill in the manipulation of the joint details. For really tight or complex joints, it is often necessary to study the form of the joint with carefully drawn large-scale layouts. Actual dimensions and form of the member ends and gusset plates may be derived from these drawings.

The truss shown in Figure 8.14 has some features that are quite common for small trusses. All member ends are connected by only two bolts, the minimum required by the specifications. This simply indicates that the minimum-sized bolt chosen has sufficient capacity to develop the forces in all members with only two bolts. At the top chord joint between the support and the peak, the top chord member is shown as being continuous (uncut) at the joint. This is quite common where the lengths of members available are greater than the joint-to-joint distances in the truss, a cost savings in member fabrication as well as connection.

If, as shown in Figure 8.14, there are only one or a few of the trusses to be used in a building, the fabrication may indeed be as shown in the illustration. However, if there are many such trusses or the truss is actually a manufactured, standardized product, it is much more likely to be fabricated employing welding for shop work and bolting only for field connections.

8.6 WELDING

In some instances welding is an alternative to bolting. Frequently, a connecting device (bearing plate, framing angle, etc.) is welded to one member in the shop and bolted to a connecting member in the field. However, in some instances, joints are fully welded, whether done in the shop or on-site. For some situations, welding is the only reasonable means to make a joint.

One advantage of welding is the ability to connect members directly, eliminating the need for intermediate devices such as gusset plates or framing angles. Another advantage is the elimination of holes that cause reduction of the area of tension members. Welding also creates very rigid joints, which is sometimes an advantage in reducing deformation due to loads in framed systems.

Electric Arc Welding

Electric arc welding is the process generally used in steel building construction. In this type of welding, an electric arc is formed between an electrode (welding rod) and the two pieces of metal to be joined. Globules of melted metal from the electrode flow into the molten seat at the joint and, when cool, solidify to join the connecting members.

The term *penetration* indicates the depth from the original surface of the base metal to the point at which fusion ceases. *Partial penetration,* the failure of the weld metal and base metal to fuse at the root of the weld, produces inferior welds. The craft skill of the welder is essential to the welding process.

Welded Connections

Several joints are shown in Figure 8.15. In general, there are three classifications of joints: *butt joints* (also called *groove welds*), *tee joints,* and *lap joints.* The joint used for a given situation depends on the magnitude of the loads, the manner in which load is applied, and the budgeted cost for construction.

A weld commonly used for structural steel in building construction is the *fillet* weld. Approximately triangular in cross section, it is formed between the two intersecting surfaces of the joined members, as shown in Figures 8.16*a* and *b.* The size of a fillet weld is the side dimension of the isosceles triangle that forms the weld cross section. The throat of a fillet weld is the distance from the hypotenuse to the peak of the triangle. The exposed surface of a fillet weld is not the plane surface shown in Figure 8.16*a;* it is usually somewhat crowned, as shown in Figure 8.16*b.* Therefore, the actual throat is usually greater than that in Figure 8.16*a* and the additional dimension is considered to be reinforcement for the weld.

Stresses in Fillet Welds

In Figure 8.16*a,* if $AB = BC$, then the throat dimension BD is $(\sqrt{2}/2)(BC)$, or $0.707(BC)$.

In other words, the throat dimension is 0.707 times the weld size. Consider a ½-in. weld—that is, a weld with dimension BC equal to one-half inch. The weld throat is 0.707(0.5) = 0.3535 in. The resistance factor for welds is 0.75. Then, if the ultimate shearing stress for the weld is 42 ksi, the design strength of a weld ϕR_n is 0.75(0.3535)(42) = 11.1 kips/in. If the ultimate shearing stress is 36 ksi, the design strength is 0.75(0.3535)(36) = 9.55 kips/in.

The permissible unit stresses used in the preceding paragraph are those for welds made with E 70 XX- and E 60 XX-type electrodes. Note that the stress in a fillet weld is considered as shear on the throat, regard-

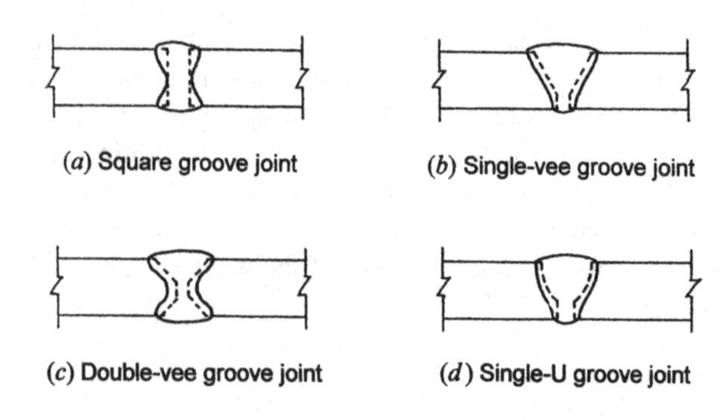
(*a*) Square groove joint (*b*) Single-vee groove joint

(*c*) Double-vee groove joint (*d*) Single-U groove joint

Butt Joints

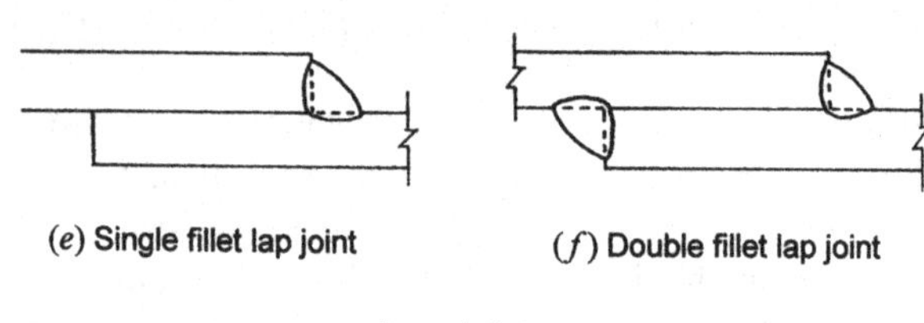
(*e*) Single fillet lap joint (*f*) Double fillet lap joint

Lap Joints

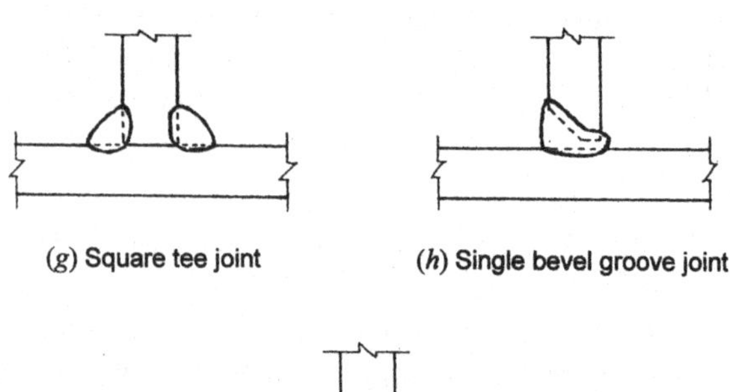
(*g*) Square tee joint (*h*) Single bevel groove joint

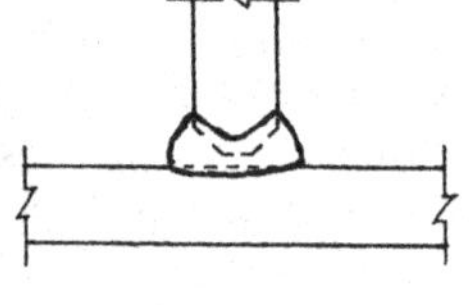
(*i*) Double bevel groove joint

Tee Joints

Figure 8.15 Common forms for welded joints.

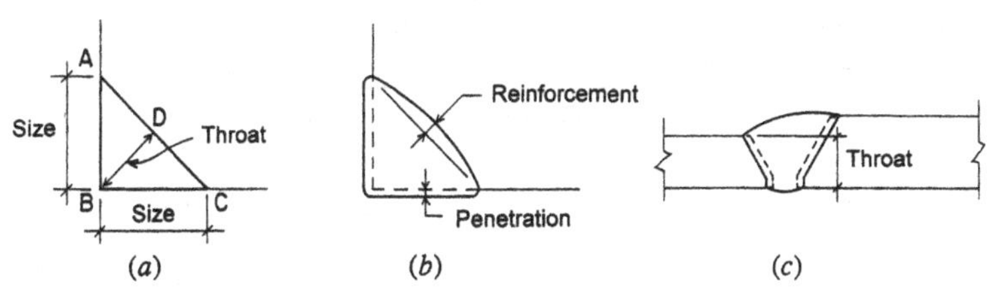

Figure 8.16 Considerations for dimensions of welds.

less of the direction of the applied load. Table 8.4 lists the design strengths of fillet welds of various sizes.

The stresses allowed for the metal of the connected parts (known as the *base metal*) apply to complete penetration groove welds that are loaded in tension or compression parallel to the axis of the connected members or are loaded in tension perpendicular to the effective throat.

Table 8.5 shows the relationship between the minimum weld size and the thickness of material in joints connected only by fillet welds. The maximum size of a fillet weld applied to the square edge of a plate or rolled shape that is ¼ in. or more in thickness should be 1⁄16 in. less than the thickness of the edge. Along edges of materials less than ¼ in. thick, the maximum size may be equal to the thickness of the material.

The effective cross-section area of butt and fillet welds is the product of the weld's effective length and the effective throat thickness. The minimum effective length of a fillet weld should not be less than four times the weld size. For starting and stopping the arc, add a distance equal to

TABLE 8.4 Design Strengths for Fillet Welds ($øR_n$)

	Design Strength (kips.in.)		Design Strength (kN/mm)		
Size of Weld (in.)	E 60 XX Electrodes	E 70 XX Electrodes	E 60 XX Electrodes	E 70 XX Electrodes	Size of Weld (mm.)
3⁄16	3.6	4.2	0.63	0.73	4.76
¼	4.8	5.6	0.84	0.97	6.35
5⁄16	6.0	7.0	1.04	1.22	7.94
⅜	7.2	8.4	1.25	1.46	9.53
½	9.5	11.1	1.67	1.95	12.7
⅝	11.9	13.9	2.09	2.40	15.9
¾	14.3	16.7	2.51	2.92	19.1

TABLE 8.5 Relation Between Material Thickness and Size of Fillet Welds

Material Thickness of the Thicker Part Joined		Minimum Sizes of Fillet Weld	
in.	mm	in.	mm
To 1/4 inclusive	To 6.35 inclusive	1/8	3.18
Over 1/4 to 1/2	Over 6.35 to 12.7	3/16	4.76
Over 1/2 to 3/4	Over 12.7 to 19.1	1/4	6.35
Over 3/4	Over 19.1	5/16	7.94

the weld size to each end of the required design length when specifying lengths to the welder.

Figure 8.17*a* represents two plates connected by fillet welds. The welds marked *A* are longitudinal (parallel to the load direction), and the weld marked *B* is a transverse weld. For the loading shown, the stress distribution in the longitudinal welds is not uniform, and the stress in the transverse weld is approximately 30 percent higher per unit length.

A transverse weld placed at the end of a member (see Figure 8.17*b*) gains strength if the weld is returned around the corner for a distance not less than twice the weld size. These end returns, called *boxing,* help to resist tearing action on the weld. Another means for producing a secure joint is to place the second weld as shown in Figure 8.17*b,* which helps prevent twisting of the connected member in the joint.

The 1/4-in. fillet weld is usually considered to be the smallest practical weld for structural steel, and the 5/16-in. weld is the most economical weld. The 5/16-in. weld is usually the largest weld that can be made in a single pass of the electrode. Larger welds must be built up with more passes, as shown in Figure 8.17*c*.

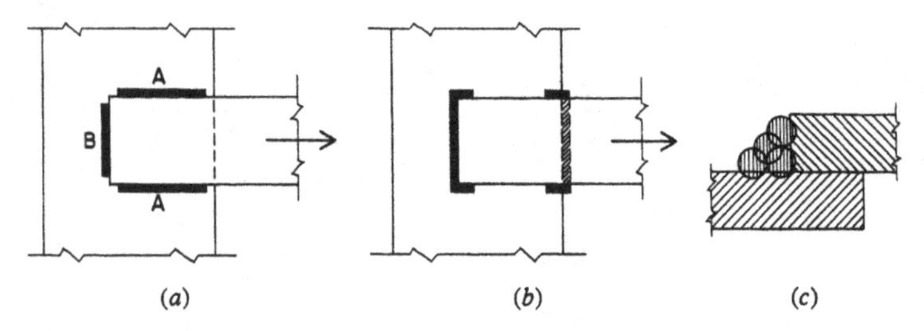

Figure 8.17 Welding of lapped steel elements.

8.7 DESIGN OF WELDED CONNECTIONS

Members to be connected must be held firmly in place during welding. It is often necessary to use temporary connecting devices during erection to achieve this condition. Although welding in the shop is often automated, field welding is almost always done by hand and usually in difficult circumstances for the welder. There is generally some greater confidence in the quality of shop welding, and designs for field welds are usually done somewhat more conservatively.

The following examples illustrate the design of some simple connections using fillet welds.

Example 2. A bar of A36 steel, 3 × 7⁄16 in. [76.2 × 11 mm] in cross section, is to be welded with E 70 XX electrodes to the back of a channel so that the full tensile strength of the bar may be developed. Determine the size and length for the fillet welds (see Figure 8.18).

Solution: The tension capacity of the steel bar is determined as

$$T_u = \phi F_y A = 0.9(36)(3 \times 0.4375) = 42.5 \text{ kips}$$

The weld must be adequate to resist this force. The maximum size is 3⁄8 in., which is 1⁄16 in. less than the bar thickness. Table 8.4 yields a value of 8.4 kips/in. for this size weld. The required weld length to develop the bar strength is thus:

$$L = \frac{42.5}{8.4} = 5.06 \text{ in.}$$

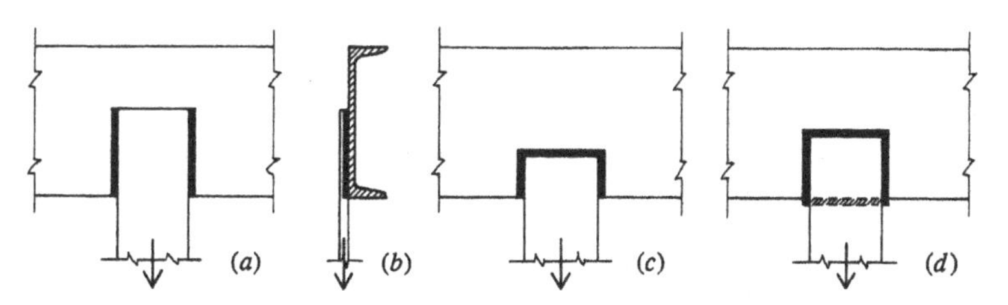

Figure 8.18 Variations of the form of a welded connection.

Figure 8.18 shows three ways to arrange this weld. In Figure 8.18*a*, the weld is divided into two equal parts. For two starts and stops, place a weld on each side of the bar equal to (5.06/2) + 0.75 = 3.28, say, 3.5 in.

The weld in Figure 8.18*c* has three parts: a 3-in.-long weld across the end of the bar and a 3-in. weld split between the two sides. Each of the side welds should be a minimum of 2 in.

Neither of the welds in Figures 8.18*a* or *c* resist twisting on the bar. To accommodate this action, the best weld is that shown in Figure 8.18*d*, where a weld is added on the back of the bar. This is definitely the most secure weld, although either of the other two forms is usually acceptable.

Example 3. Connect a 3½-×-3½-×-5⁄16-in. angle of A36 steel to a plate by fillet welds using E 70 XX electrodes. Find the weld dimensions to develop the angle's full tensile strength.

Solution: From Table A.5 in the appendix, the cross-sectional area of the angle is 2.09 in.^{2} [1348 mm^2]. The tensile capacity of the angle is

$$T_u = \phi F_y A = 0.9(36)(2.09) = 67.7 \text{ kips [301 kN]}$$

For the 5⁄16-in.-thick angle leg, the maximum recommended weld is ¼ in. From Table 8.4, the weld capacity is 5.6 kips/in. The total effective length required is thus:

$$L = \frac{67.7}{5.6} = 12.1 \text{ in. [307 mm]}$$

This total length could be divided equally between the two sides of the angle. However, assuming the tension load in the angle to coincide with its centroid, the distribution of the load is not equally divided between the two sides. Therefore, some designers prefer to proportion the lengths on the two sides so that they correspond appropriately to their positions. If this is desired, the following procedure may be used.

From Table A.5, the centroid of the angle is 0.99 in. from the back of the angle. Referring to the two weld lengths as shown in Figure 8.19, note that their lengths should be in inverse proportion to their distances from the angle centroid. Thus:

$$L_1 = \frac{2.51}{3.5}(12.1) = 8.67 \text{ in. [220 mm]}$$

and

$$L_2 = \frac{0.99}{3.5}(12.1) = 3.42 \text{ in. [87 mm]}$$

Adding for starts and stops, reasonable specified lengths are 9.25 and 4.0 in.

When angle shapes are used as tension members and are connected at their ends by fastening of only one leg, it is questionable to assume a full stress distribution across the whole angle cross section. Some designers therefore prefer to ignore the development of tension in the unconnected leg and to limit the member capacity to the force obtained by considering only the connected leg. In this case, this means the consideration of a bar that is 3½ × 5⁄16 in. For this assumption the maximum tension force is

$$T_u = 0.9(36)(3.5 \times 0.3125) = 35.4 \text{ kips [158 mm]}$$

and the total required weld length is

$$L = \frac{35.4}{5.6} = 6.32 \text{ in. [161 mm]}$$

Adding for starts and stops this means a total weld length of (6.32/2) + 0.5 = 3.66 in., say, 3.75 in., on each side of the angle leg.

Problem 8.7.A. A 4-×-4-×-½-in. angle of A36 steel is to be welded to a plate with E 70 XX electrodes to develop the full tensile strength of the angle. Using ⅜-in. fillet welds, compute the design lengths for the welds

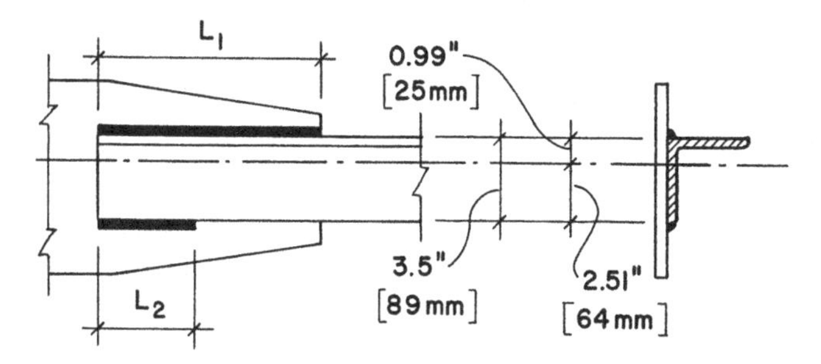

Figure 8.19 Form of the welded connection in Example 2.

on the two sides of the angle, assuming the development of tension on the full cross section of the angle.

Problem 8.7.B. Redesign the connection in Problem 8.7.A, assuming that the tension force is developed only in the connected leg of the angle.

Plug and Slot Welds

One method of connecting two overlapping plates uses a weld in a hole made in one of the two plates (see Figure 8.20). Plug and slot welds are those in which the entire area of the hole or slot receives weld metal. The maximum and minimum diameters of plug and slot welds, and the maximum length of slot welds are shown in Figure 8.20. If the plate containing the hole is not more than ⅝ in. thick, the hole should be filled with weld metal. If the plate is more than ⅝ in. thick, the weld metal should be at least one-half the thickness of the plate but not less than ⅝ in.

The stress in a plug or slot weld is considered to be shear on the area of the weld at the plane of contact of the two plates being connected. The ultimate shearing stress is 42 ksi [145 MPa] for E 70 XX electrodes, and 36 ksi [248 MPa] for E 60 XX electrodes, and ϕ is 0.75.

A somewhat similar weld consists of a continuous fillet weld at the circumference of a hole, as shown in Figure 8.20*c*. This is not a plug or slot weld and is subject to the usual requirements for fillet welds.

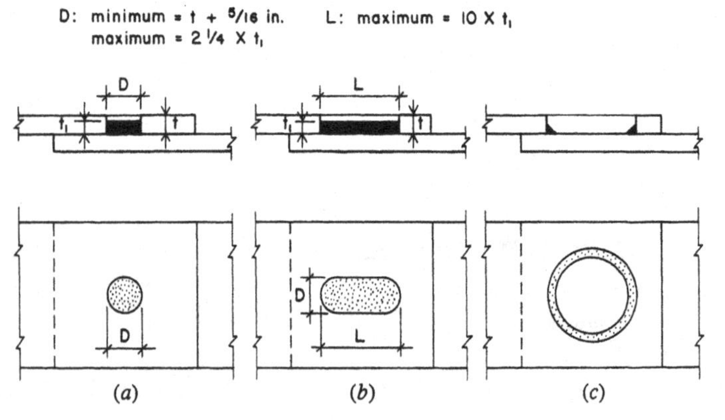

Figure 8.20 Welds placed in holes: (a) plug weld, (b) slot weld, and (c) fillet weld in a large hole.

8.8 WELDED STEEL FRAMES

Welding is presently used extensively for assembly of steel frames, both in the fabricating shop and in the field. In many cases, joints are of a classic form, deriving from the forms of connected elements and from common tasks for steel structures.

In the early development of steel structures, in the nineteenth century, most assembly was done with rivets. Both bolting with high-strength bolts and welding took over in the mid-twentieth century, and riveting has largely been discontinued. Nevertheless, the classic forms of joints that were developed for riveting are still widely used. Most fabrication in the shop is now done with welding. Bolting is largely utilized for field connections.

In many cases, from strictly a functioning point of view, joints may be alternatively developed with bolting or welding. The decision has often to do with location of the work: in the shop or in the field. Many of the joints shown as bolted in illustrations in this book could as well be welded.

In some cases, joints are developed with some assembly achieved in the fabricating shop and some in the field. A common example is the standard framed beam connection using a pair of angles. It is now common practice to attach the angles to the supported beam's ends in the shop by welding and to complete the connection to the supporting member in the field with bolts.

A common structure is the multistory, multibay steel frame for multilevel buildings. Parts of these three-dimensional frameworks are frequently developed as rigid frame bents to resist lateral forces due to wind or earthquakes. These bents have mostly been achieved in recent years by welding of the beams directly to the columns. However, recent experiences in major earthquakes have caused reconsideration of these jointing methods.

8.9 CONTROL JOINTS: DESIGN FOR SELECTED BEHAVIOR

Most connections between elements of a structure are designed to *resist* movement within the joint. The usual purpose of fastening devices, such as bolts and welds, is to hold the connected parts firmly together. A *control joint,* on the other hand, is a joint in which some form of movement is deliberately facilitated. The three most common reasons for such a joint are the following:

1. *Thermal expansion.* All materials expand and contract with changes in temperature. In large buildings, or any long building part, the total movement can be considerable. This requires some consideration for the forces that can accumulate and may damage the construction.
2. *Structural actions.* Actions of continuous and long-span structures can cause movements at their supports that can disrupt other construction. The support joint may need to accommodate some aspects of this movement with reduced resistance while providing the basic support that is mandatory.
3. *Seismic separation.* Multi-massed buildings may have discrete parts that want to move differently under the dynamic effect of an earthquake. One technique for dealing with this is to provide a tolerance for independent movement of the parts at their points of connection.

Of course, the simplest form of a control joint is no connection—that is, the non-joining of the parts. From a structural point of view, this is essentially what most seismic separation joints are. However, a common situation is where some form of connection is required that must simultaneously facilitate some form of movement. For example, it may be desirable to support a beam in a manner that does not restrict its end rotation; in other words, provide a simple support for vertical force, not a fixed one.

A common device used for bolted connections is the slotted hole, as shown in Figure 8.21. If the width of the hole is only slightly larger than the bolt diameter, the plate is reasonably restrained from moving in the direction perpendicular to the slot, even if the bolt is not tight. In the other direction, however, with the bolt in place, the plate can be moved a small distance. This device can be used to permit small movements (such as those from thermal expansion), but it is actually more frequently used for tolerance of erection inaccuracies.

There are many types and many details for control joints. The following example illustrates a situation where some kind of structural resistance is required but other resistances need to be reduced as much as possible.

The behavior of trusses requires some allowance for change in the length of the chords as loading changes—basically, as the live load comes and goes. This implies that the truss supports can facilitate some overall change in the length of the truss, which requires some change in

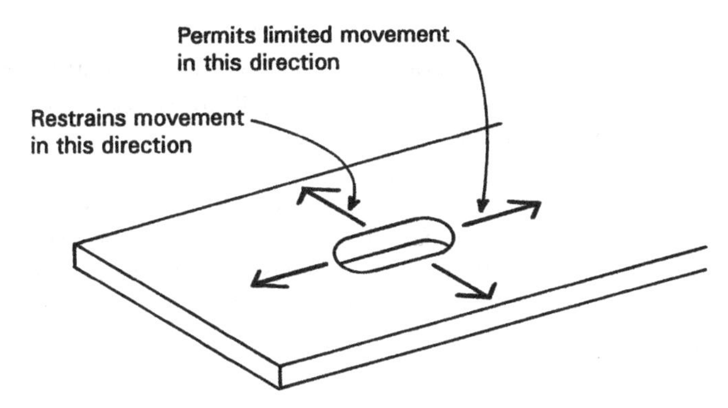

Figure 8.21 Slotted slip connection. A common means for facilitating movement in a connection in a single direction.

the actual distance between the support points. If movements are small in actual dimension and there is some possibility of nondestructive deformation in the truss-to-support connections, it may not be necessary to make any special provision for the movements. If, however, the actual dimensions of movements are large and both the support structure and the connections to it are virtually unyielding, problems will occur unless actual provision is effectively made to facilitate the movements.

A wide range in temperature can also produce considerable length change in long structures. These effects should be considered in terms of movements at the supports as well as length change. Following are two examples:

1. *Long-span steel truss, supported by masonry piers.* In this case, the movements will be considerable and the supports essentially unyielding. Special provision must be made at one or both supports for some actual dimension of movement.
2. *Trusses erected during cold weather but later subjected to warm weather or to the warmer conditions maintained in the enclosed building.* In this case, even though provisions for movement due to loading stresses may be unnecessary, the length changes due to thermal change should be considered.

A technique that is sometimes used to reduce the need for any special provision for movement at supports is to leave the support connections in a stable but nontightened condition until after the building construction is essentially completed. This allows the truss deformations resulting from the dead load to accumulate during construction so that the critical effects are limited to the deformations caused by the live loads. Where the dead load is a major part of the total design load, this is often quite effective.

When provision must be made for movement, the precise form of movements and their approximate anticipated magnitudes must be carefully determined. For support of a horizontal-spanning structure, such as

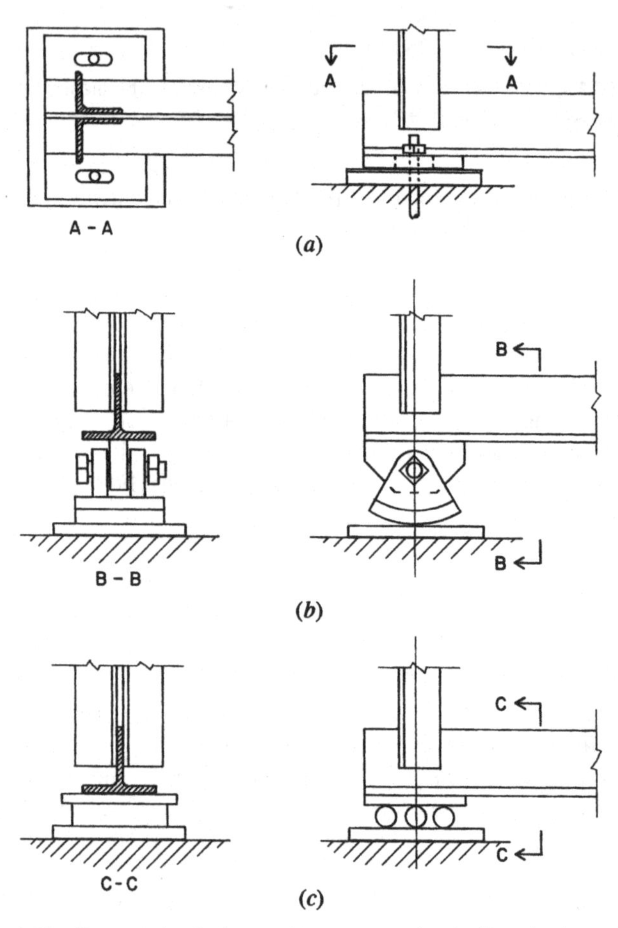

Figure 8.22 Support details for steel trusses, used to facilitate horizontal movement at the joint.

a beam or truss, the minimum structural resistance required is that in response to vertical force. Thus, the joint must restrict vertical movement while possibly tolerating both horizontal movement and rotation.

Figure 8.22 shows several details that may be used where provision for movement must be made at the truss supports. The need for these is a matter of judgment and must be considered in terms of the full development of the building construction.

In Figure 8.22*a* the method used is a slight modification of a common joint that uses an end bearing plate bolted to the support with anchor bolts embedded in masonry or concrete. In this version, a second steel plate is used; thus, one plate comes welded to the truss and the other is set in place on the top of the wall. Slotted holes (Figure 8.21) are provided in the truss end plate. These also provide for some inaccuracy in the precise positioning of the anchor bolts—a wise choice in any regard. However, if the nuts on the anchor bolts are only hand-tightened and a friction-reducing interface material is used, resistance to horizontal movement is minimal.

The method shown in Figure 8.22*b* uses a rather elaborate rocker device. An advantage gained with this joint is the very low resistance to rotation, if indeed this is a concern.

The detail in Figure 8.22*c* is a variation on a so-called *roller bearing joint.* In this case, three steel rods are used to separate the truss end plate and the wall top plate. Horizontal movement is thus only slightly resisted. However, here as in Figure 8.22*a,* there is little provision for reduction of effects of end rotation of the truss.

Because control joints are often provided because of possible effects on the surrounding or supported construction, their effectiveness must in the end be evaluated for both structural adequacy and overall value to the building.

9

GENERAL CONSIDERATIONS FOR BUILDING STRUCTURES

This chapter contains discussions of some general issues relating to design of building structures. These concerns have mostly not been addressed in the presentations in earlier chapters, but they require consideration when you are dealing with whole building design situations. Application of these materials is illustrated in the design examples in Chapter 10.

9.1 CHOICE OF BUILDING CONSTRUCTION

Materials, methods, and details of building construction vary considerably on a regional basis. Many factors affect this situation, including the effects of response to climate and regional availability of construction materials. Even in a single region, differences occur among individual buildings, based on styles of architectural design and techniques of builders. Nevertheless, at any given time there are usually a few predominant, popular methods of construction that are employed for most buildings of a given type and size. The construction methods and details shown here are reasonable, but in no way are they intended to illustrate a singular, superior style of building.

It is not possible to choose the materials and forms for a building structure without considering its integration with the general building construction. In some cases it may also be necessary to consider the elements required for various building services, such as those for piping, electrical service, lighting, communication, roof drainage, and the HVAC (heating, ventilating, and air conditioning) systems.

For multistory buildings it is necessary to accommodate the placement of stairs, elevators, and the vertical elements for various building services—particularly for air ducts between building levels. A major consideration for multistory buildings is the planning of the various levels so that they work when superimposed on top of each other. Bearing walls and columns must be supported from below.

Choice of the general structural system as well as the various individual elements of the system is typically highly dependent on the general architectural design of the building. Hopefully the two issues—structural planning and architectural planning—are dealt with simultaneously from preliminary design to final construction drawings.

9.2 STRUCTURAL DESIGN STANDARDS

Use of methods, procedures, and reference data for structural design are subject to the judgment of the designer. Many guides exist, but some individual selection is often required. Strong influences on choices include the following:

Building code requirements from the enforceable statutes relating to the location of the building.

Acceptable design standards as published by professional groups, such as the reference from the American Society of Civil Engineers (ASCE) referred to frequently in this book (Ref.1).

Recommended design standards from industry organizations, such as the AISC and ACI.

The body of work from current texts and references produced by respected authors.

Some reference is made to these sources in this book. However, much of the work is also simply presented in a manner familiar to the authors, based on their own experiences. If you pursue this subject, you are sure to encounter styles and opinions that differ from those presented here. Making one's own choices in the face of those conflicts is part of the progress of professional growth.

9.3 LOADS FOR STRUCTURAL DESIGN

Loads used for structural design must be derived primarily from enforceable building codes. However, the principal concern of codes is public health and safety. Performance of the structure for other concerns may not be adequately represented in the minimum requirements of the building code. Issues sometimes not included in code requirements are as follows:

Effects of deflection of spanning structures on nonstructural elements of the construction.

Sensations of bounciness of floors by building occupants.

Protection of structural elements from damage due to weather or normal usage.

It is quite common for professional structural designers to have situations where they use their own judgment in assigning design loads. This ordinarily means using increased loads, because the minimum loads required by codes must always be recognized.

Building codes currently stipulate both the load sources and the form of combinations to be used for design. The following loads are listed in the 2002 edition of the *ASCE Minimum Design Loads for Buildings and Other Structures* (Ref. 1), hereinafter referred to as ASCE 2002.

D = dead load

E = earthquake-induced force

L = live load, except roof load

L_r = roof live load

S = snow load

W = load due to wind pressure

Additional special loads are listed but these are the commonly occurring loads. This chapter contains descriptions of some of these loads.

9.4 DEAD LOADS

Dead load consists of the weight of the materials of which the building is constructed such as walls, partitions, columns, framing, floors, roofs, and ceilings. In the design of a beam or column, the dead load used must include an allowance for the weight of the structural member itself. Table 9.1, which lists the weights of many construction materials, may be used in the computation of dead loads. Dead loads are due to gravity and result in downward vertical forces.

Dead load is generally a permanent load, once the building construction is completed, unless remodeling or rearrangement of the construction occurs. Because of this permanent, long-time character, the dead load requires certain considerations in design, such as the following:

1. It is always included in design loading combinations, except for investigations of singular effects, such as deflections due to only live load.
2. Its long-time character has some special effects that cause sag and require reduction of design stresses in wood structures, development of long-term, continuing settlements in some soils, and production of creep effects in concrete structures.
3. Dead load contributes some unique responses, such as the stabilizing effects that resist uplift and overturn caused by wind forces.

Although we can determine weights of materials with reasonable accuracy the complexity of most building construction makes the computation of dead loads possible only on an approximate basis. This adds to other factors to make design for structural behaviors a very approximate science. As in other cases, this should not be used as an excuse for sloppiness in the computational work, but it should be recognized as a fact to temper concern for high accuracy in design computations.

9.5 BUILDING CODE REQUIREMENTS FOR STRUCTURES

Structural design of buildings is most directly controlled by building codes, which are the general basis for the granting of building permits—the legal permission required for construction. Building codes (and the permit-granting process) are administered by some unit of government: city, county, or state. Most building codes, however, are based on some model code.

TABLE 9.1 Weight of Building Construction

	psf[a]	kPa[a]
Roofs		
3-ply ready roofing (roll, composition)	1	0.05
3-ply felt and gravel	5.5	0.26
5-ply felt and gravel	6.5	0.31
Shingles: Wood	2	0.10
Asphalt	2–3	0.10–0.15
Clay tile	9–12	0.43–0.58
Concrete tile	6–10	0.29–0.48
Slate, 3 in.	10	0.48
Insulation: Fiber glass batts	0.5	0.025
Foam plastic, rigid panels	1.5	0.075
Foamed concrete, mineral aggregate	2.5/in.	0.0047/mm
Wood rafters: 2 × 6 at 24 in.	1.0	0.05
2 × 8 at 24 in.	1.4	0.07
2 × 10 at 24 in.	1.7	0.08
2 × 12 at 24 in.	2.1	0.10
Steel deck, painted: 22 gage	1.6	0.08
20 gage	2.0	0.10
Skylights: Steel frame with glass	6–10	0.29–0.48
Aluminum frame with plastic	3–6	0.15–0.29
Plywood or softwood board sheathing	3.0/in.	0.0057/mm
Ceilings		
Suspended steel channels	1	0.05
Lath: Steel mesh	0.5	0.025
Gypsum board, ½ in.	2	0.10
Fiber tile	1	0.05
Drywall, gyspum board, ½ in.	2.5	0.12
Plaster: Gypsum	5	0.24
Cement	8.5	0.41
Suspended lighting and HVAC, average	3	0.15
Floors		
Hardwood, ½ in.	2.5	0.12
Vinyl tile	1.5	0.07
Ceramic tile: ¾ in.	10	0.48
Thin-set	5	0.24
Fiberboard underlay, 0.625 in.	3	0.15
Carpet and pad, average	3	0.15
Timber deck	2.5/in.	0.0047/mm
Steel deck, stone concrete fill, average	35–40	1.68–1.92
Concrete slab deck, stone aggregate	12.5/in.	0.024/mm
Lightweight concrete fill	8.0/in.	0.015/mm

(continued)

TABLE 9.1 (*Continued*)

	psf[a]	kPa[a]
Floors (Continued)		
Wood joists: 2 × 8 at 16 in.	2.1	0.10
2 × 10 at 16 in.	2.6	0.13
2 × 12 at 16 in.	3.2	0.16
Walls		
2 × 4 studs at 16 in., average	2	0.10
Steel studs at 16 in., average	4	0.20
Lath. plaster—see *Ceilings*		
Drywall, gypsum board, ½ in.	2.5	0.10
Stucco, on paper and wire backup	10	0.48
Windows, average, frame + glazing:		
Small pane, wood or metal frame	5	0.24
Large pane, wood or metal frame	8	0.38
Increase for double glazing	2–3	0.10–0.15
Curtain wall, manufactured units	10–15	0.48–0.72
Brick veneer, 4 in., mortar joints	40	1.92
½ in., mastic-adhered	10	0.48
Concrete block:		
Lightweight, unreinforced, 4 in.	20	0.96
6 in.	25	1.20
8 in.	30	1.44
Heavy, reinforced, grouted, 6 in.	45	2.15
8 in.	60	2.87
12 in.	85	4.07

[a]Average weight per square foot of surface, except as noted.

Values given as /in. or /mm are to be multiplied by actual thickness of material.

Model codes are more similar than different, and are in turn largely derived from the same basic data and standard reference sources, including many industry standards. In the several model codes and many city, county, and state codes, however, there are some items that reflect particular regional concerns. With respect to control of structures, all codes have materials (all essentially the same) that relate to the following issues:

1. *Minimum required live loads.* All building codes have tables that provide required values to be used for live loads. Tables 9.2 and 9.3 contain some loads as specified in ASCE 2002 (Ref. 1).
2. *Wind loads.* These are highly regional in character with respect to concern for local windstorm conditions. Model codes provide data with variability on the basis of geographic zones.

3. *Seismic (earthquake) effects.* These are also regional, with predominant concerns in the western states. This data, including recommended investigations, is subject to quite frequent modification, because the area of study responds to ongoing research and experience.
4. *Load duration.* Loads or design stresses are often modified on the basis of the time span of the load, varying from the life of the structure for dead load to a few seconds for a wind gust or a single major seismic shock. Safety factors are frequently adjusted on this basis. Some applications are illustrated in the work in the design examples.
5. *Load combinations.* Formerly mostly left to the discretion of designers, these are now quite commonly stipulated in codes, mostly because of the increasing use of ultimate strength design and the use of factored loads.
6. *Design data for types of structures.* These deal with basic materials (wood, steel, concrete, masonry, etc.), specific structures (rigid frames, towers, balconies, pole structures, etc.) and special problems (foundations, retaining walls, stairs, etc.). Industry-wide standards and common practices are generally recognized, but local codes may reflect particular local experience or attitudes. Minimal structural safety is the general basis, and some specified limits may result in questionably adequate performances (bouncy floors, cracked plaster, etc.).
7. *Fire resistance.* For the structure, there are two basic concerns, both of which produce limits for the construction. The first concern is for structural collapse or significant structural loss. The second concern is for containment of the fire to control its spread. These concerns produce limits on the choice of materials (e.g., combustible or noncombustible) and some details of the construction (cover on reinforcement in concrete, fire insulation for steel beams, etc.).

The work in the design examples in Chapter 10 is based largely on criteria from ASCE 2002 (Ref. 1).

9.6 LIVE LOADS

Live loads technically include all the nonpermanent loadings that can occur, in addition to the dead loads. However, the term as commonly used

TABLE 9.2 Minimum Floor Live Loads

Building Occupancy or Use	Uniformly Distributed Load (psf)	Concentrated Load (lb)
Apartments and Hotels		
Private rooms and corridors serving them	40	
Public rooms and corridors serving them	100	
Dwellings, One- and Two-Family		
Uninhabitable attics without storage	10	
Uninhabitable attics with storage	20	
Habitable attics and sleeping rooms	30	
All other areas except stairs and balconies	40	
Office Buildings		
Offices	50	2000
Lobbies and first-floor corridors	100	2000
Corridors above first floor	80	2000
Stores		
Retail		
First floor	100	1000
Upper floors	75	1000
Wholesale, all floors	125	1000

Source: ASCE 2002 (Ref. 1), used with permission of the publishers, American Society of Civil Engineers.

TABLE 9.3 Live Load Element Factor, K_{LL}

Element	K_{LL}
Interior columns	4
Exterior columns without cantilever slabs	4
Edge columns with cantilever slabs	3
Corner columns with cantilever slabs	2
Edge beams without cantilever slabs	2
Interior beams	2
All other members not identified above	1

Source: ASCE 2002 (Ref. 1), used with permission of the publishers, American Society of Civil Engineers.

usually refers only to the vertical gravity loadings on roof and floor surfaces. These loads occur in combination with the dead loads but are generally random in character and must be dealt with as potential contributors to various loading combinations, as discussed in Section 9.8.

Roof Loads

In addition to the dead loads they support, roofs are designed for a uniformly distributed live load. The minimum specified live load accounts for general loadings that occur during construction and maintenance of the roof. For special conditions, such as heavy snowfalls, additional loadings are specified.

The minimum roof live load in psf is specified in ASCE 2002 (Ref. 1) in the form of an equation, as follows:

$$L_r = 20\,R_1\,R_2 \text{ in which } 12 \leq L_r \leq 20$$

In the equation, R_1 is a reduction factor based on the tributary area supported by the structural member being designed (designated as A_t and quantified in ft²) and is determined as follows:

$$\begin{aligned} R_1 &= 1, \text{ for } A_t \geq 200 \text{ ft}^2 \\ &= 1.2 - 0.001\,A_{t,} \text{ for } 200 \text{ ft}^2 < A_t < 600 \text{ ft}^2 \\ &= 0.6, \text{ for } A_t \geq 600 \text{ ft}^2 \end{aligned}$$

Reduction factor R_2 accounts for the slope of a pitched roof and is determined as follows:

$$\begin{aligned} R_2 &= 1, \text{ for } F \leq 4 \\ &= 1.2 - 0.05\,F, \text{ for } 4 < F < 12 \\ &= 0.6, \text{ for } F \geq 12 \end{aligned}$$

The quantity F in the equations for R_2 is the number of inches of rise per ft for a pitched roof (for example: $F = 12$ indicates a rise of 12 in., or an angle of 45°).

The design standard also provides data for roof surfaces that are arched or domed and for special loadings for snow or water accumulation. Roof surfaces must also be designed for wind pressures on the roof surface, both upward and downward. A special situation that must be considered is that of a roof with a low dead load and a significant upward wind load (outward suction) that exceeds the dead load.

Although the term *flat roof* is often used, there is generally no such thing; all roofs must be designed for some water drainage. The minimum required pitch is usually ¼ in./ft, or a slope of approximately 1:50. With roof surfaces that are close to flat, a potential problem is that of *ponding,* a phenomenon in which the weight of the water on the surface causes deflection of the supporting structure, which in turn allows for more water accumulation (in a pond), causing more deflection, and so on, resulting in a progressive collapse condition.

Floor Live Loads

The live load on a floor represents the probable effects created by the occupancy. It includes the weights of human occupants, furniture, equipment, stored materials, and so on. All building codes provide minimum live loads to be used in the design of buildings for various occupancies. Because there is a lack of uniformity among different codes in specifying live loads, the local code should always be used. Table 9.2 contains a sample of values for floor live loads as given in ASCE 2002 (Ref. 1) and commonly specified by building codes.

Although expressed as uniform loads, code-required values are usually established large enough to account for ordinary concentrations that occur. For offices, parking garages, and some other occupancies, codes often require the consideration of a specified concentrated load as well as the distributed loading. This required concentrated load is listed in Table 9.2 for the appropriate occupancies.

Where buildings are to contain heavy machinery, stored materials, or other contents of unusual weight, these must be provided for individually in the design of the structure.

When structural framing members support large areas, most codes allow some reduction in the total live load to be used for design. These reductions, in the case of roof loads, are incorporated in the formulas for roof loads given previously. The following is the method given in ASCE 2002 (Ref. 1) for determining the reduction permitted for beams, trusses, or columns that support large floor areas.

The design live load on a member may be reduced in accordance with the formula:

$$L = L_0\left(0.25 + \frac{15}{\sqrt{K_{LL}A_T}}\right)$$

where L = reduced design live load per square foot of area supported by the member

L_0 = unreduced live load supported by the member

K_{LL} = live load element factor (see Table 9.3)

A_T = tributary area supported by the member

L shall not be less than $0.50L_0$ for members supporting one floor and L shall not be less than $0.40L_0$ for members supporting two or more floors.

In office buildings and certain other building types, partitions may not be permanently fixed in location but may be erected or moved from one position to another in accordance with the requirements of the occupants. To provide for this flexibility, it is customary to require an allowance of 15 to 20 psf, which is usually added to other dead loads.

9.7 LATERAL LOADS (WIND AND EARTHQUAKE)

As used in building design, the term *lateral load* is usually applied to the effects of wind and earthquakes, because they induce horizontal forces on stationary structures. From experience and research, design criteria and methods in this area are continuously refined, with recommended practices being presented through the various model building codes.

Space limitations do not permit a complete discussion of the topic of lateral loads and design for their resistance. The following discussion summarizes some of the criteria for design in ASCE 2002 (Ref. 1). Examples of application of these criteria are given in the design examples of building structural design in Chapter 10. For a more extensive discussion, refer to *Simplified Building Design for Wind and Earthquake Forces* (Ref. 6).

Wind

Where wind is a regional problem, local codes are often developed in response to local conditions. Complete design for wind effects on buildings includes a large number of both architectural and structural concerns. The following is a discussion of some of the requirements from ASCE 2002 (Ref. 1).

Basic Wind Speed. This is the maximum wind speed (or velocity) to be used for specific locations. It is based on recorded wind histories and adjusted for some statistical likelihood of occurrence. For the United States, recommended minimum wind speeds are taken from maps provided in the ASCE standard. As a reference point, the speeds are those recorded at the standard measuring position of 10 m (approximately 33 ft) above the ground surface.

Wind Exposure. This refers to the conditions of the terrain surrounding the building site. The ASCE standard uses three categories, labeled B, C, and D. Qualifications for categories are based on the form and size of wind-shielding objects within specified distances around the building.

Simplified Design Wind Pressure (p_s). This is the basic reference equivalent static pressure based on the critical wind speed and is determined as follows:

$$p_s = \lambda I p_{S30}$$

where $\lambda\lambda$ = adjustment factor for building height and exposure
I = importance factor
p_{S30} = simplified design wind pressure for exposure B, at height of 30 ft, and for $I = 1.0$

The importance factor for ordinary circumstances of building occupancy is 1.0. For other buildings, factors are given for facilities that involve hazard to a large number of people, for facilities considered to be essential during emergencies (such as windstorms), and for buildings with hazardous contents.

The design wind pressure may be positive (inward) or negative (outward, suction) on any given surface. Both the sign and the value for the pressure are given in the design standard. Individual building surfaces, or parts thereof, must be designed for these pressures.

Design Methods. Two methods are described in the Code for the application of wind pressures.

Method 1 (simplified procedure). This method is permitted to be used for relatively small, low-rise buildings of simple symmetrical shape. It is the method described here and used for the examples in Chapter 10.

Method 2 (analytical procedure). This method is much more complex and is prescribed to be used for buildings that do not fit the limitations described for Method 1.

Uplift. Uplift may occur as a general effect, involving the entire roof or even the whole building. It may also occur as a local phenomenon such as that generated by the overturning moment on a single shear wall.

Overturning Moment. Most codes require that the ratio of the dead load resisting moment (called the restoring moment, stabilizing moment, etc.) to the overturning moment be 1.5 or greater. When this is not the case, uplift effects must be resisted by anchorage capable of developing the excess overturning moment. Overturning may be a critical problem for the whole building, as in the case of relatively tall and slender tower structures. For buildings braced by individual shear walls, trussed bents, and rigid-frame bents, overturning is investigated for the individual bracing units.

Drift. Drift refers to the horizontal deflection of the structure caused by lateral loads. Code criteria for drift are usually limited to requirements for the drift of a single story (horizontal movement of one level with respect to the next above or below). As in other situations involving structural deformations, effects on the building construction must be considered; thus, the detailing of curtain walls or interior partitions may affect limits on drift.

Special Problems. The general design criteria given in most codes are applicable to ordinary buildings. More thorough investigation is recommended (and sometimes required) for special circumstances such as the following:

Tall buildings. These are critical with regard to their height dimension as well as the overall size and number of occupants inferred. Local wind speeds and unusual wind phenomena at upper elevations must be considered.

Flexible structures. These may be affected in a variety of ways, including vibration or flutter as well as simple magnitude of movements.

Unusual shapes. Open structures, structures with large overhangs or other projections, and any building with a complex shape should be carefully studied for the special wind effects that may occur. Wind tunnel testing may be advised or even required by some codes.

Earthquakes

During an earthquake, a building is shaken up and down and back and forth. The back-and-forth (horizontal) movements are typically more violent and tend to produce major destabilizing effects on buildings; thus, structural design for earthquakes is mostly done in terms of considerations for horizontal (called lateral) forces. The lateral forces are actually generated by the weight of the building—or, more specifically, by the mass of the building that represents both an inertial resistance to movement and a source for kinetic energy once the building is actually in motion. In the simplified procedures of the equivalent static force method, the building structure is considered to be loaded by a set of horizontal forces consisting of some fraction of the building weight. An analogy would be to visualize the building as being rotated vertically 90° to form a cantilever beam, with the ground as the fixed end and with a load consisting of the building weight.

In general, design for the horizontal force effects of earthquakes is quite similar to design for the horizontal force effects of wind. The same basic types of lateral bracing (shear walls, trussed bents, rigid frames, etc.) are used to resist both force effects. There are indeed some significant differences, but in the main, a system of bracing that is developed for wind bracing will most likely serve reasonably well for earthquake resistance as well.

Because of its considerably more complex criteria and procedures, we have chosen not to illustrate the design for earthquake effects in the examples in this book. Nevertheless, the development of elements and systems for the lateral bracing of the building in the design examples here is quite applicable in general to situations where earthquakes are a predominant concern. For structural investigation, the principal difference is in the determination of the loads and their distribution in the building. Another major difference is in the true dynamic effects, critical wind force being usually represented by a single, major, one-direction punch from a gust, whereas earthquakes represent rapid back-and-forth, reversing-direction actions. However, once the dynamic effects are translated into equivalent static forces, design concerns for the bracing systems are very similar, involving considerations for shear, overturning, horizontal sliding, and so on.

For a detailed explanation of earthquake effects and illustrations of the investigation by the equivalent static force method refer to *Simplified Building Design for Wind and Earthquake Forces* (Ref. 6).

9.8 LOAD COMBINATIONS AND FACTORS

The various types of load sources, as described in the preceding section, must be individually considered for quantification. However, for design work the possible combination of loads must be also be considered. Using the appropriate combinations, we must determine the design load for individual structural elements. The first step in finding the design load is to establish the critical combinations of load for the individual element. Using ASCE 2002 (Ref. 1) as a reference, consider the following combinations:

1.4(Dead load)

1.2(Dead load) + 1.6(Live load) + 0.5(Roof load)

1.2(Dead load) + 1.6(Roof load) + Live load or 0.8(Wind load)

1.2(Dead load) + 1.6(Wind load) + Live load + 0.5(Roof load)

1.2(Dead load) + 1.0(Earthquake load) + Live load + 0.2(Snow load)

0.9(Dead load) + 1.0(Earthquake load) or 1.6(Wind load)

9.9 DETERMINATION OF DESIGN LOADS

The following example demonstrates the process of determination of loading for individual structural elements. Additional examples are presented in the building design cases in Chapter 10.

Figure 9.1 shows the plan layout for the framed structure of a multistory building. The vertical structure consists of columns and the horizontal floor structure of a deck and beam system. The repeating plan unit of 24 × 32 ft is called a *column bay.* Assuming lateral bracing of the building to be achieved by other structural elements, the columns and beams shown here will be designed for dead load and live load only.

The load to be carried by each element of the structure is defined by the unit loads for dead load and live load and the *load periphery* (also

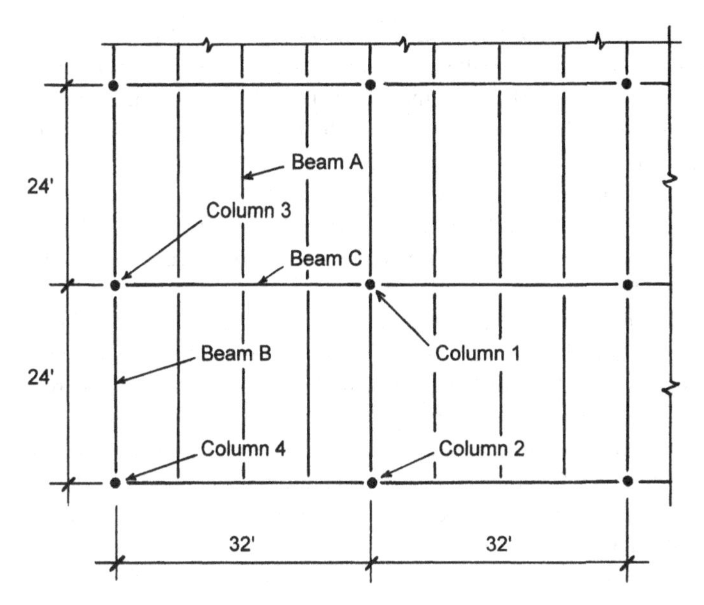

Figure 9.1 Reference for determination of peripheral distributed loads.

called *tributary area*) for the individual elements. The load periphery for an element is established by the layout and dimensions of the framing system. With reference to the labeled elements in Figure 9.1, the load peripheries are as follows:

Beam A: $8 \times 24 = 192$ ft^2
Beam B: $4 \times 24 = 96$ ft^2
Beam C: $24 \times 24 = 576$ ft^2 (Note that beam C carries only three of the four beams per bay of the system; the fourth is carried directly by the columns.)
Column 1: $24 \times 32 = 768$ ft^2
Column 2: $12 \times 32 = 384$ ft^2
Column 3: $16 \times 24 = 384$ ft^2
Column 4: $12 \times 16 = 192$ ft^2

For each of these elements, the unit dead load and unit live load from the floor is multiplied by the floor areas computed for the individual elements. Any possible live-load reduction (as described in Section 9.6) is made for the individual elements based on their load periphery area.

Additional dead load for the elements consists of the dead weight of the elements themselves. For the columns and beams at the building edge, another additional dead load consists of the portion of the exterior wall construction supported by the elements. Thus, column 2 carries an area of the exterior wall defined by the multiple of the story height times 32 ft. Column 3 carries 24 ft of wall, and column 4 carries 28 ft of wall (12 + 16).

The column loads are determined by the indicated supported floor, to which is added the weight of the columns. For an individual story column, this would be added to loads supported above this level—from the roof and any upper levels of floor.

The loads as described are used in the defined combinations described in Section 9.8. If any of these elements are involved in the development of the lateral bracing structure, the appropriate wind or earthquake loads are also added.

Floor live loads may be reduced by the method described in Section 9.6. Reductions are based on the tributary area supported and the number of levels supported by members.

Computations of design loads using the process described here are given for the building design cases in Chapter 10.

9.10 STRUCTURAL PLANNING

Planning a structure requires the ability to perform two major tasks. The first is the logical arranging of the structure itself, regarding its geometric form, its actual dimensions and proportions, and the ordering of the elements for basic stability and reasonable interaction. All of these issues must be faced, whether the building is simple or complex, small or large, of ordinary construction or totally unique. Spanning beams must be supported and have depths adequate for the spans, horizontal thrusts of arches must be resolved, columns above should be centered over columns below, and so on.

The second major task in structural planning is the development of the relationships between the structure and the building in general. The building plan must be "seen" as a structural plan. The two may not be quite the same, but they must fit together. "Seeing" the structural plan (or possibly alternative plans) inherent in a particular architectural plan is a major task for designers of building structures.

Hopefully, architectural planning and structural planning are done interactively, not one after the other. The more the architect knows about the structural problems and the structural designer (if another person) knows about architectural problems, the more likely it is possible that an interactive design development may occur.

Although each individual building offers a unique situation if all of the variables are considered, the majority of building design problems are highly repetitious. The problems usually have many alternative solutions, each with its own set of pluses and minuses in terms of various points of comparison. Choice of the final design involves the comparative evaluation of known alternatives and the eventual selection of one.

The word *selection* may seem to imply that all the possible solutions are known in advance, not allowing for the possibility of a new solution. The more common the problem, the more this may be virtually true. However, the continual advance of science and technology and the fertile imagination of designers make new solutions an ever-present possibility, even for the most common problems. When the problem is truly a new one in terms of a new building use, a jump in scale, or a new performance situation, there is a real need for innovation. Usually, however, when new solutions to old problems are presented, their merits must be compared to established previous solutions in order to justify them. In its broadest context, the selection process includes the consideration of all possible alternatives: those well known, those new and unproven, and those only imagined.

9.11 BUILDING SYSTEMS INTEGRATION

Good structural design requires integration of the structure into the whole physical system of the building. It is necessary to realize the potential influences of structural design decisions on the general architectural design and on the development of the systems for power, lighting, thermal control, ventilation, water supply, waste handling, vertical transportation, firefighting, and so on. The most popular structural systems have become so in many cases largely because of their ability to accommodate the other subsystems of the building and to facilitate popular architectural forms and details.

9.12 ECONOMICS

Dealing with dollar cost is a very difficult, but necessary, part of structural design. For the structure itself, the bottom-line cost is the delivered cost of the finished structure, usually measured in units of dollars per square foot of the building. For individual components, such as a single wall, units may be used in other forms. The individual cost factors or components, such as cost of materials, labor, transportation, installation, testing, and inspection, must be aggregated to produce a single unit cost for the entire structure.

Designing for control of the cost of the structure is only one aspect of the design problem, however. The more meaningful cost is that for the entire building construction. It is possible that certain cost-saving efforts applied to the structure may result in increases of cost of other parts of the construction. A common example is that of the floor structure for multistory buildings. Efficiency of floor beams occurs with the generous provision of beam depth in proportion to the span. However, adding inches to beam depths with the unchanging need for dimensions required for floor and ceiling construction and installation of ducts and lighting elements means increasing the floor-to-floor distance and the overall height of the building. The resulting increases in cost for the added building skin, interior walls, elevators, piping, ducts, stairs, and so on, may well offset the small savings in cost of the beams. The really effective cost-reducing structure is often one that produces major savings of nonstructural costs, in some cases at the expense of less structural efficiency.

Real costs can only be determined by those who deliver the completed construction. Estimates of cost are most reliable in the form of actual offers or bids for the construction work. The further the cost estimator is from the actual requirement to deliver the goods, the more speculative the estimate. Designers, unless they are in the actual employ of the builder, must base any cost estimates on educated guesswork deriving from some comparison with similar work recently done in the same region. This kind of guessing must be adjusted for the most recent developments in terms of the local markets, competitiveness of builders and suppliers, and the general state of the economy. Then the four best guesses are placed in a hat and one is drawn out.

Serious cost estimating requires a lot of training and experience and an ongoing source of reliable, timely information. For major projects, various sources are available in the form of publications or computer databases.

The following are some general rules for efforts that can be made in the structural design work in order to have an overall general cost-saving attitude:

1. Reduction of material volume is usually a means of reducing cost. However, unit prices for different grades must be noted. Higher grades of steel or wood may be proportionally more expensive than the higher stress values they represent; more volume of cheaper material may be less expensive.
2. Use of standard, commonly stocked products is usually a cost savings, as special sizes or shapes may be premium priced. Wood 2 × 3 studs may be higher in price than 2 × 4 studs, because the 2 × 4 is so widely used and bought in large quantities.
3. Reduction in the complexity of systems is usually a cost savings. Simplicity in purchasing, handling, managing of inventory, and so on will be reflected in lower bids as builders anticipate simpler tasks. Use of the fewest number of different grades of materials, sizes of fasteners, and other such variables is as important as the fewest number of different parts. This is especially true for any assemblage done on the building site; large inventories may not be a problem in a factory but usually are on a restricted site.
4. Cost reduction is usually achieved when materials, products, and construction methods are highly familiar to local builders and construction workers. If real alternatives exist, choice of the "usual" one is the best course.
5. Do not guess at cost factors; use real experience, yours or others. Costs vary locally, by job size and over time. Keep up-to-date with cost information.
6. In general, labor cost is greater than material cost. Labor for building forms, installing reinforcement, pouring, and finishing concrete surfaces is *the* major cost factor for site-poured concrete. Savings in these areas are much more significant than saving of material volume.
7. For investment buildings speed of construction may be a major advantage. However, getting the structure up fast is not a true advantage unless the other aspects of the construction can take advantage of the time gained. Steel frames often go up quickly, only to stand around and rust while the rest of the work catches up.

10

BUILDING STRUCTURES: DESIGN EXAMPLES

This chapter contains examples of the design of structural systems for buildings. The buildings selected for design are not intended as examples of good architectural design, but rather have been selected to create a range of common situations in order to demonstrate the use of various structural components. Design of individual elements of the structural systems is based on the materials presented in earlier chapters. The purpose here is to show a broader context of design work by dealing with the whole structure and with the building in general.

10.1 BUILDING ONE

Figure 10.1 shows a one-story, box-shaped building intended for commercial occupancy. As the detail section (Figure 10.1*d*) shows, the exterior walls are principally of reinforced masonry with concrete blocks (concrete masonry units, or CMUs), and the roof structure consists of light steel trusses supporting a formed sheet steel deck. This is the struc-

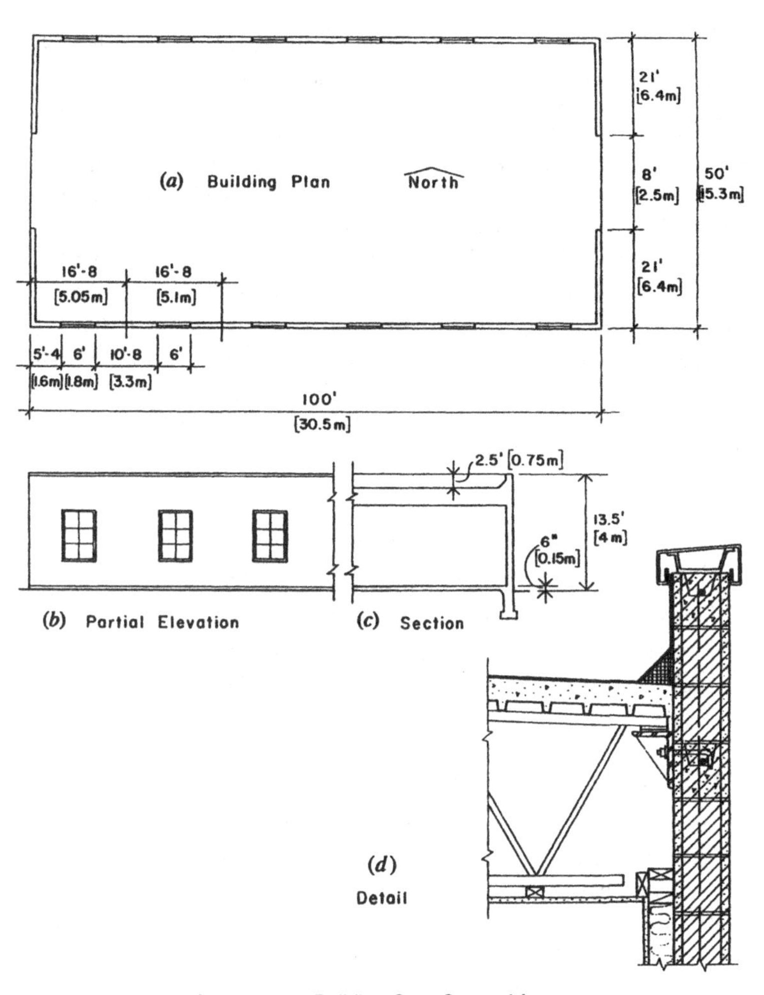

Figure 10.1 Building One: General form.

ture that is discussed first, although other options for the construction, using various steel elements, will also be discussed.

General Considerations

The following data is assumed for design:

Roof dead load = 15 psf, not including the weight of the structure
Roof live load = 20 psf, reducible for large supported areas

The section in Figure 10.1*d* indicates that the wall continues above the top of the roof to create a parapet, and that the steel trusses are supported at the wall face. The span of the joists is thus established as approximately 48 ft, which is used for their design.

The section in Figure 10.1*c* indicates that the building has a generally flat, horizontal roof surface. As the construction section shows, the roof deck is placed directly on top of the trusses and the ceiling is supported by direct attachment to the bottom of the trusses. For reasonable drainage of the roof surface, a slope of at least ¼ inch per foot (2 percent) must be provided, although the following work assumes a constant depth of the trusses for design purposes.

Construction consists of:

K-series open-web steel joists. (See Section 3.9.)
Reinforced hollow concrete masonry construction.
Formed sheet steel deck. (See Table 3.7.)
Deck surfaced with lightweight insulating concrete fill.
Multiple-ply, hot-mopped, felt and gravel roofing.
Suspended ceiling with gypsum drywall.

Design of the Roof Structure

Spacing of the open-web joists must be coordinated with the selection of the roof deck and the details for construction of the ceiling. For a trial design, a spacing of 4 ft is assumed. From Table 3.7, with deck units typically achieving three spans or more, the lightest deck in the table (22 gage) may be used. Choice of the deck configuration (rib width) depends on the type of materials placed on top of the deck and the means used to attach the deck to the supports.

Adding the weight of the deck to the other roof dead load produces a total dead load of 17 psf for the superimposed load on the joists. As illustrated in Section 3.9, the design for a K-series joist is as follows:

$$\text{Joist dead load} = 4(17) = 68 \text{ lb/ft (not including joist)}$$

$$\text{Joist live load} = 4(20) = 80 \text{ lb/ft}$$

$$\text{Total factored load} = 1.2(68) + 1.6(80) = 82 + 128 = 210 \text{ lb/ft} + \text{the joist weight}$$

For the 48-ft span, the following alternative choices are obtained from Table 3.4:

24K9 at 12.0 plf, total factored load = 1.2(12 + 68) + 128 = 96 + 128 = 224 lb/ft (less than the table value of 313 lb/ft

26K5 at 10.6 plf, total factored load = 1.2(68 + 10.6) + 128 = 94 + 128 = 222 lb/ft (less than the table value of 233 lb/ft)

Live-load capacity for L/360 deflection exceeds the requirement for both of these choices.

Although the 26K5 is the lightest permissible choice, there may be compelling reasons for using a deeper joist. For example, if the ceiling is directly attached to the bottoms of the joists, a deeper joist will provide more space for passage of building service elements. Deflection will also be reduced if a deeper joist is used. Pushing the live-load deflection to the limit means a deflection of $(1/360)(48 \times 12) = 1.9$ in. Although this may not be critical for the roof surface, it can present problems for the underside of the structure, involving sag of ceilings or difficulties with nonstructural walls built up to the ceiling.

Choice of a 30K7 at 12.3 plf results in considerably less deflection at a small premium in additional weight.

Note that Table 3.4 is abridged from a larger table in the reference, and there are therefore many more choices for joist sizes. The example here is meant only to indicate the process for use of such references.

Specifications for open-web joists give requirements for end-support details and lateral bracing (see Ref. 4). If the 30K7 is used for the 48-ft span, for example, four rows of bridging are required.

Although the masonry walls are not designed for this example, note that the support indicated for the joists in Figure 10.1*d* results in an eccentric load on the wall. This induces bending in the wall, which may be objectionable. An alternative detail for the roof-to-wall joint is shown in Figure 10.2, in which the joists sit directly on the wall with the joist top chord extending to form a short cantilever. This is a common detail, and the reference supplies data and suggested details for this construction.

Alternative Roof Structure with Interior Columns

If a clear spanning roof structure is not required for this building, it may be possible to use some interior columns and a framing system for the roof with quite modest spans. Figure 10.3*a* shows a framing plan for a

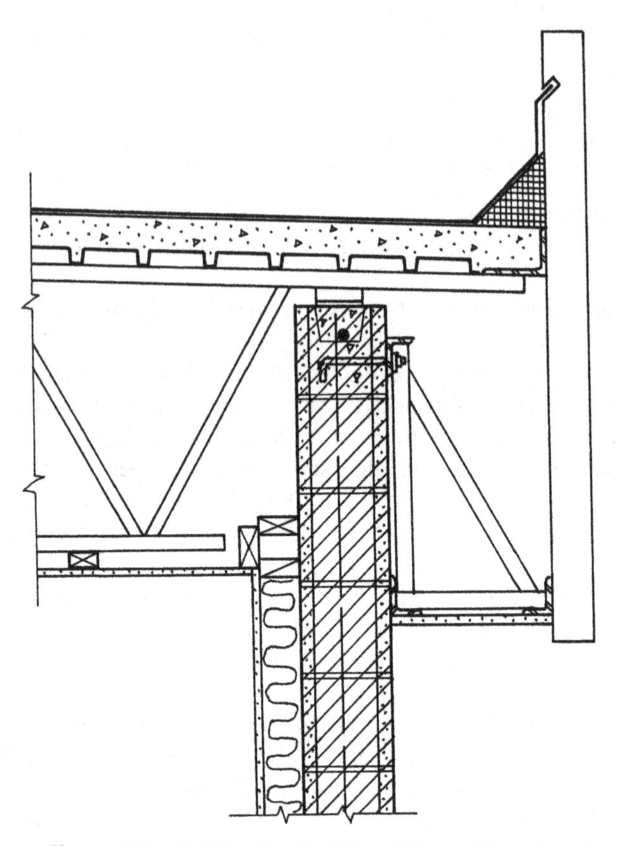

Figure 10.2 Building One: Variation of the roof-to-wall joint.

system that uses ten columns at 16 ft, 8 in. on center in each direction. Although short-span joists may be used with this system, it would also be possible to use a longer-span deck, as indicated on the plan. This span exceeds the capability of the deck with 1.5-in. ribs, but decks with deeper ribs are available.

A second possible framing arrangement is shown in Figure 10.3*b,* in which the deck spans the other direction and only eight columns are used. This arrangement allows for wider column spacing, which increases the beam spans but eliminates 60 percent of the interior columns and their footings—a major cost savings.

Beams in continuous rows can sometimes be made to simulate a continuous beam action without the need for moment-resistive connections.

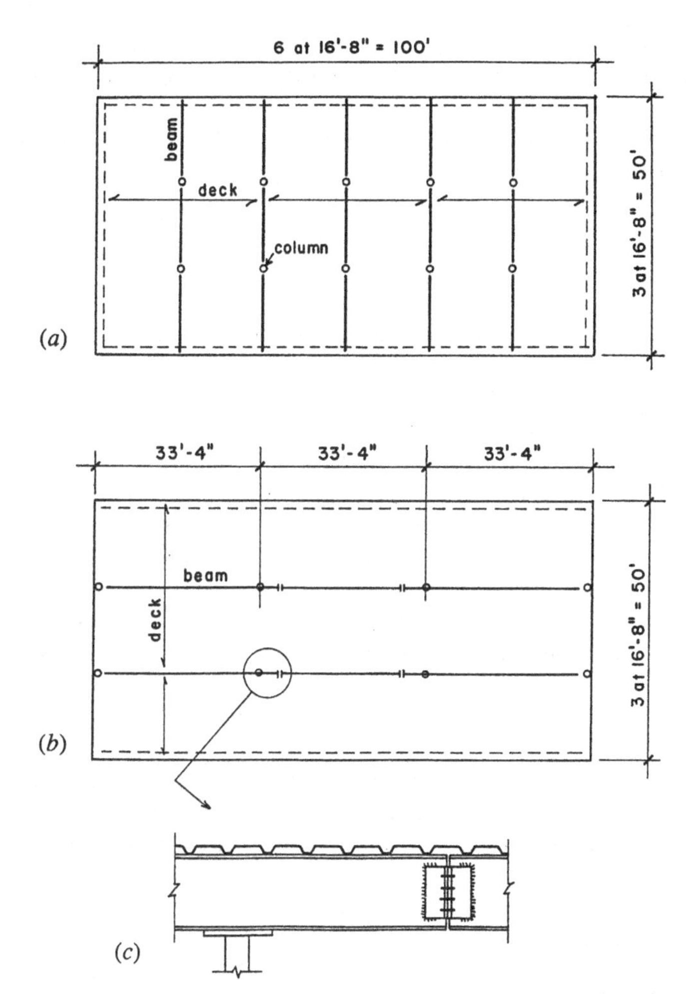

Figure 10.3 Building One: Options for roof framing plan and interior column locations.

Use of beam splice joints off the columns, as shown in Figure 10.3*c*, allows for relatively simple connections but some advantages of the continuous beam. A principal gain thus achieved is a reduction in deflections.

For the beam in Figure 10.3*b*, assuming a slightly heavier deck, an approximate dead load of 20 psf will result in a beam factored load of

$$w = 16.67[1.2(20) + 1.6(16)] = 827 \text{ plf} + \text{the beam, say } 900 \text{ plf}$$

Note: The beam periphery of 33.3 × 16.67 = 555 ft^2 qualifies the beam for a roof live-load reduction, indicating the use of 16 psf as discussed in Section 9.6.

The simple beam bending moment for the 33.3-ft span with the factored load is

$$M_u = \frac{wL^2}{8} = \frac{0.900 \times (33.3)^2}{8} = 125 \text{ kip-ft}$$

The required moment resistance of the beam is therefore

$$M_n = 125/0.9 = 139 \text{ kip-ft}$$

From Table 3.1 the lightest W-shape beam permitted is a W 14 × 34. From Figure 3.11, observe that the total load deflection will be approximately L/240, which is usually not critical for roof structures. Furthermore, the live-load deflection will be less than one-half this amount, which is quite a modest value. It is assumed that the continuous connection of the deck to the beam top flange is adequate to consider the beam to have continuous lateral support, permitting the use of a resisting moment of M_p.

Also possible is the use of the beam framing indicated in Figure 10.3*b* with a continuous beam having pinned connections off the columns. This will reduce both the maximum bending moment and the deflection for the beam, and most likely permit a slightly lighter beam.

If the three-span beam is constructed with three simple-spanning segments, the detail at the top of the column will be as shown in Figure 10.4. Although this may be possible, the detail shown in Figure 10.3*c*, with the beam-to-beam connection off the column, is a better framing detail. The investigation shown in Figure 10.5 is made with a beam splice 4 ft from the column. The center portion of the three-span beam thus becomes a simple span of 25.33 ft. The end reactions for the center portion become loads on the ends of the extensions of the outer portions. The resulting solution for the beam reactions, shear, and bending moments is shown in the figure.

The maximum factored bending moment determined in Figure 10.5 is 100 kip-ft, or approximately 80 percent of that for the simple span beam. This requires a beam with a resisting moment (M_p) of 100/0.9 = 111 kip-ft. Reference to Table 3.1 indicates the possibility of reducing the shape to a W 16 × 26.

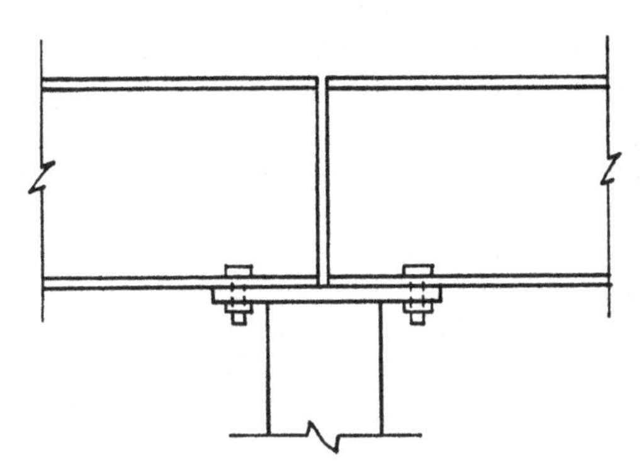

Figure 10.4 Framing detail at the top of the steel column with simple beam action.

The total load on the beam is the approximate load on the column. Thus, the column factored load is $0.900 \times 33.3 = 30$ kips. The required design strength of the column is therefore $30(1/0.85) = 35.3$ kips. Assuming an unbraced column height of 10 ft, the following choices may be found for the column:

From Table 4.4, a 3-in. pipe (nominal size, standard weight)

From Table 4.5, an HHS 3-in.-square tube, with 3/16-in.-thick wall

Design for Lateral Loads

To enable this size and form of building to resist wind or earthquake forces, a *box system* is typically used, consisting of shear walls and a horizontal diaphragm roof. If planning of solid walls does not provide sufficient potential for development of shear walls, a braced frame (trussed) or a rigid frame in the wall planes can be used.

The following must be considered in the design of the building structure for wind:

1. Inward and outward pressure on exterior building surfaces, causing bending of the wall studs and an addition to the gravity loads on roofs
2. Total lateral (horizontal) force on the building, requiring bracing by the roof diaphragm and the shear walls
3. Uplift on the roof, requiring anchorage of the roof structure to its supports

4. Total effect of uplift and lateral forces, possibly resulting in overturn (toppling) of the entire building

Uplift on the roof depends on the roof shape and the height above ground. For this low, flat-roofed building, the ASCE standard (Ref. 1) requires an uplift pressure of 10.7 psf. In this case, the uplift pressure does not exceed the roof dead weight of 16 psf, so anchorage of the roof construction is not required. However, common use of metal framing devices for light-wood-frame construction provides an anchorage with considerable resistance.

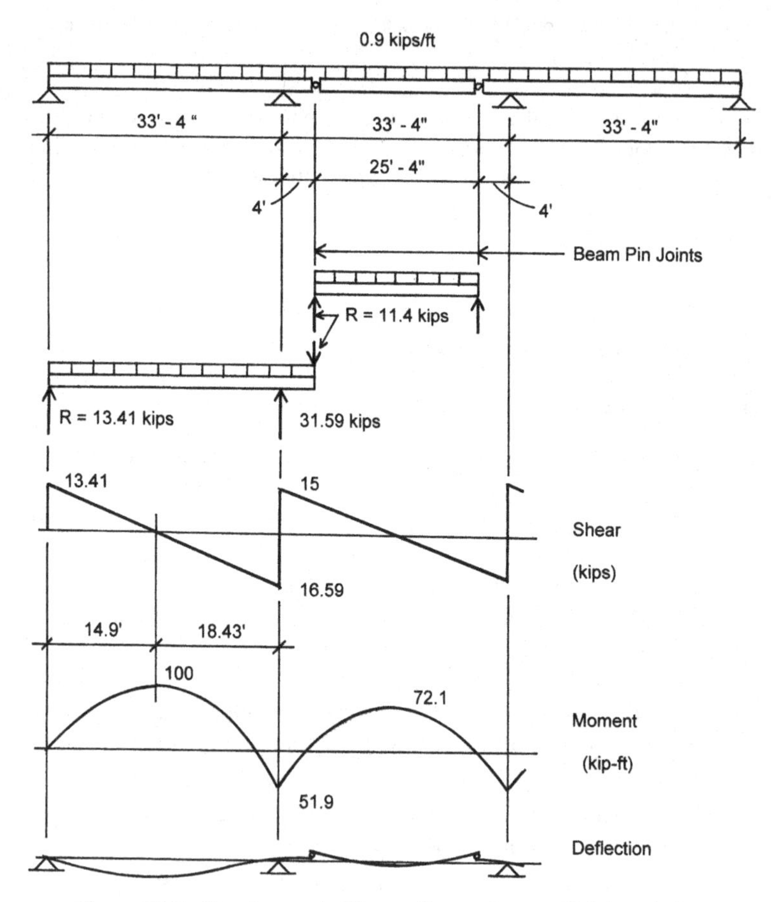

Figure 10.5 Development of the continuous beam with internal pins.

Overturning of the building is not likely critical for a building with this squat profile (50 ft wide by only 13.5 ft high). Overturn of the whole building is usually more critical for towerlike building forms or for extremely light construction. Of separate concern is the overturn of individual bracing elements; in this case, the individual shear walls.

Wind Force on the Bracing System

The building's bracing system must be investigated for horizontal force in the two principal orientations: east–west and north–south. If the building is not symmetrical, the force must be investigated in each direction on each building axis: east, west, north, and south.

The horizontal wind force on the north and south walls of the building is shown in Figure 10.6. This force is generated by a combination of positive (direct, inward) pressure on the windward side and negative (suction, outward) on the lee side of the building. The pressures shown as Case 1 in Figure 10.6 are obtained from data in the ASCE 2003 (Ref. 1) chapter on wind loads. (See discussion of wind loads in Chapter 9.) The single pressures shown in the figure are intended to account for the combination of positive and suction pressures. The ASCE standard provides for two zones of pressure—a general one and a small special increased area of pressure at one end. The values shown in Figure 10.6 for these pressures are derived by considering a critical wind velocity of 90 mph and an exposure condition B, as described in the standard.

The range for the increased pressure in Case 1 is defined by the dimension a and the height of the windward wall. The value of a is established as 10 percent of the least plan dimension of the building or 40 percent of the wall height, whichever is smaller, but not less than 3 ft. For this example, a is determined as 10 percent of 50 ft, or 5 ft. The distance for the pressure of 12.8 psf in Case 1 is thus $2(a) = 10$ ft.

The design standard also requires that the bracing system be designed for a minimum pressure of 10 psf on the entire area of the wall. This sets up two cases (Case 1 and Case 2 in Figure B.1 in Appendix B) that must be considered. Because the concern for the design is the generation of maximum effect on the roof diaphragm and the end shear walls, the critical conditions may be determined by considering the development of end reaction forces and maximum shear for an analogous beam subjected to the two loadings. This analysis is shown in Figure 10.7, from which it is apparent that the critical concern for the end shear walls and the maximum effect in the roof diaphragm is derived from Case 2 in Figure 10.6.

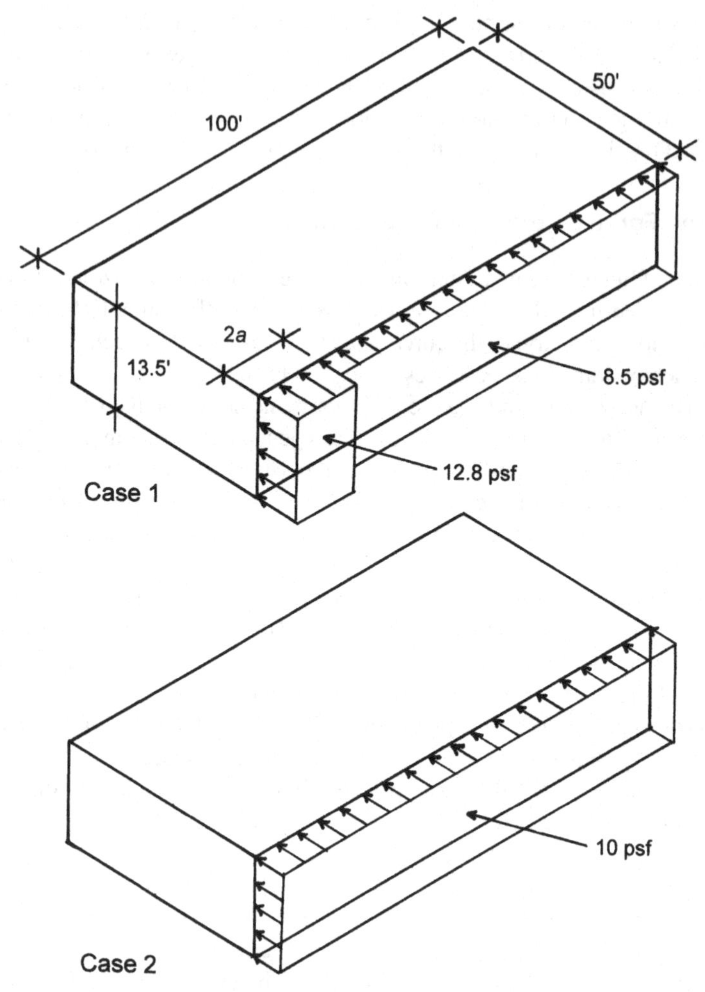

Figure 10.6 Building One: Wind pressure on the south wall, ASCE 2002 (Ref. 1).

The actions of the horizontal wind force resisting system in this regard are illustrated in Figure 10.8. The initial force comes from wind pressure on the building's vertical sides. The wall spans vertically to resist this uniformly distributed load, as shown in Figure 10.8*a*. Assuming the wall function to be as shown in Figure 10.8*a*, the north–south wind force delivered to the roof edge is determined as

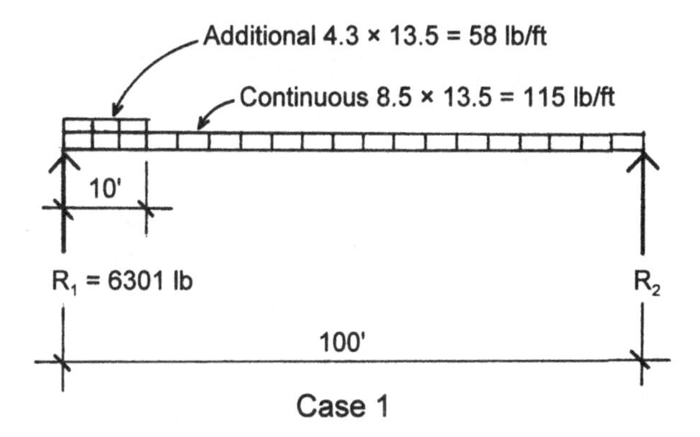

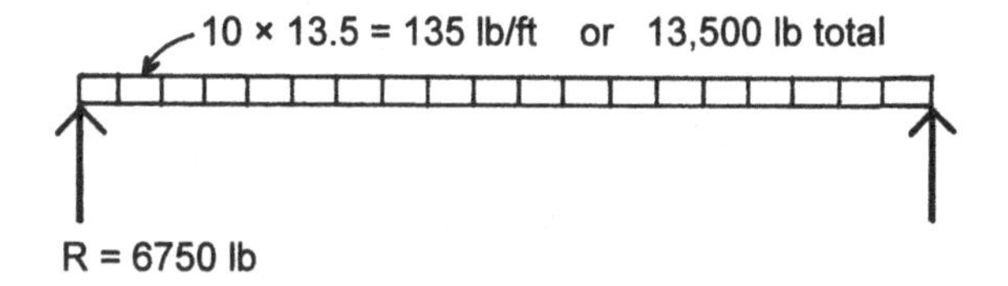

Case 2

Figure 10.7 Building One: Resultant wind forces on the end shear walls.

$$\text{Total } W = (10\text{psf})(100 \times 13.5) = 13{,}500 \text{ lb}$$

$$\text{Roof edge } W = 13{,}500 \times \frac{6.75}{11} = 8284 \text{ lb}$$

In resisting this load, the roof functions as a spanning member supported by the shear walls at the east and west ends of the building. The investigation of the diaphragm as a 100-ft simple span beam with uniformly distributed loading is shown in Figure 10.9. The end reaction and maximum diaphragm shear force is found as

$$R = V = \frac{8284}{2} = 4142 \text{ lb}$$

which produces a maximum unit shear in the 50-ft-wide diaphragm of

$$v = \frac{\text{Shear force}}{\text{Roof width}} = \frac{4142}{50} = 82.8 \text{ lb/ft}$$

The moment diagram shown in Figure 10.9 indicates a maximum value of 104 kip-ft at the center of the span. This moment is used to determine the maximum force in the diaphragm chord at the roof edges. The force must be developed in both compression and tension as the wind direction reverses.

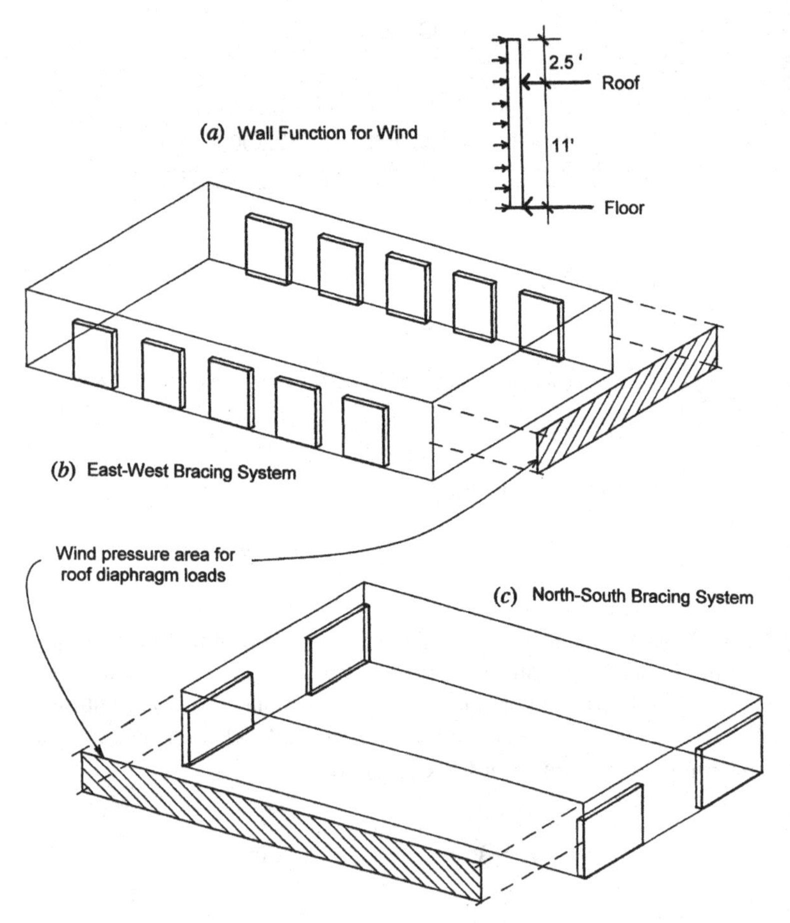

Figure 10.8 Wall bracing functions and wind pressures for design.

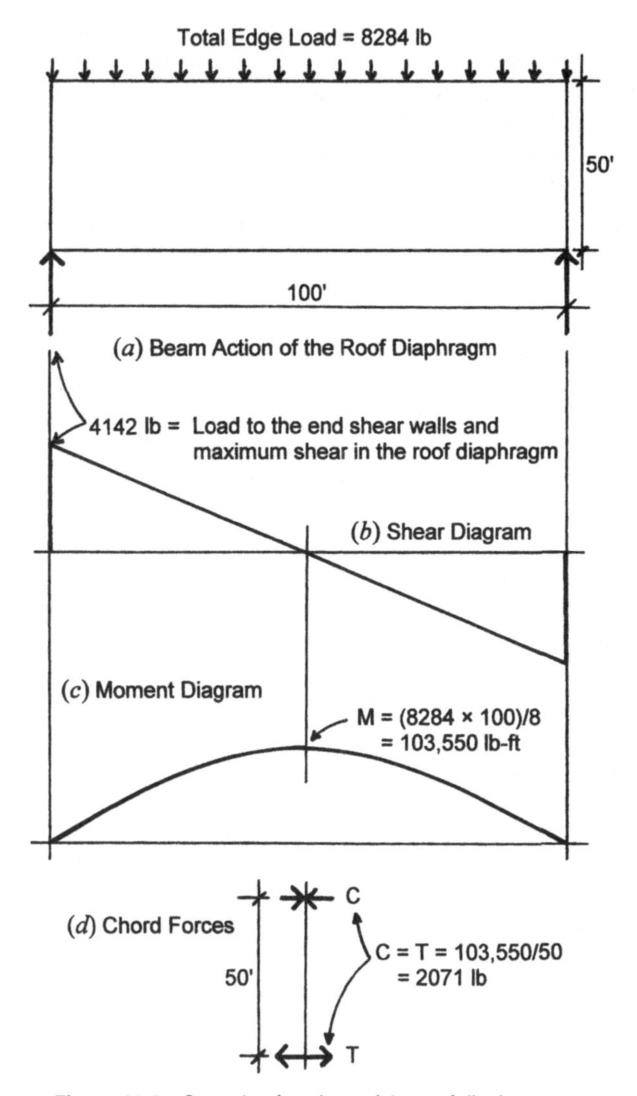

Figure 10.9 Spanning functions of the roof diaphragm.

Designing steel decks as horizontal diaphragms is routine, done with data supplied by deck manufacturers—individual deck products and connecting elements have rated capacities. Elements of the steel frame are used as diaphragm chords, collectors, ties, and drag struts. For example, Building One's masonry walls could be used as shear walls, whereas the lightest steel deck would suffice for diaphragm action if adequately attached to the framing. In fact, the minimum construction permitted by codes and industry standards is adequate for anything other than major windstorm conditions.

Alternative Truss Roof

If a gabled (double-sloped) roof form is desirable for Building One, a possible roof structure is shown in Figure 10.10. The building profile shown in Figure 10.10*a* is developed with a series of trusses, spaced at plan intervals, as shown for the beam-and-column rows in Figure 10.3*a*. The truss form is shown in Figure 10.10*b*. An investigation for a unit loading on this truss was done in Section 7.4, and the results displayed in Figure 7.13. The true unit loading for the truss is derived from the form of construction and is approximately eleven times the unit load used in the example. This accounts for the values of the internal forces in the members as displayed in Figure 10.10*e*.

The detail in Figure 10.10*d* shows the use of double-angle members with joints developed with gusset plates. The top chord is extended to form the cantilevered edge of the roof. For clarity of the structure, the detail shows only the major structural elements. Additional construction would be required to develop the roofing, ceiling, and soffit.

In trusses of this size, it is common to extend the chords without joints for as long as possible. Available lengths depend on the sizes of members and the usual lengths in stock by local fabricators. Figure 10.10*c* shows a possible layout that creates a two-piece top chord and a two-piece bottom chord. The longer top chord piece is thus 36 ft plus the overhang, which may be difficult to obtain if the angles are small.

The roof construction illustrated in Figure 10.10*d* shows the use of a long-span steel deck that bears directly on top of the top chord of the trusses. This option simplifies the framing by eliminating the need for intermediate framing between the trusses. For the truss spacing of 16-ft 8-in. as shown in Figure 10.3*a,* the deck will be quite light and this is a feasible system. However, the direct bearing of the deck adds a spanning function to the top chord, and the chords must be considerably heavier to work for this added task.

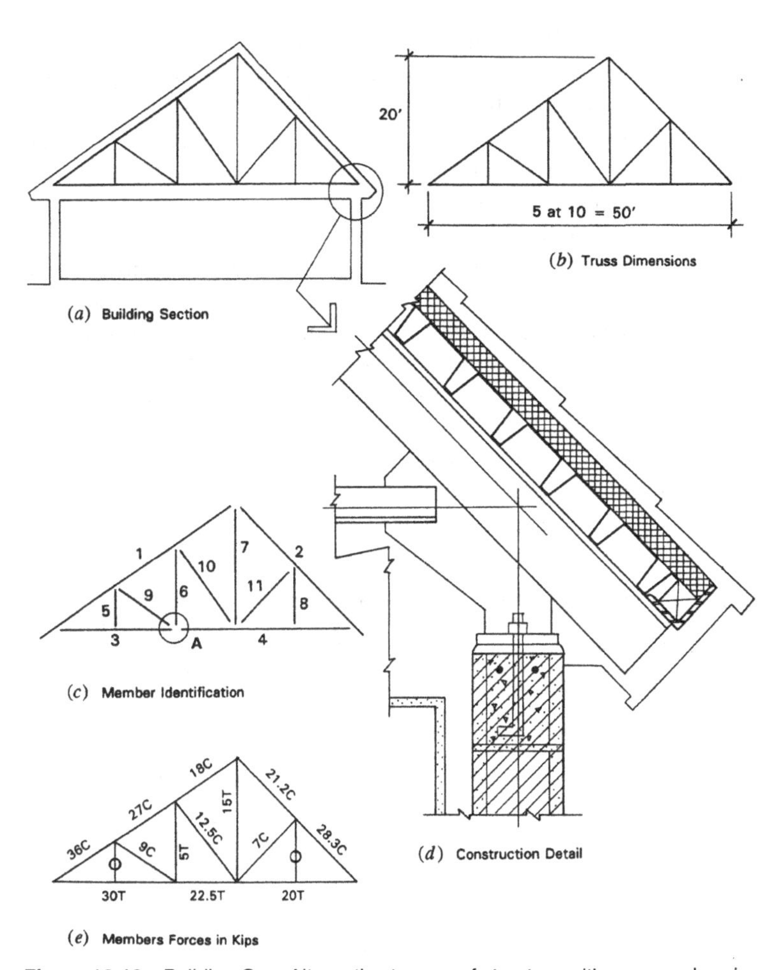

Figure 10.10 Building One: Alternative truss roof structure with masonry bearing walls.

The loading condition for the truss as shown in Section 7.4 indicates concentrated forces of 1000 lb each at the top chord joints. *Note:* It is typical procedure to assume this form of loading, even though the actual load is distributed along the top chord (roof load) and the bottom chord (ceiling load). If the total factored loading consisting of live load, roof dead load, ceiling dead load, and truss weight is approximately 66 psf, the single joint load is

$$P = (66)(10)(16.67) = 11{,}000 \text{ lb}.$$

This is 11 times the load in the truss in Section 7.4, so the internal forces for the gravity loading will be 11 times those shown in Figure 7.13. These values are shown here in Figure 10.10*e*.

Table 10.1 summarizes the design of the truss members, using all double angles. Selection of tension members reflects a desire for welded joints and a minimum angle leg ⅜ in. thick, while compression members come from the AISC tables.

Various forms may be used for the members and the joints of this truss. The loading and span is quite modest here, so the truss members will be quite small and joints will have minimum forces. A common form for this truss would use tee shapes for the top and bottom chords and double angles for interior members with the angles welded directly to the tee stems (see Figure 8.13*b*). Bolted connections are possible but probably not practical for this size truss.

For light trusses, a minimum size member can be derived from the layout, dimensions, magnitude of forces, and joint design. For example,

TABLE 10.1 Design of the Truss Members

No.	Force (kips)	Length (ft)	Member Choice
1	40 C	12	Combined compression and bending member 6 × 4 × ½
2	31 C	14.2	Max. L/r = 200, min. r = 0.85 6 × 4 × ½
3	33 T	10	Max. L/r = 300, min. r = 0.4 3 × 2½ × ⅜
4	25 T	10	3 × 2½ × ⅜
5	0	6.67	2½ × 2½ × ⅜
6	6 T	13.33	Max. L/r = 300, min. r = 0.53 in. 2½ × 2½ × ⅜
7	17 T	20	Max. L/r = 300, min. r = 0.8 3½ × 3½ × ⅜
8	0	10	2½ × 2½ × ⅜
9	10 C	12	2½ × 2½ × ⅜
10	14 C	16.67	Max. L/r = 200, min. r = 1.0 3½ × 2½ × ⅜
11	8 C	14.2	Max. L/r = 200, min. r = 0.85 3 × 2½ × ⅜

angle legs may be just wide enough to accommodate bolts or thick enough to accommodate minimum size fillet welds. Minimum L/r ratios for members are another source for minimum criteria. Minimum design sometimes results in a poor truss design, with the combination of form, size, and proposed construction type not meshing well.

Truss chords may consist of structural tees with most gusset plates eliminated, except possibly at the supports. If the truss is too large for transporting in one piece, a scheme must be developed for splicing some joints in the field.

To derive an approximate design for the top chord, consider the following combined function equation:

$$\frac{P_u}{\phi_c P_n} + \frac{M_{ux}}{\phi_b M_{nx}} \leq 1.0$$

For which the following values are used

$$P_u = 40 \text{ kips}, \ \phi_c P_n = 216 \text{ kips, Ratio} = 40/216 = 0.185$$

$$M_{ux} = 16.5 \text{ kip-ft}, \ \phi_b M_{nx} = 27.7 \text{ kip-ft, Ratio} = 16.5/27.7 = 0.596$$

$$\text{Sum of the ratios} = 0.781 < 0$$

These computations indicate that the member choices are reasonable, although the AISC Specifications require a more elaborate investigation.

A possible form for a bolted truss with gusset plates is shown in Figure 8.14. Of course, with bolted joints, the net section of the tension members must be considered for design selection.

Developing a complete roof construction entails the design of a roof deck system, and possibly the support of a ceiling system. For closely spaced trusses, a deck may span from truss to truss and be directly supported by the truss top chords. When truss spacing is greater, a purlin system may be used, with the purlins supporting the deck and in turn being supported at truss joints—which would eliminate the bending in the top chords.

Building One: Gable-Form Roof with Welded Steel Bent

Figure 10.11 shows a scheme for the development of a gable-form roof profile using a welded steel bent structure. The bent is fabricated from steel plates that are welded to form I-shaped members. The two bent

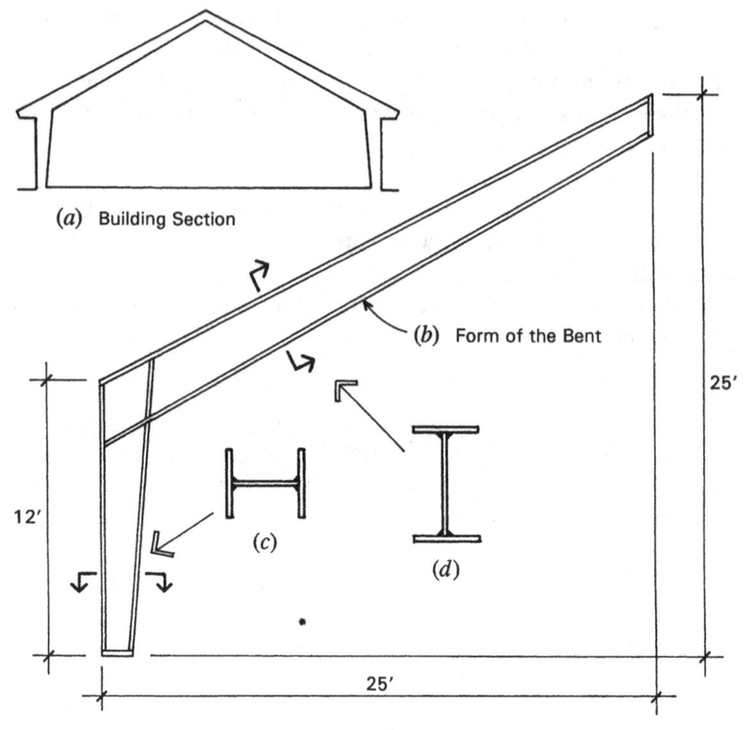

Figure 10.11 Building One: Alternative structure with three-hinged steel bents.

halves are connected at the roof peak to develop a three-hinged structure that is commonly executed in laminated wood and precast concrete, as well as with welded steel members. With the pinned supports and the pin at the peak, this is a structure that can be analyzed by static methods alone.

The complete investigation and design of this structure will not be shown here, although examples can be found in many references. Note that there will be a horizontal outward push at the supports, as in the case of an arch. This must be resolved by a tie across the bottom of the structure or by a thrust-resisting abutment foundation.

10.2 BUILDING TWO

Figure 10.12 shows a partial framing plan for the roof structure of a one-story industrial building. The system as shown, with 48-ft square bays, is

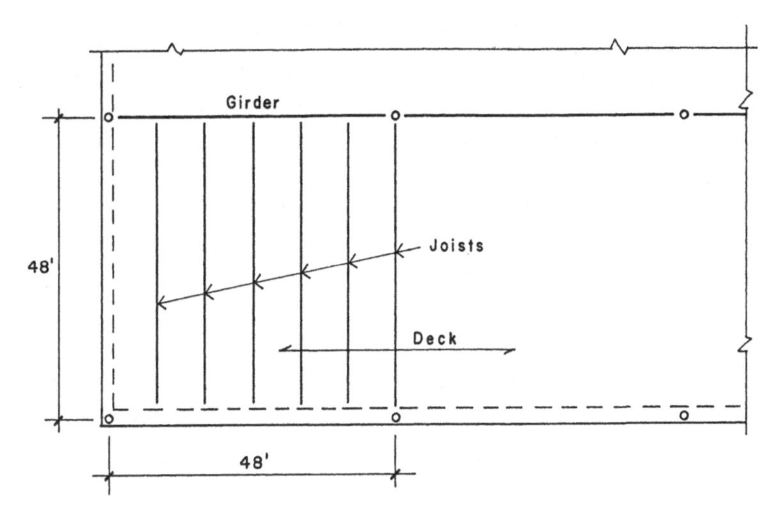

Figure 10.12 Building Two: Partial roof framing plan.

repeated a number of times in each direction. The same live load and dead load given for Building One will be used.

The plan in Figure 10.12 indicates a series of girders supported by columns, a series of joists perpendicular to the girders, and a roof deck supported by the joists. Selection of the joist spacing is affected by the following:

Load on a single joist. Based on the span and the load on a single joist, there is some feasible range for the spacing of individual manufactured joists. Upper limits for spacing relate to strength of joists; lower limits relate to economy in terms of the number of joists used.

Selected deck. Plywood panels have quite short-span potential. Steel decks cover a wide range, depending on the deck form and rib depth. Proprietary products have specific limits, obtainable from the manufacturers.

Point loads on girders. This relates to the lateral bracing of the girders and to truss panel points if the girders are trusses.

A very wide range of possible combinations exists for the choice of the deck, joists, girders, and columns for this structure. Two common situations are shown here.

Alternative One: Joists and W-Shape Girders

For this system, joists are placed at 4-ft centers and are supported by W-shape girders. For this span and loading, the choices for the joists are the same as for Building One. The remainder of the discussion for this system deals with the W-shape girders.

Because of the large area supported by a single girder, the roof live load drops to 12 psf (see Section 9.6). Assuming a total dead load with the joists of 20 psf, the average linear load on a girder is found to be

$$w_u = [1.2(20) + 1.6(12)] \times 48 = 2074 \text{ lb/ft, or } 2.074 \text{ kips/ft}$$

In fact, the joist loads constitute point loads at 4 ft on the girder, but the difference in the maximum bending moment is quite small if the load is considered to be uniformly distributed. Adding some additional dead load for the assumed weight of the girder, a reasonable approximation for design is a total load of 2.2 kips/ft. For simple beam action, the maximum bending moment is

$$M = \frac{wL^2}{8} = \frac{2.2(48)^2}{8} = 633.6 \text{ kip-ft}$$

If the simple beam is used, Table 3.1 indicates the least weight shape to be a W 30 × 90 of A36 steel.

It is possible to consider the use of the scheme with the beam joints off of the columns, as was done for Building One. If a similar reduction in the maximum bending moment of 20 percent can be achieved, the moment drops to (0.80)(633.6) = 507 kip-ft, and the lightest shape from Table 3.1 is a W 24 × 76.

For either of these choices for the girder, deflection should be investigated, including effects on the drainage of the large flat roof area, as well as recommended limits for maximum live load and total load deflections. Lateral bracing of the girder should also be considered, as discussed in Sections 3.5 and 3.13.

For the interior column, the total load is 2.2(48) = 105.6 kips. Assuming a height of 20 ft and a K of 1, some choices are as follows:

From Table 4.2: W 8 × 31 of A36 steel

From Table 4.4: 8-in.-diameter standard weight pipe, 6-in.-diameter XS pipe

From Table 4.5: square tube, 8 × 8 × 3/16, 7 × 7 × 1/4, 6 × 6 × 5/16

Column selection may relate to the development of the framing connection to the girder or to other factors involving the general construction.

Alternative Two: Joists and Joist Girders

A variation on the preceding example is to use trusses in place of the W-shape girders. Manufacturers of open-web joists have a product for this use, called a *joist girder,* as described in Section 3.9. This product is typically designed by the manufacturer, so the work required is limited to determination of the truss span, loading, and depth. The truss form (profile) will be determined based on the joist spacing and the manufacturer's standard details.

The spaced point loads of the joists determine a horizontal module for the girder. This dimension must be matched with the truss depth so that the truss panels do not approach the extremes shown in Figure 10.13. A rule of thumb suggested by the industry is that the depth of the truss in inches should approximate the truss span in feet. This puts the span-to-depth ratio in a reasonable range (10 to 15). For this example, a depth of 48 inches will produce square truss panels and a reasonable span-to-depth ratio of 12.

With the 4-ft joist spacing, there are 12 panels in the truss and the unit load at the truss–girder panel point on the top chord is equal to the total load on one joist. Using the loading determined previously, the panel point load is $2.074 \times 4 = 8.296$, say 8.3 kips. As described in Section 3.9, the joist girder is thus designated as

48G12N8.3

indicating a 48-in.-deep truss girder with 12 panels (actually 11 joists) and a single joist load of 8.3 kips.

Options for the columns for this structure would generally be the same as those determined for the scheme with the W-shape girders. However, the framing between the trusses and columns creates some different considerations for the choice of the column shape.

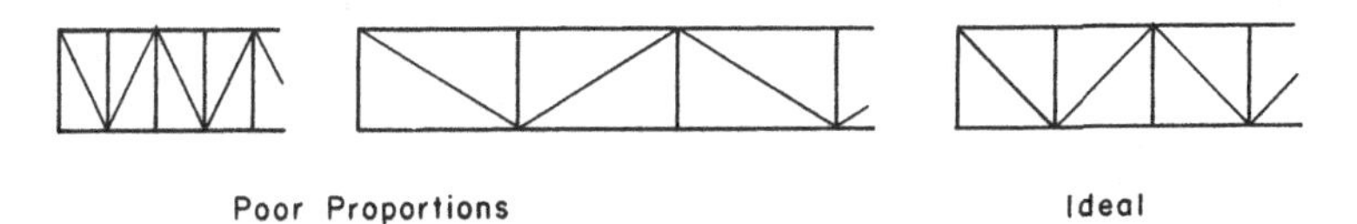

Figure 10.13 Development of form for the truss girder.

This basic system (open-web joists and joist girders) is also used for an alternative floor structure for Building Three.

10.3 BUILDING THREE

This is a modest-size office building, generally qualified as being low rise (see Figure 10.14). In this category there is a considerable range of choice for the construction, although in a particular place, at a particular time, a few popular forms of construction tend to dominate the field.

General Considerations

Some modular planning is usually required for this type of building, involving the coordination of dimensions for spacing of columns, window mullions, and interior partitions in the building plan. This modular coordination may also be extended to development of ceiling construction, lighting, ceiling HVAC elements, and the systems for access to electric power, phones, and other signal wiring systems. There is no single magic number for this modular system; all dimensions between 3 and 5 ft have been used and strongly advocated by various designers. Selection of a particular proprietary system for the curtain wall, interior modular partitioning, or an integrated ceiling system may establish a reference dimension.

For buildings built as investment properties, with speculative occupancies that may vary over the life of the building, it is usually desirable to accommodate future redevelopment of the building interior with some ease. For the basic construction, this means a design with as few permanent structural elements as possible. At a bare minimum, what is usually required is the construction of the major structure (columns, floors, and roof), the exterior walls, and the interior walls that enclose stairs, elevators, restrooms, and risers for building services. Everything else should be nonstructural or demountable in nature, if possible.

Spacing of columns on the building interior should be as wide as possible, basically to reduce the number of freestanding columns in the rented portion of the building plan. A column-free interior may be possible if the distance from a central core (grouped permanent elements) to the outside walls is not too far for a single span. Spacing of columns at the building perimeter does not affect this issue, so additional columns are sometimes used at this location to reduce their size for gravity loading or to develop a stiffer perimeter rigid frame system for lateral loads.

The space between the underside of suspended ceilings and the top of floor or roof structures above must typically contain many elements besides those of the basic construction. This usually represents a situation

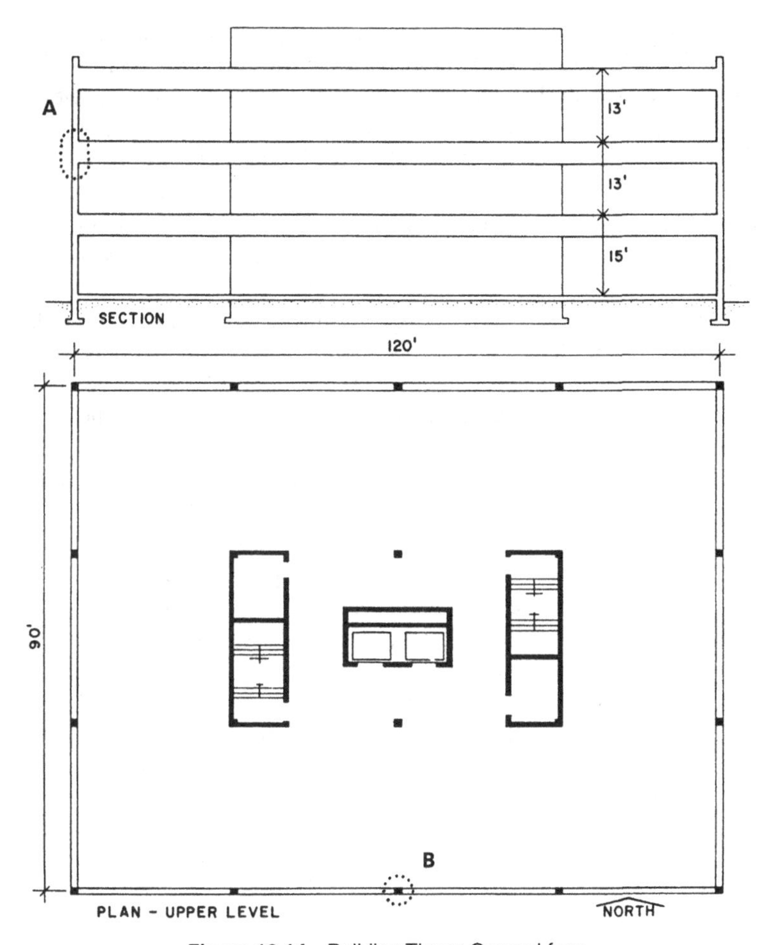

Figure 10.14 Building Three: General form.

requiring major coordination for the integration of the space needs for the elements of the structural, HVAC, electrical, communication, lighting, and firefighting systems.

A major design decision that must often be made very early in the design process is the overall dimension of the space required for this collection of elements. Depth permitted for the spanning structure and the general level-to-level vertical building height will be established—and not easy to change later, if the detailed design of any of the enclosed systems indicates a need for more space.

Generous provision of the space for building elements makes the work of the designers of the various other building subsystems easier, but the overall effects on the building design must be considered. Extra height for the exterior walls, stairs, elevators, and service risers all result in additional cost, making tight control of the level-to-level distance very important.

A major architectural design issue for this building is the choice of a basic form of the construction of the exterior walls. For the column-framed structure, two elements must be integrated: the columns and the nonstructural infill wall. The basic form of the construction as shown in Figure 10.15 involves the incorporation of the columns into the wall, with windows developed in horizontal strips between the columns. With the exterior column and spandrel covers developing a general continuous surface, the window units are thus developed as "punched" holes in the wall.

The windows in this example do not exist as parts of a continuous curtain wall system. They are essentially single individual units, placed in and supported by the general wall system. The curtain wall is developed as a stud-and-surfacing system not unlike the typical light wood stud wall system in character. The studs in this case are light-gage steel, the exterior covering is a system of metal-faced sandwich panel units, and the interior covering, where required, is gypsum drywall, attached to the metal studs with screws.

Detailing of the wall construction (as shown in detail A of Figure 10.15) results in a considerable interstitial void space. Although taken up partly with insulation materials, this space may easily contain elements for the electrical system or other services. In cold climates, a perimeter hot-water-heating system would most likely be used, and it could be incorporated in the wall space shown here.

Design Criteria The following are used for the design work:

Design codes. ASCE 2002 Standard (Ref.1) and 1997 UBC (Ref. 2)

Live loads.

Roof: 20 psf, reducible as described in Section 9.6.

Floor: From Table 9.1, 50 psf minimum for office areas, 100 psf for lobbies and corridors, 20 psf for movable partitions

Wind. Map speed of 90 mph, exposure B

Assumed construction loads.

Floor finish: 5 psf

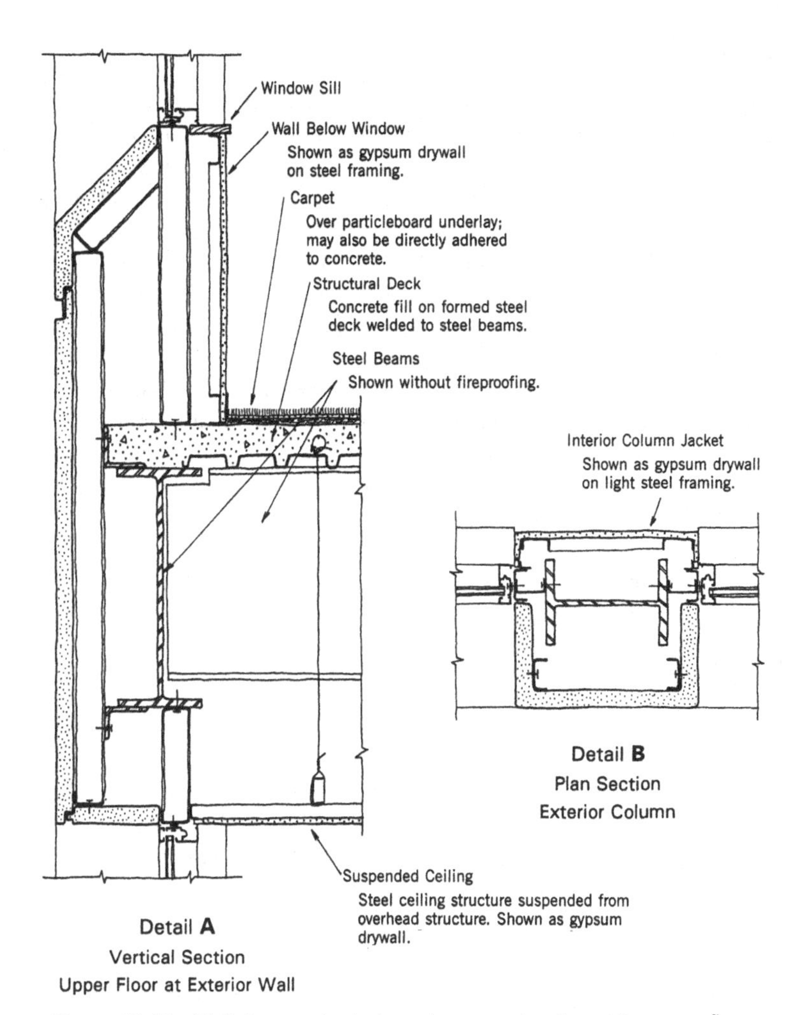

Figure 10.15 Wall, floor, and exterior column construction at the upper floors.

Ceilings, lights, ducts: 15 psf

Walls (average surface weight):

Interior, permanent: 15 psf

Exterior curtain wall: 25 psf

Steel for rolled shapes. ASTM A36, $F_y = 36$ ksi

Structural Alternatives

Structural options for this example are considerable, including possibly the light wood frame if the total floor area and zoning requirements permit its use. Certainly, many steel frame, concrete frame, and masonry bearing wall systems are feasible. Choice of the structural elements will depend mostly on the desired plan form, the form of window arrangements, and the clear spans required for the building interior.

At this height and taller, the basic structure must usually be steel, reinforced concrete, or masonry. Options illustrated here include versions in steel and concrete.

Design of the structural system must take into account both gravity and lateral loads. Gravity requires developing horizontal spanning systems for the roof and upper floors and the stacking of vertical supporting elements. The most common choices for the general lateral bracing system are the following (see Figure 10.16).

Core shear wall system (Figure 10.16*a*). Use of solid walls around core elements (stairs, elevators, restrooms, duct shafts) produce a very rigid vertical structure; the rest of the construction may lean on this rigid core.

Truss-braced core. Similar to the shear wall core; trussed bents replace solid walls.

Perimeter shear walls (Figure 10.16*b*). Turns the building into a tube-like structure; walls may be structurally continuous and pierced by holes for windows and doors, or may be built as individual, linked piers between vertical strips of openings.

Mixed exterior and interior shear walls or trussed bents. For some building plans the perimeter or core systems may not be feasible, requiring use of some mixture of walls and/or trussed bents.

Full rigid frame bent system (Figure 10.16*c*). Uses all the available bents described by vertical planes of columns and beams.

Perimeter rigid frame bents (Figure 10.16*d*). Uses only the columns and spandrel beams in the exterior wall planes, resulting in only two bents in each direction for this building plan.

In the right circumstances, any of these systems may be acceptable for this size building. Each has some advantages and disadvantages from both structural and architectural design points of view.

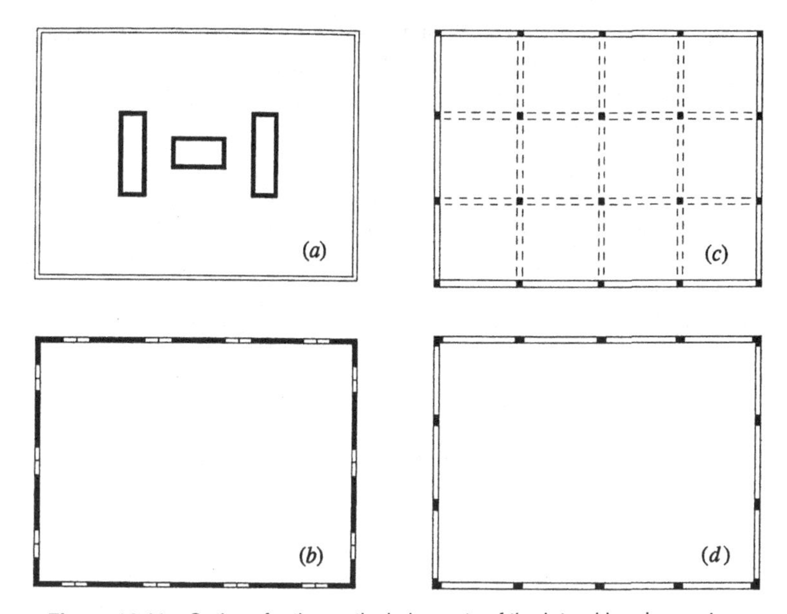

Figure 10.16 Options for the vertical elements of the lateral bracing system.

Presented here are schemes for use of three lateral bracing systems: a truss-braced core, a rigid frame bent, and multistory shear walls. For the horizontal roof and floor structures, several schemes are also presented.

Design of the Steel Structure

Figure 10.17 shows a partial plan of a framing system for the typical upper floor that uses rolled steel beams spaced at a module related to the column spacing. As shown, the beams are 7.5 ft on center and the beams that are not on the column lines are supported by column line girders. Thus, three-fourths of the beams are supported by the girders and the remainder are supported directly by the columns. The beams in turn support a one-way spanning deck.

Within this basic system, there are a number of variables.

Beam spacing. Affects the deck span and the beam loading.

Deck. A variety available, as discussed later.

Beam/column relationship in plan. As shown, permits possible development of vertical bents in both directions.

Column orientation. The W shape has a strong axis and accommodates framing differently in different directions.

Fire protection. Various means, as related to codes and general building construction.

These issues and others are treated in the following discussions.

Inspection of the framing plan in Figure 10.17 reveals a few common elements of the system as well as several special beams required at the building core. The discussions that follow are limited to treatments of the common elements—that is the members labeled "Beam" and "Girder" in Figure 10.17.

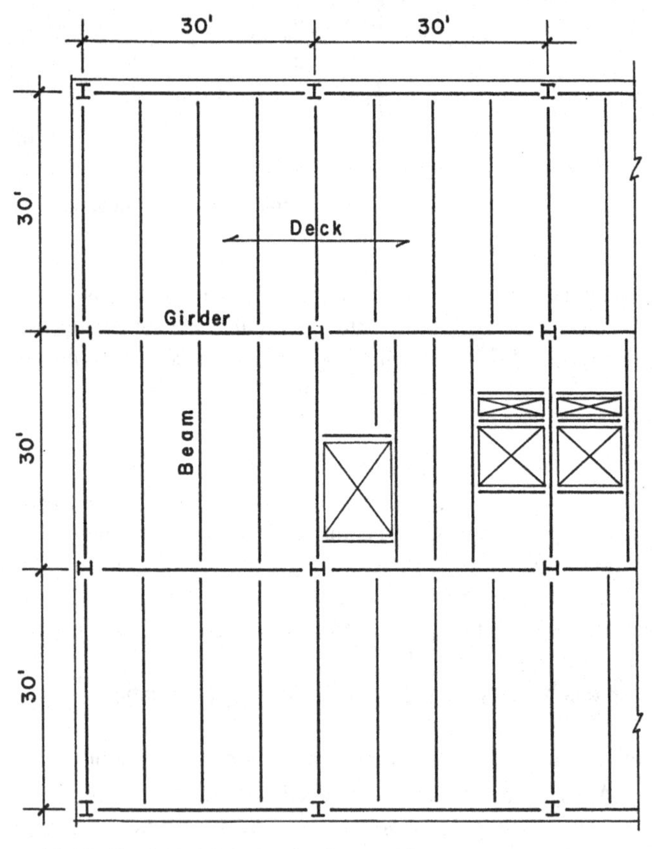

Figure 10.17 Partial framing plan for the steel floor structure at the upper levels.

For the design of the speculative rental building, we must assume that different plan arrangements of the floors are possible. Thus, it is not completely possible to predict where there will be offices and where there will be corridors, each of which require different live loads. It is therefore not uncommon to design for some combinations of loading for the general system that relates to this problem. For the design work here, the following will be used:

For the deck: Live load = 100 psf

For the beams: Live load = 80 psf, with 20 psf added to dead load for movable partitions

For girders and columns: Live load = 50 psf, with 20 psf added to dead load

The Structural Deck Several options are possible for the floor deck. In addition to structural concerns, which include gravity loading and diaphragm action for lateral loads, consideration must be given to fire protection for the steel; to the accommodation of wiring, piping, and ducts; and to attachment of finish floor, roofing, and ceiling constructions. For office buildings, there are often networks for electrical power and communication that must be built into the ceiling, wall, and floor constructions.

If the structural floor deck is a concrete slab, either site-cast or precast, there is usually a nonstructural fill placed on top of the structural slab; power and communication networks may be buried in this fill. If a steel deck is used, closed cells of the formed sheet steel deck units may be used for some wiring, although this is no longer a common practice.

For this example, the selected deck is a steel deck with 1.5-in.-deep ribs, on top of which is cast a lightweight concrete fill with a minimum depth of 2.5 in. over the steel units. The unit average dead weight of this deck depends on the thickness of the sheet steel, the profile of the deck folds, and the unit density of the concrete fill. For this example, it is assumed that the average weight is 30 psf. Adding to this the assumed weight of the floor finish and suspended items, the total dead load for the deck design is thus 50 psf.

Although industry standards exist for these decks (see Ref. 5), data for deck design should be obtained from deck manufacturers.

The Common Beam As shown in Figure 10.17, this beam spans 30 ft and carries a load strip that is 7.5 ft wide. The total peripheral load support area for the beam is thus 7.5 × 30 = 225 ft². This allows for a reduced live load as follows (see Section 9.6):

$$L = L_0\left(0.25 + \frac{15}{\sqrt{K_{LL}A_T}}\right) = 80\left(0.25 + \frac{15}{\sqrt{2 \times 225}}\right) = 77\text{psf}$$

The beam loading is thus:

Live load = 7.5(77) = 578 lb/lineal ft (or plf)
Dead load = 7.5(50 + 20) = 525 plf + beam weight, say 560 plf
Total factored unit load = 1.2(560) + 1.6(578) = 672 +925
= 1597 plf
Total supported factored load = 1.597(30) = 48 kips

Assuming that the welding of the steel deck to the top flange of the beam provides almost continuous lateral bracing, the beam may be selected on the basis of flexural failure. For this load and span, Table 3.2 yields the following possible choices: W 16 × 45, W 18 × 40, or W 21 × 44. Actual choice may be affected by various considerations. For example, the table used does not incorporate concerns for deflection or lateral bracing. The deeper shape will obviously produce the least deflection, although in this case the live-load deflection for the 16-in. shape is within the usual limit (see Figure 3.11). This beam becomes the typical member, with other beams being designed for special circumstances, including the column-line beams, the spandrels, and so on.

The Common Girder Figure 10.18 shows the loading condition for the girder, as generated only by the supported beams. Although this ignores the effect of the weight of the girder as a uniformly distributed load, it is reasonable for use in an approximate design, because the girder weight is a minor loading.

Note that the girder carries three beams and thus has a total load periphery of 3(225) = 675 ft². The reduced live load is thus (see Section 9.6):

$$L = (80 \times 675)\left(0.25 + \frac{15}{\sqrt{2 \times 675}}\right) = 35{,}100 \text{ lb, or } 35.1 \text{ kips}$$

The unit beam load for design of the girder is determined as follows:

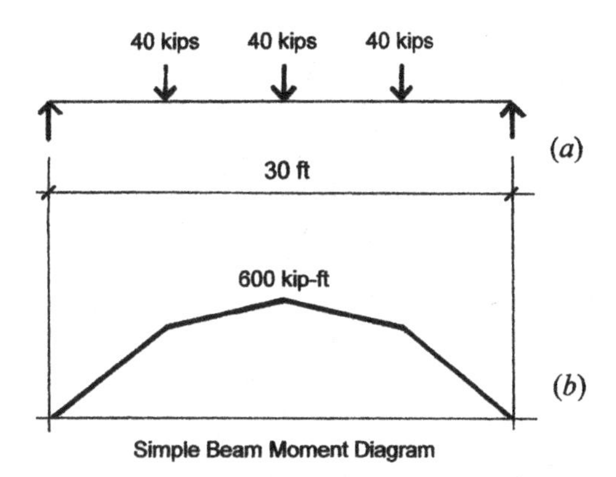

Figure 10.18 Loading condition for gravity load on the girder: (a) loads from the supported beams; (b) form of the simple beam moment diagram.

Live load = 35.1/3 = 11.7 kips

Dead load = 0.560(30) = 16.8 kips

Factored unit load = 1.2(16.8) + 1.6(11.7) = 38.88 kips

To account for beam weights use 40 kips

From Figure 10.18, maximum moment is 600 kip-ft

Selection of a member for this situation may be made using various data sources. Because this member is laterally braced at only 7.5-ft intervals, attention must be paid to this point. The maximum moment together with the laterally unbraced length can be used in Table 3.1 to determine acceptable choices. Possibilities are W 18 ×106, W 21 ×101, W 24 ×94, W 27 ×94 or W 30 × 90. The deeper members will have less deflection and will allow greater room for building service elements in the enclosed floor/ceiling space. However, shallower beams may reduce the required story height, resulting in cost savings.

Computation for deflections may be performed with formulas that recognize the true form of loading. However, approximate deflection values may be found using an equivalent load derived from the maximum moment, as discussed in Section 3.4. For this example, the equivalent uniform load (EUL) is obtained as follows:

$$M = \frac{WL}{8} = 600 \text{ kip-ft}$$

$$W = \frac{8M}{L} = \frac{8 \times 600}{30} = 160 \text{ kips}$$

This hypothetical uniformly distributed load may be used with the formula for deflection of a simple beam (see Figure B.1) to find an approximate deflection. However, for a quick check, Figure 3.11 indicates that for this span all the previous choice options will have a total load deflection of less than 1/240 of the span.

Although deflection of individual elements should be investigated, there are wider issues regarding deflection, such as the following:

Bounciness of floors. This involves the stiffness and the fundamental period of spanning elements and may relate to the deck and/or the beams. In general, use of the static deflection limits usually ensures a reasonable lack of bounce, but just about anything that increases stiffness improves the situation.

Transfer of load to nonstructural walls. With the building construction completed, live-load deflections of the structure may result in bearing of spanning members on nonstructural construction. Reducing deflections of the structure will help for this, but some special details may be required for attachment between the structure and the nonstructural construction.

Deflection during construction. The deflection of the girders plus the deflection of the beams adds up to a cumulative deflection at the center of a column bay. This may be critical for live load, but can also create problems during construction. If the steel beams and steel deck are installed dead flat, then construction added later will cause deflection from the flat condition. In this example, that would include the concrete fill, which can cause a considerable deflection at the center of the column bay. One response is to camber (bow upward) the beams by bending them in the fabricating shop so that they deflect to a flat position under the dead load.

Column Design for Gravity Loads Design of the steel columns must include considerations for both gravity and lateral loads. Gravity loads for individual columns are based on the column's *periphery,* which is usually defined as the area of supported surface on each level supported. Loads are actually delivered to the columns by the beams and girders, but the peripheral area (also called the *tributary area*) is used for load tabulation and determination of live-load reductions.

If beams are rigidly attached to columns with moment-resistive connections—as is done in development of rigid frame bents—then gravity loads will also cause bending moments and shears in the columns. Otherwise, the gravity loads are essentially considered only as axial compressive loads.

Involvement of the columns in development of resistance to lateral loads depends on the form of the lateral bracing system. If trussed bents are used, some columns will function as chords in the vertically cantilevered trussed bents, which will add some compressive forces and possibly cause some reversals with net tension in the columns. If columns are parts of rigid frame bents, the same chord actions will be involved, but the columns will also be subject to bending moments and shears from the rigid frame lateral actions.

Whatever the lateral force actions may do, the columns must also work for gravity load effects alone. In this part this investigation is made and designs are completed without reference to lateral loads. This yields some reference selections, which can then be modified (but not reduced) when the lateral resistive system is designed. Later discussions in this chapter present designs for both a trussed bent system and a rigid frame system.

There are several different cases for the columns, due to the framing arrangements and column locations. For a complete design of all columns it would be necessary to tabulate the loading for each different case. For illustration purposes here, tabulation is shown for a hypothetical interior column. The interior column illustrated assumes a general periphery of 900 sq ft of general roof or floor area. Actually, the floor plan in Figure 10.14 shows that all the interior columns are within the core area, so there is no such column. However, the tabulation yields a column that is general for the interior condition, and can be used for approximate selection. As will be shown later, all the interior columns are involved in the lateral force systems, so this also yields a takeoff size selection for the design for lateral forces.

Table 10.2 is a common form of tabulation used to determine the column loads. For the interior columns, the table assumes the existence of a rooftop structure (penthouse) above the core, thus creating a fourth story for these columns.

Table 10.2 is organized to facilitate the following determinations:

1. Dead load on the periphery at each level, determined by multiplying the area by an assumed average dead load per sq ft. Loads de-

termined in the process of design of the horizontal structure may be used for this estimate.

2. Live load on the periphery areas.
3. The reduced live load to be used at each story, based on the total supported periphery areas above that story.
4. Other dead loads directly supported, such as the column weight and any permanent walls within the load periphery.
5. The total load collected at each level.
6. A design load for each story, using the total accumulation from all levels supported.

For the entries in Table 10.2, the following assumptions were made:

Roof unit live load = 20 psf (reducible)

Roof dead load = 40 psf (estimated, based on the similar floor construction)

Penthouse floor live load = 100 psf (for equipment, average)

TABLE 10.2 Service Load Tabulation for the Interior Column

		Load Tabulation (lb)	
Level Supported	Load Source and Computation	Dead Load	Live Load
Penthouse roof	Live load, not reduced = 20 psf × 225		4500
225 ft^2	Dead load = 40 psf × 225	9000	
Building roof	Live load, not reduced = 20 psf × 675		13,500
675 ft^2	Dead load = 40 psf × 675	27,000	
Penthouse floor	Live load = 100 psf × 225		22,500
225 ft^2	Dead load = 50 psf × 225	11,250	
	Story loads + loads from above	47,250	40,500
	Reduced live load (50%)		20,250
Third floor	Live load = 50 psf × 900		45,000
900 ft^2	Dead load = 70 psf × 900	63,000	
	Story loads + loads from above	110,250	85,500
	Reduced live load (50%)		42,750
Second floor	Story loads	63,000	45,000
(same as third)	Story loads + loads from above	173,250	130,500
	Reduced live load (50%)		65,250

Penthouse floor dead load = 50 psf
Floor live load = 50 psf (reducible)
Floor dead load = 70 psf (including partitions)

Table 10.3 summarizes the design for the four-story column. For the pin-connected frame, a K factor of 1.0 is assumed and the full story heights are used as the unbraced column lengths.

Although column loads in the upper stories are quite low, and some small column sizes would be adequate for the loads, a minimum size of 10 in. is maintained for the W shapes for two reasons.

The first consideration involves the form of the horizontal framing members and the type of connections between the columns and the horizontal framing. All the H-shaped columns must usually facilitate framing in both directions, with beams connected both to column flanges and webs. With standard framing connections for field bolting to the columns, a minimum beam depth and flange width are required for practical installation of the connecting angles and bolts.

The second consideration involves the problem of achieving splices in the multistory column. If the building is too tall for a single-piece column, a splice must be used somewhere, and the stacking of one column piece on top of another to achieve a splice is made much easier if the two pieces are of the same nominal size group.

TABLE 10.3 Design of the Interior Column

Design Loads for Each Story (lb)	Possible Choices
Penthouse, unbraced height = 13 ft $P_u = \phi P_n = 1.2(9{,}000) + 1.6(4{,}500) = 10{,}800 + 7200$ $= 18{,}000$ lb	W 8 × 24
Third Story, unbraced height = 13 ft $P_u = \phi P_n = 1.2(47{,}250) + 1.6(20{,}250) = 58{,}700 + 32{,}400$ $= 91{,}100$ lb	W 8 × 24, W 10 × 33
Second Story, unbraced height = 13 ft $P_u = \phi P_n = 1.2(110{,}250) + 1.6(42{,}750) = 132{,}300 + 68{,}400$ $= 200{,}700$ lb	W 8 ×31, W 10 × 45, W 12 × 45
First Story, unbraced height = 15 ft $P_u = \phi P_n = 1.2(173{,}250) + 1.6(65{,}250) = 207{,}900 + 104{,}400$ $= 312{,}300$ lb	W 8 ×58, W 10 × 54, W 12 × 53, W 14 × 53

Add to this a possible additional concern relating to the problem of handling long pieces of steel during transportation to the site and erection of the frame. The smaller the member's cross section, the shorter the piece that is feasible to handle.

For all of these reasons, a minimum column is often considered to be the W 10 × 33, which is the lightest shape in the group that has an 8-in.-wide flange. It is assumed that a splice occurs at 3 ft above the second-floor level (a convenient, waist-high distance for the erection crew), making two column pieces approximately 18 and 23 ft long. These lengths are readily available and quite easy to handle with the 10-in. nominal shape. On the basis of these assumptions, a possible choice would be for a W 10 × 33 for the penthouse and the third-story column and a W 10 × 54 for the lower two stories.

Alternative Floor Construction with Trusses

A framing plan for the upper floor of Building Three is shown in Figure 10.19, indicating the use of open-web steel joists and joist girders. Although this construction might be extended to the core and the exterior spandrels, it is also possible to retain the use of rolled shapes for these purposes. There are various possibilities for the development of lateral bracing with this scheme; one is the use of the same trussed core bents that were previously designed. Although somewhat more applicable to longer spans and lighter loads, this system is reasonably applicable to this situation as well.

One potential advantage of using the all-truss framing for the horizontal structure is the higher degree of freedom of passage of building service elements within the enclosed space between ceilings and the supported structure above. A disadvantage is the usual necessity for greater depth of the structure, adding to building height—a problem that increases with the number of stories.

Design of the Open-Web Joists General concerns and basic design for open-web joists are presented in Section 3.9. Using the data for this example, a joist design is as follows:

Joists at 3 ft on center, span of 30 ft

Dead load = 3(70) = 210 lb/ft not including joists

Live load = 3(100) = 300 lb/ft not reduced. This is a high live load but it permits location of a corridor anywhere on the plan and also reduces deflection and bounciness.

Total factored load = 1.2(210) + 1.6(300) = 252 + 480 = 732 lb/ft

Referring to Table 3.4, note that choices may be considered for any joist that will carry the total load of 732 lb/ft and a live load of 300 lb/ft on a span of 30 ft. The following choices are possible:

24K9 at 12 lb/ft, permitted load = 807 − 1.2(12) = 793 lb/ft
26K9, stronger than 24K9 and only 0.2 lb/ft heavier
28K8, stronger than 24K9 and only 0.7 lb/ft heavier
30K7, stronger than 24K9 and only 0.3 lb/ft heavier

All of the joists listed are economically equivalent. Choice would be made considering details of the general building construction. A shallower joist depth means a shorter story height and less overall building height. A deeper joist yields more open space in the floor construction for ducts, wiring, piping, and so on, and also means less deflection and less floor bounce.

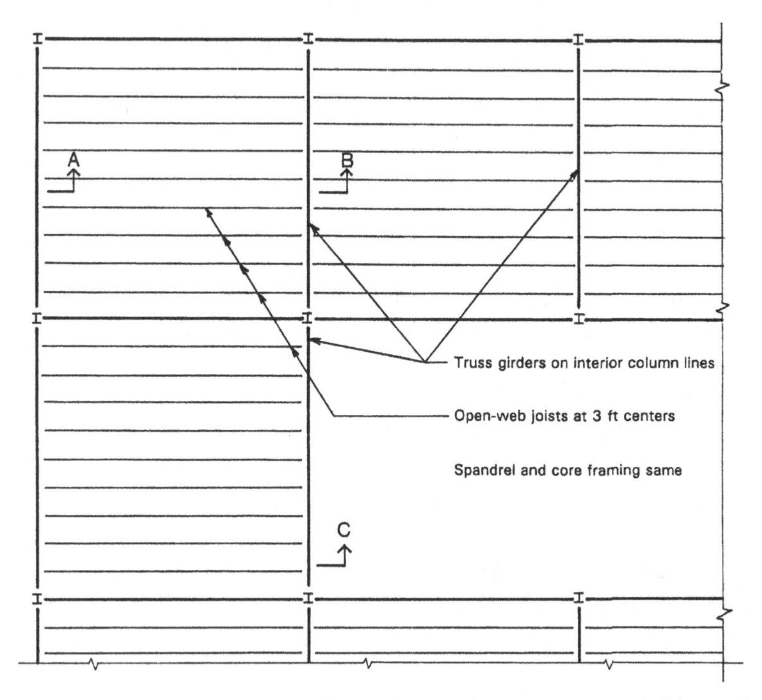

Figure 10.19 Building Three: Alternate floor framing using open-web joists and joist girders.

Design of the Joist Girders Joist girders are also discussed in Section 3.9. Both the joists and the girders are likely to be supplied and erected by a single contractor. Although there are industry standards (see Ref. 4), consult the specific manufacturer for data regarding design and construction details for these products.

The pattern of the joist girder members is somewhat fixed and relates to the spacing of the supported joists. To achieve a reasonable proportion for the panel units of the truss, the dimension for the depth of the girder should be approximately the same as that for the joist spacing.

Considerations for design for the truss girder are as follows:

The assumed depth of the girder is 3 ft, which should be considered a *minimum* depth for this span (L/10). Any additional depth possible will reduce the amount of steel and also improve deflection responses. However, for floor construction in multistory buildings, this dimension is hard to bargain for.

Use a live-load reduction of 40 percent maximum with a live load of 50 psf. Thus, live load from one joist = $(3 \times 30)(0.6 \times 50) = 2700$ lb or 2.7 kips.

For dead load add a partition load of 20 psf to the construction load of 40 psf. Thus, dead load = $(3 \times 30)(60) = 5400$ lb + joist weight of 12 lb/ft $\times$ 30 = 360 lb, total = 5400 + 360 = 5760 lb or 5.76 kips

Total factored load = 1.2(5.76) + 1.6(2.7) = 6.91 + 4.32) = 11.23 kips

Figure 10.20 shows a possible form for the joist girder. For this form and the computed data the joist specification is as follows:

36G = girder depth of 3 ft
10N = ten spaces between the joists
11.23K = design factored joist load
Complete specification is thus 36G10N11.23K

Construction Details for the Truss Structure Figure 10.21 shows some details for construction of the trussed system. The deck shown here is the same as that for the scheme with W-shape framing, although the shorter span may allow use of a lighter sheet steel deck. However, the deck must also be used for diaphragm action, which may limit the reduction.

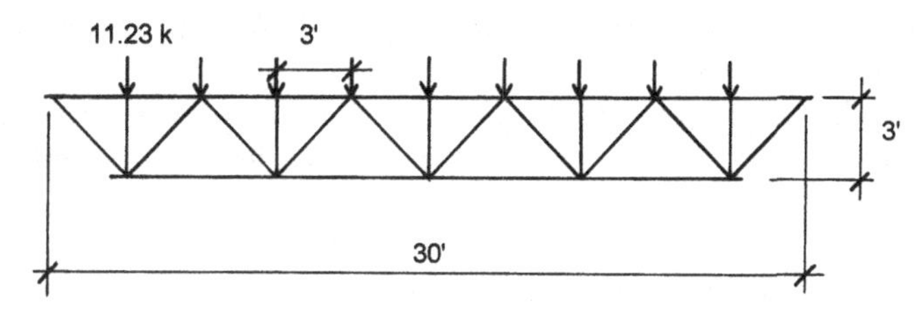

Figure 10.20 Form and data for the joist girder.

Adding to the problem of overall height for this structure is the detail at the joist support, in which the joists must sit on top of the supporting members, whereas in the all W-shape system the beams and girders have their tops level.

With the closely spaced joists, ceiling construction may be directly supported by the bottom chords of the joists. This may be a reason for selection of the joist depth. However, it is also possible to suspend the ceiling

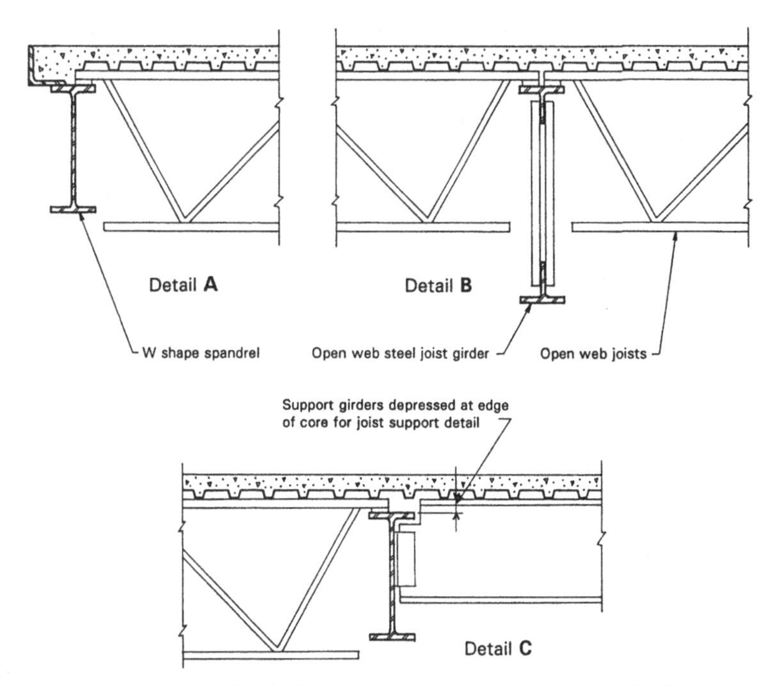

Figure 10.21 Details for the floor system with open-web joists and joist girders. For location of details see the framing plan in Figure 10.19.

from the deck, as is generally required for the all-W-shape structure with widely spaced beams.

Another issue here is the usual necessity to use a fire-resistive ceiling construction, because it is not feasible to encase the joists or girders in fireproofing material.

Design of the Trussed Bent for Wind

Figure 10.22 shows a partial framing plan for the core area, indicating the placement of some additional columns off the 30-ft grid. These columns are used together with the regular columns and some of the horizontal framing to define a series of vertical bents for the development of the trussed bracing system shown in Figure 10.23. With relatively slender diagonal members, it is assumed that the X-bracing behaves as if the tension diagonals function alone. There are thus considered to be four vertical, cantilevered, determinate trusses that brace the building in each direction.

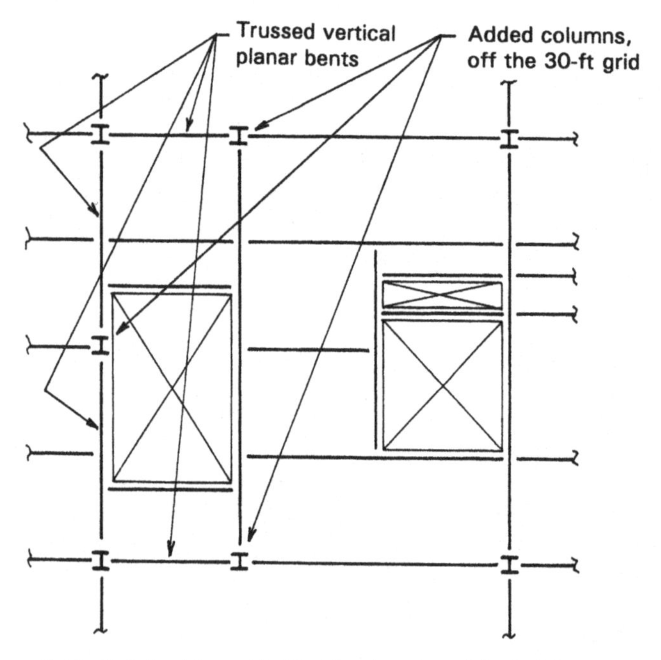

Figure 10.22 Building Three: Modified framing plan of the building core for development of the trussed bracing bents.

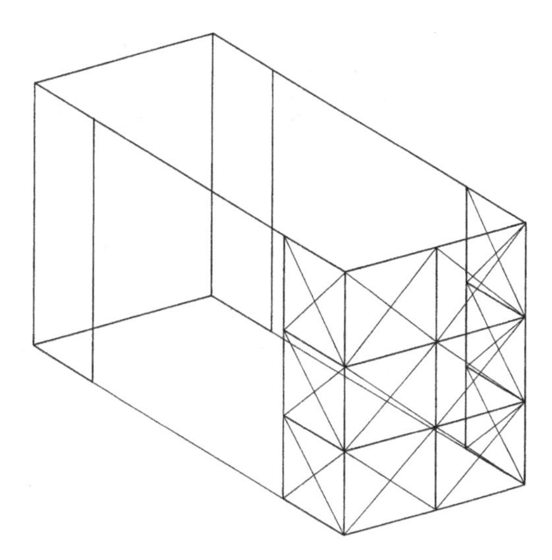

Figure 10.23 General form of the trussed bent bracing system.

With the symmetrical building exterior form and the symmetrically placed core bracing, this is a reasonable system for use in conjunction with the horizontal roof and upper floor structures to develop resistance to horizontal forces due to wind. The work that follows illustrates the design process, using criteria for wind loading from ASCE 2002 (Ref. 1).

For the total wind force on the building, we will assume a base pressure of 15 psf, adjusted for height as described in ASCE 2002 (Ref. 1). The design pressures and their zones of application are shown in Figure 10.24.

For investigation of the lateral bracing system, the design wind pressures on the outside wall surface are distributed as edge loadings to the roof and floor diaphragms. These are shown as the forces H_1, H_2, and H_3 in Figure 10.24. The horizontal forces are next shown as loadings to one of the vertical truss bents in Figure 10.25*a*. For the bent loads the total forces per bent are determined by multiplying the unit edge diaphragm load by the building width and dividing by the number of bracing bents for load in that direction. The bent loads are thus:

$$H_1 = (165.5)(92)/4 = 3807 \text{ lb}$$

$$H_2 = (199.5)(92)/4 = 4589 \text{ lb}$$

$$H_3 = (210)(92)/4 = 4830 \text{ lb}$$

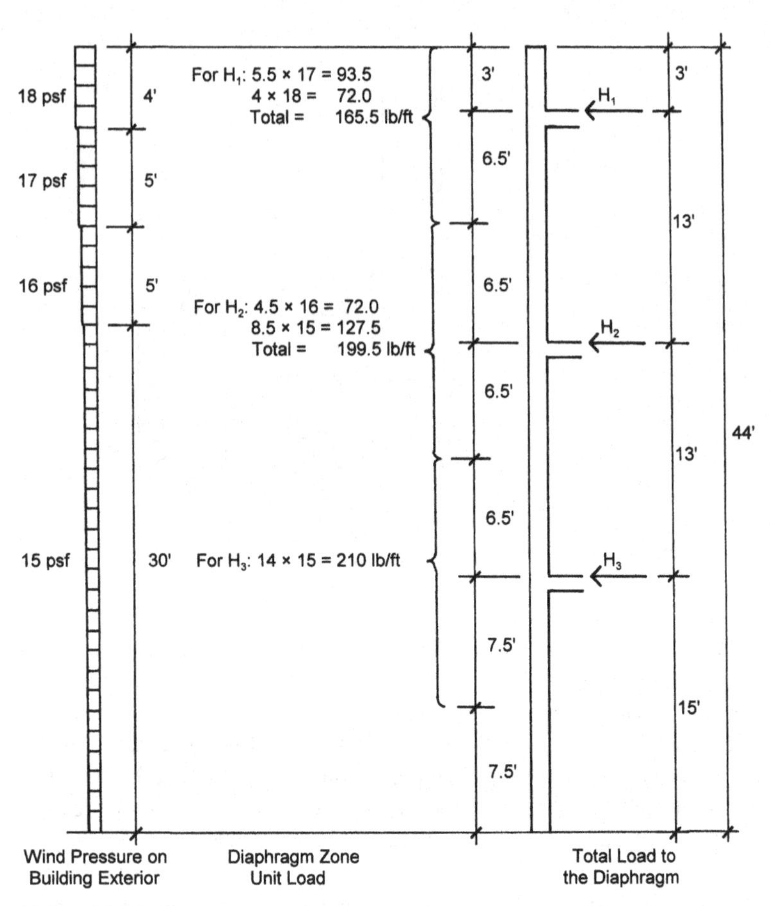

Figure 10.24 Building Three: Development of the wind loads transferred to the roof and upper-level floor diaphragms. Design wind pressures on the building exterior from ASCE 2002 (Ref. 1), assuming a base wind pressure of 15 psf. Diaphragm zones are defined by column mid-height points. Total loads on the diaphragms are found by multiplying the diaphragm zone unit load per foot by the building width.

The truss loading, together with the reaction forces at the supports, are shown in Figure 10.25*b*. The internal forces in the truss members resulting from this loading are shown in Figure 10.25*c*, with force values in pounds and sense indicated by C for compression and T for tension.

The forces in the diagonals may be used to design tension members, using the factored load combination that includes wind (see Section 9.8).

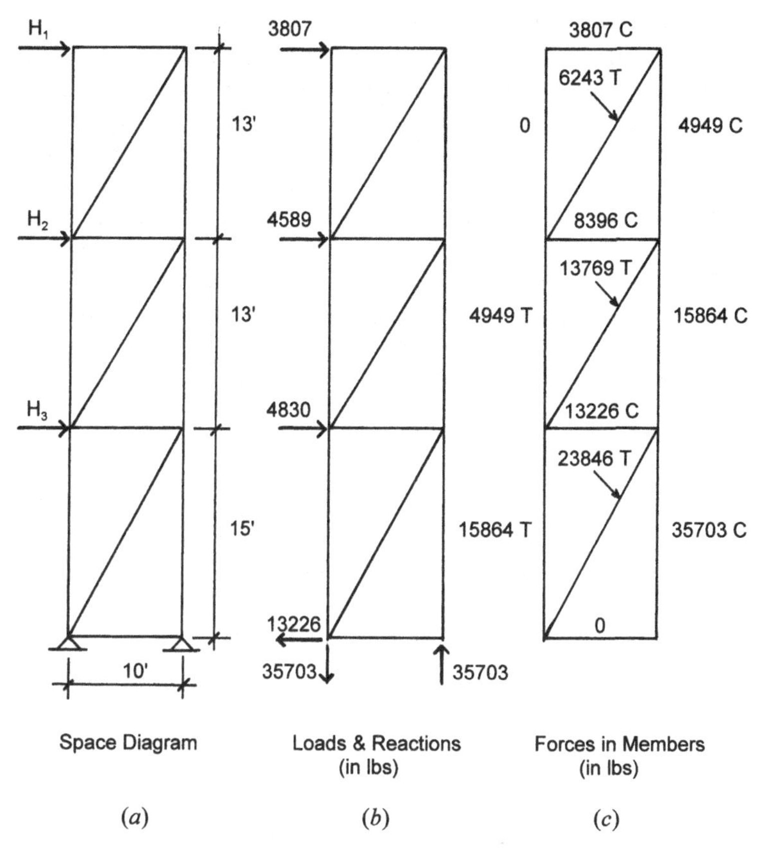

Figure 10.25 Investigation of one of the trussed bents.

The compression forces in the columns may be added to the gravity loads to see if this load combination is critical for the column design. The uplift tension force at the column should be compared with the dead load to see if the column base needs to be designed for a tension anchorage force.

The horizontal forces should be added to the beams in the core framing and an investigation should be done for the combined bending and compression. Because beams are often weak on their minor axis (y-axis), it may be practical to add some framing members at right angles to these beams to brace them against lateral buckling.

Design of the diagonals and their connections to the beam and column frame must be developed with consideration of the form of the elements and some consideration for the wall construction in which they are imbedded. Figure 10.26 shows some possible details for the diagonals and the connections. A detail problem that must be solved is that of the crossing of the two diagonals at the middle of the bent. If we use double angles for the diagonals (a common truss form), the splice joint shown in Figure 10.26 is necessary. An option is to use either single angles or channel shapes for the diagonals, allowing the members to pass each other back-to-back at the center. The latter choice, however, involves some degree of eccentricity in the members and connections and a single shear load on the bolts, so it is not advisable if load magnitudes are high. For the tension member, a recommended minimum slenderness is represented by an L/r ratio of 300.

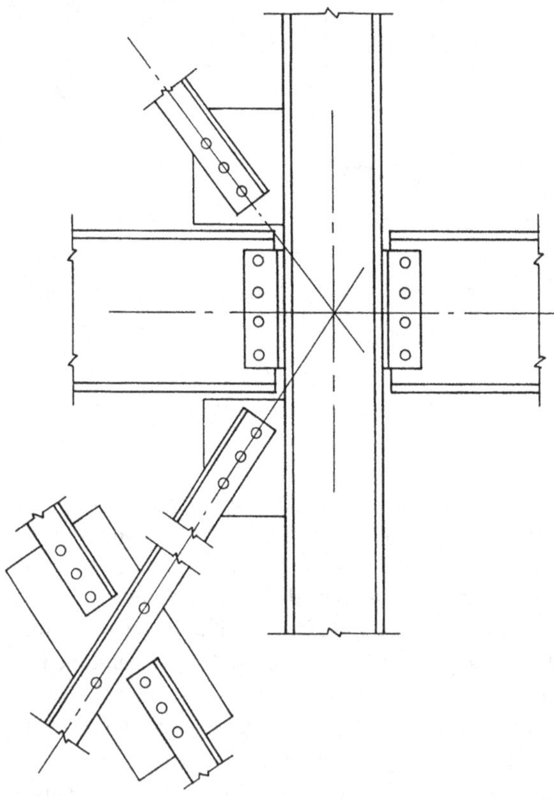

Figure 10.26 Details of the trussed bent construction with bolted joints.

Considerations for a Steel Rigid Frame

The general nature of rigid frames is discussed in Section 5.1. A critical concern for multistory, multiple-bay frames is the lateral strength and stiffness of columns. Because the building must be developed to resist lateral forces in all directions, it becomes necessary in many cases to consider the shear and bending resistance of columns in two directions (north–south and east–west, for example). This presents a problem for W-shape columns, because they have considerably greater resistance on their major (x-x) axis versus their minor (y-y) axis. Orientation of W-shape columns in plan thus sometimes becomes a major consideration in structural planning.

Figure 10.27*a* shows a possible plan arrangement for column orientation for Building Three, relating to the development of two major bracing bents in the east–west direction and five shorter and less stiff bents in the north–south direction. The two stiff bents may well be approximately equal in resistance to the five shorter bents, giving the building a reasonably symmetrical response in the two directions.

Figure 10.27*b* shows a plan arrangement for columns designed to produce approximately symmetrical bents on the building perimeter. The form of such perimeter bracing is shown in Figure 10.28.

One advantage of perimeter bracing is the potential for using deeper (and thus stiffer) spandrel beams, because the restriction on depth that applies for interior beams does not exist at the exterior wall plane. Another possibility is to increase the number of columns at the exterior, as shown in Figure 10.27*c*—a possibility that does not compromise the building interior space. With deeper spandrels and closely spaced exterior columns, a very stiff perimeter bent is possible. In fact, such a bent may have very little flexing in the members, and its behavior approaches that of a pierced wall, rather than a flexible frame.

At the expense of requiring much stronger (and heavier and/or larger) columns and expensive moment-resistive connections, the rigid frame bracing offers architectural planning advantages with the elimination of solid shear walls or truss diagonals in the walls. However, the lateral deflection (drift) of the frames must be carefully controlled, especially with regard to damage to nonstructural parts of the construction.

Considerations for a Masonry Wall Structure

An option for the construction of Building Three involves the use of structural masonry for development of the exterior walls. The walls are

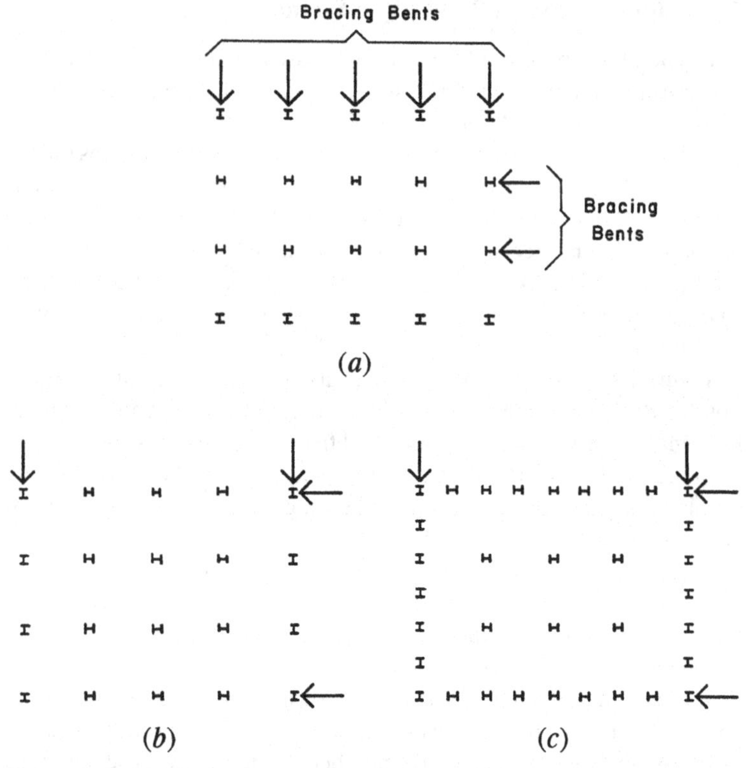

Figure 10.27 Building Three: Optional arrangements for the steel W-shape columns for development of rigid frame bents.

used for both vertical bearing loads and lateral shear wall functions. The choice of forms of masonry and details for the construction depend very much on regional considerations (climate, codes, local construction practices, etc.) and on the general architectural design. Major differences occur due to variations in the range in outdoor temperature extremes and the specific critical concerns for lateral forces.

General Considerations Figure 10.29 shows a partial elevation of the masonry wall structure and a partial framing plan of the upper floors. The wood construction shown here is questionably acceptable for fire codes. The example is presented only to demonstrate the general form of the construction.

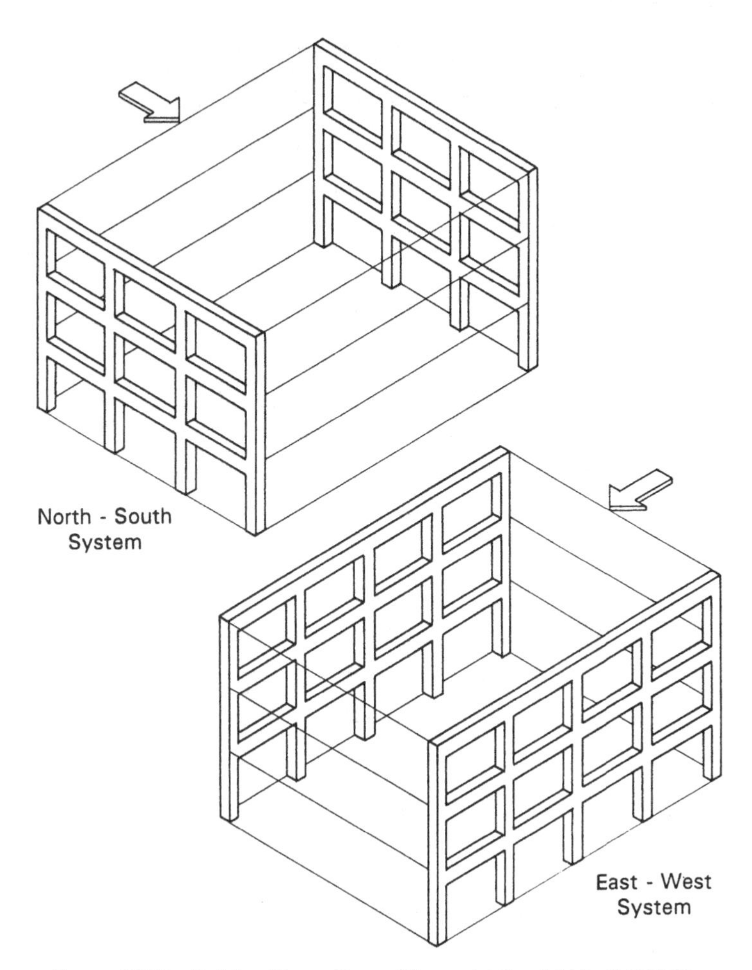

Figure 10.28 Building Three: Form of the perimeter rigid bent system.

Plan dimensions for structures using CMUs (concrete blocks) must be developed so as to relate to the modular sizes of typical CMUs. A few standard sizes are widely used, but individual manufacturers often have some special units or will accommodate requests for special shapes or sizes. However, whereas solid brick or stone units can be cut to produce precise, non-modular dimensions, the hollow CMUs generally cannot. Thus, the dimensions for the CMU structure itself must be carefully de-

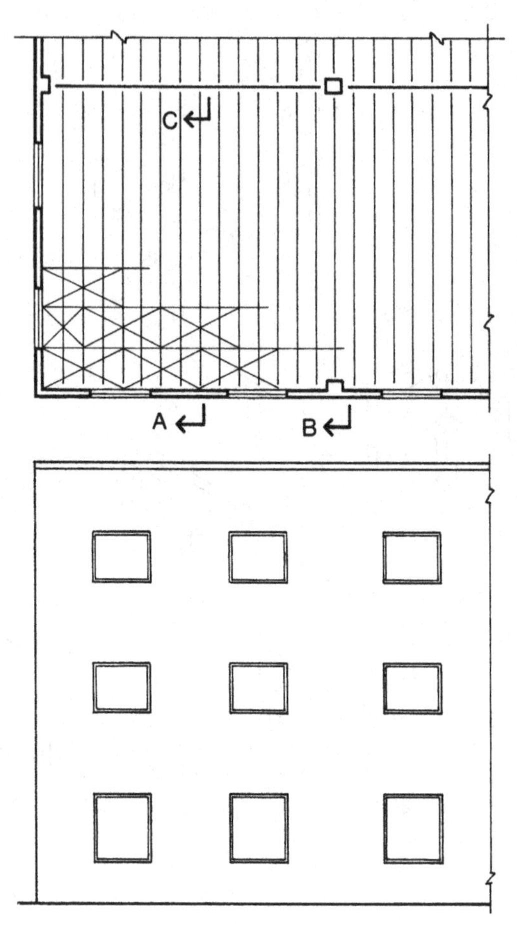

Figure 10.29 Building Three: Partial framing plan for the upper floor and partial elevation for the masonry wall.

veloped to have wall intersections, corners, ends, tops, and openings for windows and doors fall on the fixed modules.

There are various forms of CMU construction. The one shown here is that widely used where either windstorm or earthquake risk is high. This is described as *reinforced masonry* and is produced to generally emulate reinforced concrete construction, with tensile forces resisted by steel reinforcement that is grouted into the hollow voids in the block construction. The construction takes the general form shown in Figure 10.30.

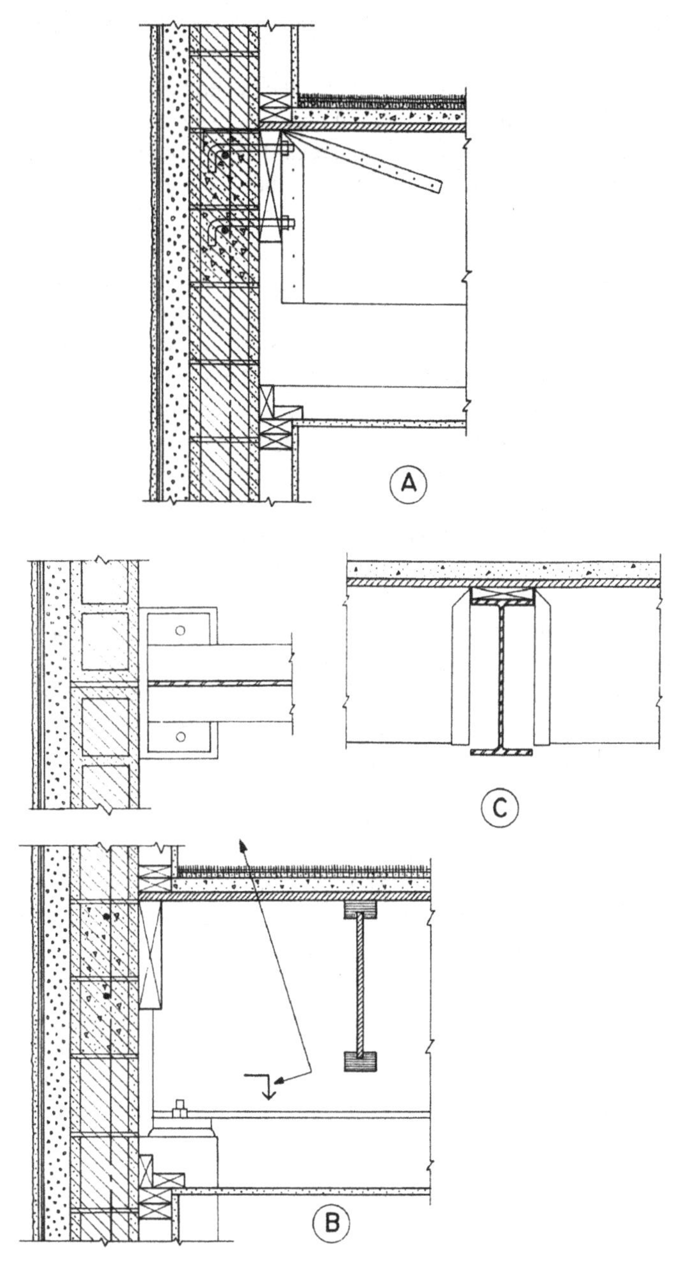

Figure 10.30 Details of the upper-level floor and exterior masonry wall construction.

Another consideration to be made for the general construction is that involving the relation of the structural masonry to the complete architectural development of the construction, regarding interior and exterior finishes, insulation, incorporation of wiring, and so on.

The Typical Floor The floor framing system here uses column-line girders that support fabricated joists and a plywood deck. The girders could be glued laminated timber but are shown here as rolled steel shapes. Supports for the steel girders consist of steel columns on the interior and masonry pilaster columns at the exterior walls.

As shown in the details in Figure 10.30, the exterior masonry walls are used for direct support of the deck and the joists, through ledgers bolted and anchored to the interior wall face. With the plywood deck also serving as a horizontal diaphragm for lateral loads, the load transfers for both gravity and lateral forces must be carefully developed in the details for this construction.

The development of the girder support at the exterior wall is a bit tricky with the pilaster columns, which must not only provide support for the individual girders at each level, but also maintain the vertical continuity of the column load from story to story. There are various options for the development of this detail. Detail C in Figure 10.30 indicates the use of a wide pilaster that virtually straddles the relatively narrow girder. The void created by the girder is thus essentially ignored with the two outer halves of the pilaster bypassing it for a real vertical continuity.

Because the masonry structure in this scheme is used only for the exterior walls, the construction at the building core is free to be developed by any of the general methods shown for other schemes. If the steel girders and steel columns are used here, it is likely that a general steel framing system might be used for most of the core framing.

Attached only to its top, the supported construction does not provide very good lateral support for the steel girder in resistance to torsional buckling. It is advisable, therefore, to use a steel shape that is not too weak on its *y*-axis, generally indicating a critical concern for lateral unsupported length.

The Masonry Walls Buildings much taller than this have been achieved with structural masonry, so the feasibility of the system is well demonstrated. The vertical loads increase in lower stories, so it is expected that some increases in structural capacity will be achieved in lower portions of the walls. The two general means for increasing wall strength are to use thicker CMUs or to increase the amount of core grouting and reinforcement.

It is possible that the usual minimum structure—relating to code minimum requirements for the construction—may be sufficient for the top story walls, with increases made in steps for lower walls. Without increasing the CMU size, there is considerable range between the minimum and the feasible maximum potential for a wall.

It is common to use fully grouted walls (all cores filled) for CMU shear walls. In this scheme that would technically involve using fully grouted construction for *all* of the exterior walls; which might likely rule against the economic feasibility of this scheme. Adding this to the concerns for thermal movements in the long walls might indicate the wisdom of using some control joints to define individual wall segments.

Design for Lateral Forces A common solution for lateral bracing is the use of an entire masonry wall as a shear wall, with openings considered as producing the effect of a very stiff rigid frame. As for gravity loads, the total lateral shear force increases in lower stories. Thus, it is also possible to consider the use of the potential range for a wall from minimum construction (defining a minimum structural capacity) to the maximum possible strength with all voids grouted and some feasible upper limit for reinforcement.

The basic approach here is to design the required wall for each story, using the total shear at that story. In the end, however, the individual story designs must be coordinated for the continuity of the construction. However, it is also possible that the construction itself could be significantly altered in each story, if it fits with architectural design considerations.

Construction Details There are many concerns for the proper detailing of the masonry construction to fulfill the shear wall functions. There are also many concerns for proper detailing to achieve the force transfers between the horizontal framing and the walls. The general framing plan for the upper floor is shown in Figure 10.29. The location of the details discussed here is indicated by the section marks on that plan.

Detail A. This shows the general exterior wall construction and the framing of the floor joists at the exterior wall. The wood ledger is used for vertical support of the joists, which are hung from steel framing devices fastened to the ledger. The plywood deck is nailed directly to the ledger to transfer its horizontal diaphragm loads to the wall.

Outward forces on the wall must be resisted by anchorage directly between the wall and the joists. Ordinary hardware elements can be used for this, although the exact details depend on the type of forces (wind or seismic), their magnitude, the details of the joists, and the details of the wall construction. The anchor shown in the detail is really only symbolic.

General development of the construction here shows the use of a concrete fill on top of the floor deck, furred out wall surfacing with batt insulation on the interior wall side, and a ceiling suspended from the joists.

Detail B. This shows the section and plan details at the joint between the girder and the pilaster. The pilaster unavoidably creates a lump on the inside of the wall in this scheme. However, it is also possible to move the pilaster to the outside of the wall.

Detail C. This shows the use of the steel beam for support of the joists and the deck. After the wood lumber piece is bolted to the top of the steel beam, the attachment of the joists and the deck become essentially the same as it would be with a timber girder.

10.4 BUILDING FOUR

Figure 10.31 shows the form of Building Three, as adapted for mill construction. Mill construction is a historic form developed in the seventeenth and eighteenth centuries consisting of heavy masonry walls with an interior framed structure for the floors and roofs. The frames were originally timber, and then later also iron and steel.

As shown here, the interior is developed with a full-building-high open space in the center, extending from the first (ground) floor level to the clear-spanning roof trusses. Upper floors are donut-shaped in plan with a single row of interior columns on each side. An open hall is formed on a cantilevered balcony that rings the open space.

Steel Floor and Columns

Figure 10.32 shows a partial framing plan for the upper floor for Building Four. Interior columns are placed at 16-ft centers, matching the location of structural columns in the exterior masonry construction. These columns support steel girders that rest on the interior columns, cantilevering past them to support the 8-ft-wide balcony. Perpendicular to these are steel beams that support a steel deck with concrete fill.

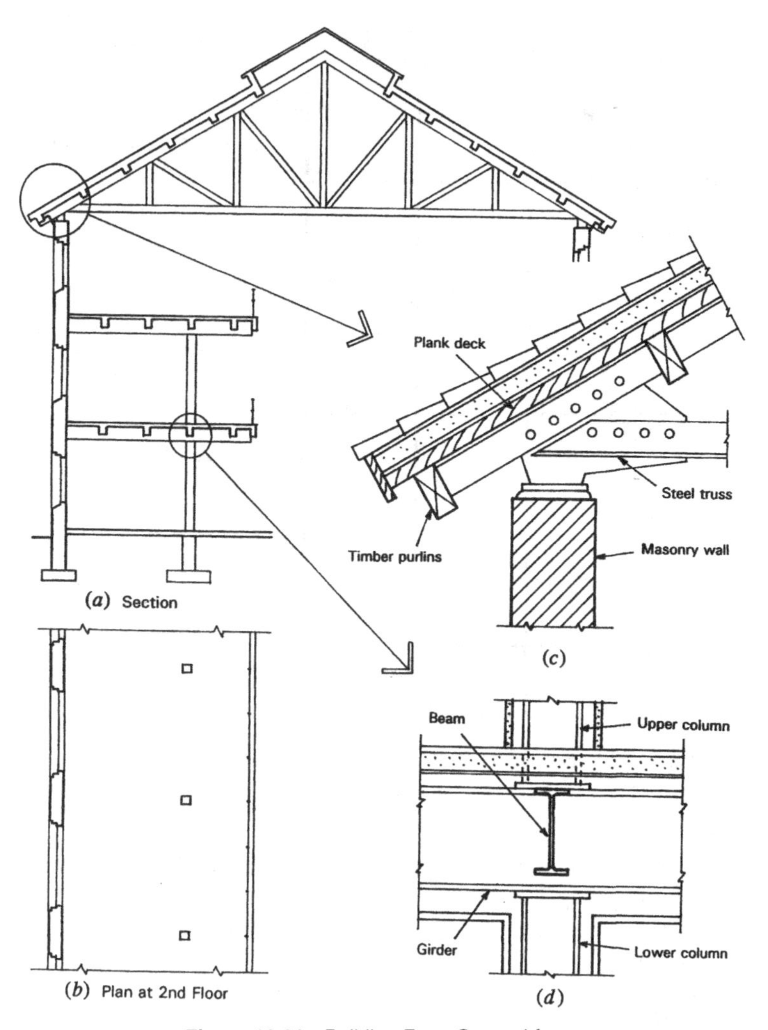

Figure 10.31 Building Four: General form.

As with Building Three, the steel beam and deck system would probably be designed for at least a 100-psf live load for the entire floor, giving some options for future rearrangement and use of space. Thus, in the building plan, the only permanent interior structural elements would be the steel columns—in addition, of course, to the necessary enclosure for stairs, elevators, restrooms, and vertical service risers. The following de-

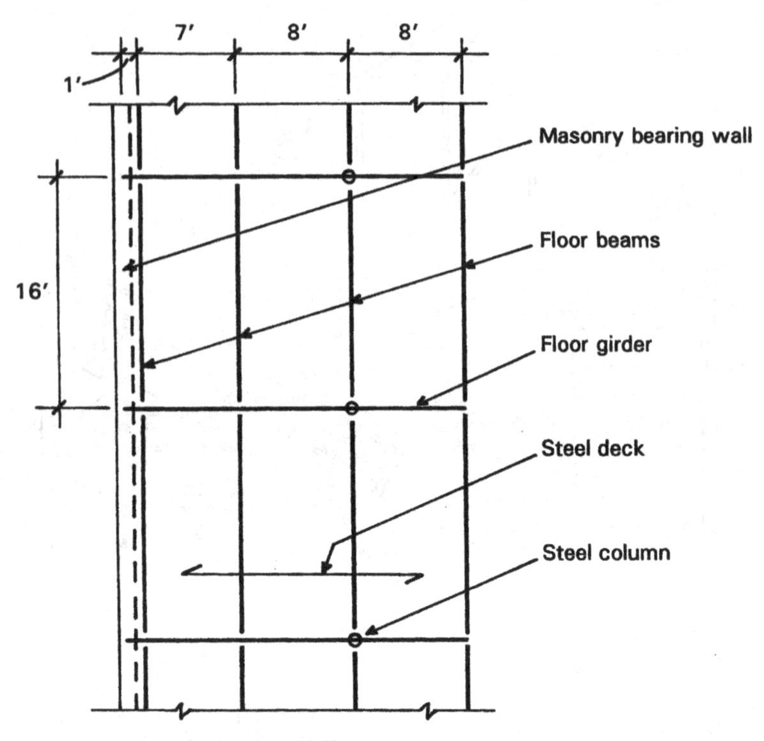

Figure 10.32 Building Four: Partial framing plan for the upper-level floor.

sign illustrations therefore use a reducible live load of 100 psf. As determined for Building Three, a total dead load for the floor is assumed to be 60 psf, including the weights of all the construction except for the steel girders and columns.

A possible choice for the deck is the same one used for Building Three, consisting of a sheet steel unit with 1.5-in.-high ribs and a concrete fill of 2.5 in. over the deck units. Units are securely attached to each other to develop a continuous deck and are welded to the top of the steel beams sufficiently to develop necessary diaphragm action for lateral loads.

Two types of beams are shown in the plan in Figure 10.32. The interior beams carry a full 8-ft-wide strip of the deck. At the exterior wall and at the edge of the balcony are beams that carry a one-half-wide (4-ft wide) strip of deck. The beam at the inside face of the exterior wall is used to prevent the wall from carrying the deck directly; thus, the only load transfer to the exterior wall is through the girder. W shapes are used for the interior beams and C, or channel, shapes for the edge beams; the latter for simplicity in the construction details.

For the interior beam, the ultimate loading for the floor is determined as

$$w_u = 1.2(60) + 1.6(100) = 232 \text{ psf}$$

and the linear load on the beam is

$$w_u = 8(232) = 1860 \text{ plf, or } 1.86 \text{ kips/ft}$$

For simple beam action, the maximum bending moment is

$$M_u = \frac{wL^2}{8} = \frac{1.86(16)^2}{8} = 59.5 \text{ kip-ft}$$

Table 3.1 indicates the possibility for a rather modest shape for this short span. Choices include W 12 × 16 and W 10 × 19. The deeper beams are lighter in weight but have very thin webs, so the details for end connections to the girders should be considered.

Using a yield stress of 36 ksi and an ultimate moment of 29.75 kip-ft, the required plastic section modulus for the channels is

$$Z = \frac{M_u}{\phi_b F_y} = \frac{29.75 \times 12}{0.9(36)} = 11.0 \text{ in.}^3$$

The AISC Manual (Ref. 3) indicates a possible choice is for a C 10 × 15.3, with a Z value of 15.9 in.^{3} Although carrying less load, this member is approximately the same weight as the W shapes for the interior beam, attesting to the structural efficiency of the W shapes.

Actual choices for these beams would be coordinated with the development of the general construction for the floor and ceiling, which must among other things provide fire protection for the steel structure.

For the girder, assume a live load of 80 psf. The load of one beam on the girder is thus:

$$P_u = (8 \times 16)\{(1.2 \times 60) + (1.6 \times 80)\} = 25{,}600 \text{ lb, or } 25.6 \text{ kips}$$

Adding a bit for the beam and girder weights, consider a design load of 26.5 kips. The girder loading and the resulting shear and moment values are thus as shown in Figure 10.33. Assuming a plastic yield failure, the lightest shape for this load is a W 16 × 26. However, lateral bracing of 8 ft is critical for this shape, so a heavier shape is indicated. The AISC Manual shows an adequate choice to be a W 16 × 36.

The critical bending moment for the girder is at the cantilever end. The cantilever is also the most critical for both static and dynamic (bounce) deflections. The 16-in.-deep shape may be adequate for static deflection and bending moment, but any extra depth will further stiffen the cantilever and reduce the possible bounciness of the cantilevered balcony. (If the lunch-hour crowd wants to use it as a running track, it probably needs at least a 27-in.-deep member.)

For approximate sizes for the columns, assume loads of 60 kips at the second story and 120 kips at the first story. Possible choices are for a

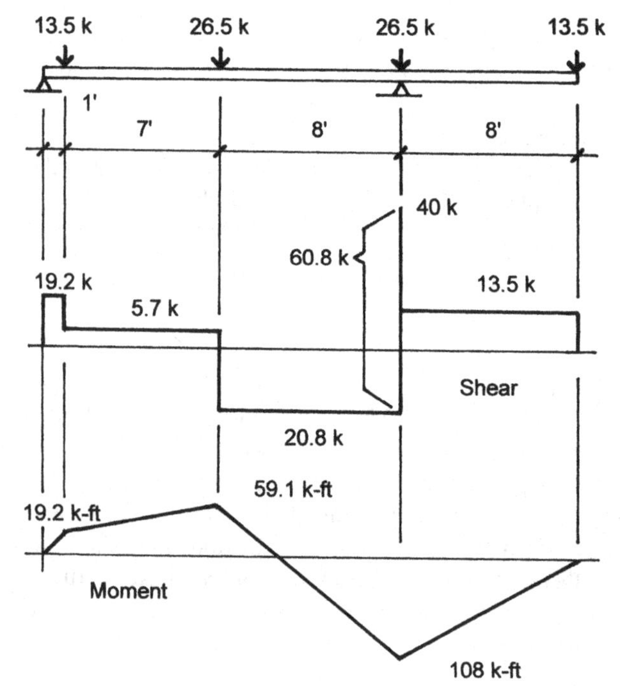

Figure 10.33 Investigation of the girder.

4-in. standard-weight pipe at the second story and a 6-in. standard pipe at the first story. Other shapes may also be used, such as a tube or a W shape.

A special construction detail problem for this structure is the generation of the cantilevered girders at the multistory columns at the second-floor level. Steel columns are ordinarily made continuous, with beams framing into their sides. For the detail as shown in Figure 10.31*d*, the girders are continuous and the columns are interrupted. This is a possibility for the relatively short three-story building, but it does require some extra consideration for the vertical compression in the girder web. (See discussion in Section 3.12.)

To prevent both bending in the girder flanges and vertical buckling of the girder web, a possible detail is the addition of a thick plate into the open U-shaped area defined by the girder flanges and web on each side of the girder. The individual story-high column segments are then fitted with bearing plates at their tops and bottoms, which are bolted to the girders to hold them in place during the construction period.

The Steel Roof Truss

As shown in Figure 10.31, the roof structure uses 60-ft-span trusses to provide a column-free interior on the top floor. The detail in Figure 10.31*c* indicates the use of a steel truss with top chords formed from structural tees (cut from W shapes) and interior members formed with double angles. Joints are formed mostly by welding the angle ends directly to the webs of the tee chords. Special joints are required to facilitate erection at the supports as well as at any interior field splice locations if the trusses cannot be shipped to the site in one piece.

Figure 10.34 shows gravity load values for the top chord joints with a unit loading. The values found for internal forces in the truss members with this loading can simply be multiplied by the true unit loading as determined from the final construction details.

The roof surface is generated by a system of purlins that span between trusses and support a steel deck. To avoid excessively heavy top chords, the purlins should deliver their end reaction loads to the top chord joints of the truss. This makes for a bit of a busy traffic problem of connections at this location, but it can probably be handled by experienced steel detailers. Placing the purlins on top of the chords would simplify the intersection problem, but at the expense of introducing a rollover stability problem for the tilted purlins.

Figure 10.35 shows wind load values based on criteria from the UBC.

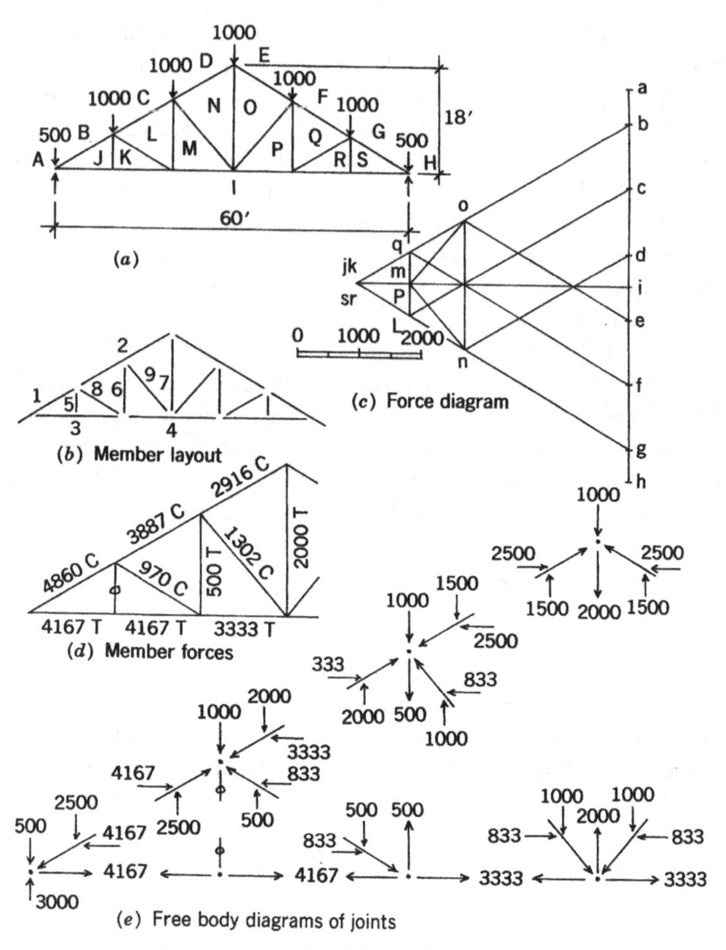

Figure 10.34 Investigation of the roof truss for gravity loading.

The form of loading is based on the roof slope and a minimum horizontal wind pressure of 20 psf at the roof level.

The results of the investigations in Figures 10.34 and 10.35 are summarized in Table 10.4. With the sloped roof surface, the load combinations likely to be critical for design consideration are as follows:

1.2 times dead load plus 1.6 times live load

Dead load plus wind load when the forces in the member are of the same sign (tension or compression)

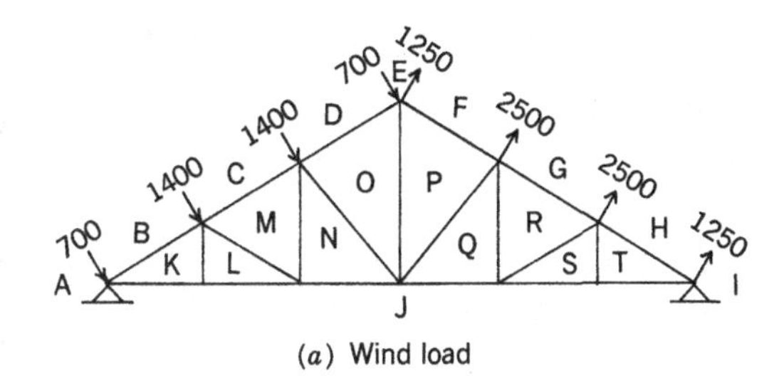

(*a*) Wind load

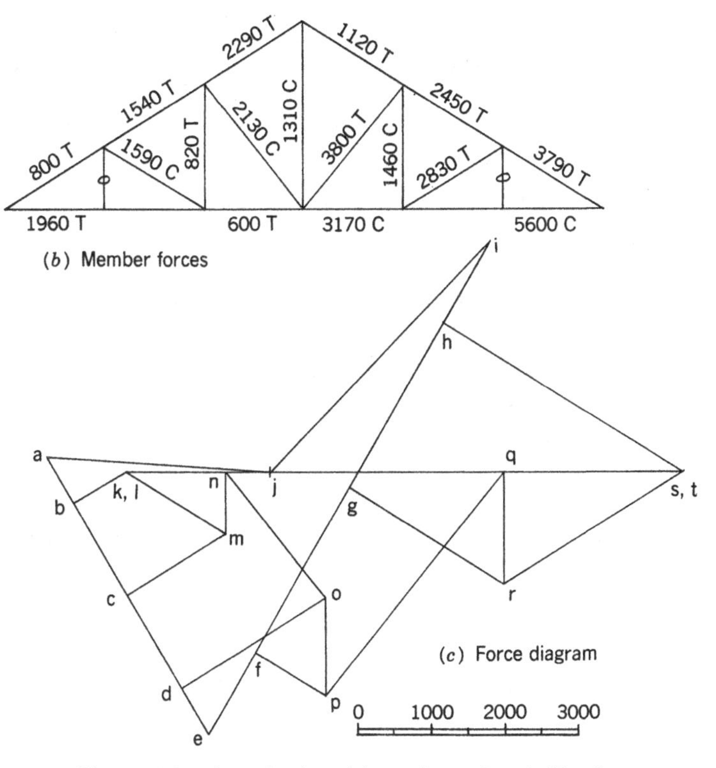

(*b*) Member forces

(*c*) Force diagram

Figure 10.35 Investigation of the roof truss for wind loading.

Dead load plus wind load when the net is a reversal of the sign of the internal force produced by gravity load alone

A complete analysis here will show that the wind load is not a critical concern for this structure, and the truss members and connections may be designed for the gravity loads alone.

TABLE 10.4 Design Forces for the Truss (in Pounds)

Member (see Figure 10.31)	Unit Gravity Load	Dead Load (3.6 × Unit)	Live Load (1.8 × Unit)	Wind Loan	1.2(LL) + 1.6(LL)
1	4860 C	17,496 C	8748 C	3790 T	35,000 C
2	3887 C	13,994 C	6997 C	2450 T	28,000 C
3	4167 T	15,000 T	7500 T	1960 T/5600 C	30,000 T
4	3333 T	12,000 T	6000 T	3170 C	24,000 T
5	(Zero force for all loadings)				
6	500 T	1800 T	900 T	820 T/1460 C	3600 T
7	2000 T	7200 T	3600 T	1310 C	14,400 T
8	970 C	3492 C	1746 C	1590 C/2830 T	7000 C
9	1302 C	4688 C	2344 C	2130 C/3800 T	9375 C

A summary of the design of the truss members for the loads in Table 10.4 is shown in Table 10.5. Selection of members must relate to several concerns, including the following:

1. Minimum thickness for interior angle members, based on the use of fillet welds for the truss joints (see Section 8.7)
2. Minimum sizes of angle pairs, based on maximum permitted L/r ratios: 120 for compression members and 200 for tension members
3. Layout of truss joints to avoid eccentricity of forces at the joints and to provide sufficient dimension for necessary welds

With members tentatively selected, joints must be drawn to scale for the layouts to verify the feasibility of the construction. On a first try, this may well indicate that something does not work and some reconsideration of the basic system is required—for instance, a closer spacing of trusses to reduce the truss loading, a reconsideration of the truss member shapes or the joint forms, or using a manufactured truss adapted for this situation by the truss producers and their designers. Such drawings save a lot of design time and ensure a smooth erection process.

Design for Lateral Forces

With the construction as shown in Figure 10.31, the vertical bracing system used for this building would most likely be the exterior structural masonry walls. Participation of the steel structure would consist of using the floor and roof construction for horizontal diaphragm functions. Although

TABLE 10.5 Design of the Truss Members

No.	Force (kips)	Length (ft)	Choice (All Double-Angles)
1	35 C	11.7	Combined bending and compression member $6 \times 4 \times \frac{1}{2}$
2	28 C	11.7	$6 \times 4 \times \frac{1}{2}$
3	30 T	10	Max. $L/r = 300$, $r = 0.4$ $3 \times 2\frac{1}{2} \times \frac{3}{8}$
4	24 T	10	$3 \times 2\frac{1}{2} \times \frac{3}{8}$
5	0	6	$2\frac{1}{2} \times 2\frac{1}{2} \times \frac{3}{8}$
6	3.6 T	12	Max. $L/r = 300$, $r = 0.48$ $2\frac{1}{2} \times 2\frac{1}{2} \times \frac{3}{8}$
7	14.4 T	18	Max. $L/r = 300$, $r = 0.72$ $2\frac{1}{2} \times 2\frac{1}{2} \times \frac{3}{8}$
8	7 C	11.7	Max. $L/r = 200$, $r = 0.7$ $3\frac{1}{2} \times 2\frac{1}{2} \times \frac{3}{8}$
9	4.4 C	15.6	Max. $L/r = 200$, $r = 0.94$ $3 \times 2\frac{1}{2} \times \frac{3}{8}$

the sloping roof deck might be used for this purpose, it is more likely that the bracing would be more effectively developed by using horizontal-plane trussing at the level of the bottom chords of the trusses.

The floor structures at the second- and third-floor levels consist of donut-shaped surfaces, as shown in Figure 10.36. This structure has a range of behaviors as a horizontal diaphragm, based on the relative size of the holes. It is difficult to establish clear lines of differentiation between the cases, although the extreme cases are reasonably certain, as illustrated in Figure 10.37.

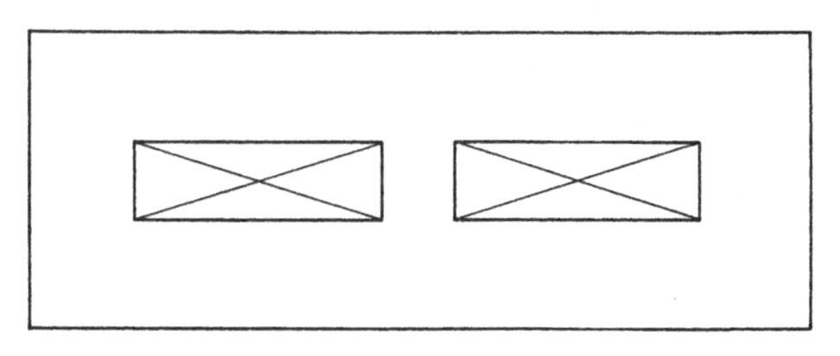

Figure 10.36 Plan of the upper-level floor diaphragm.

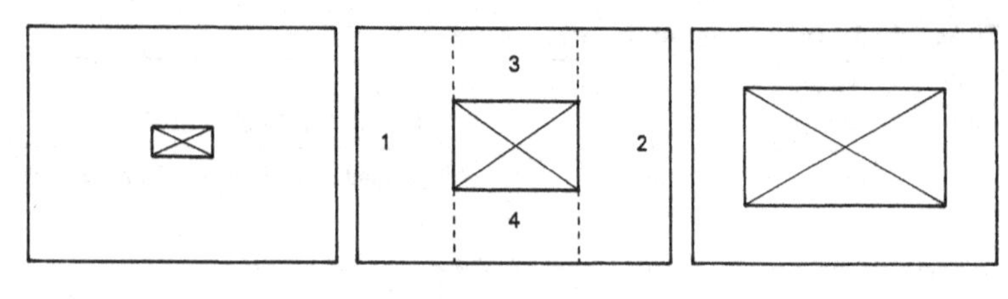

Figure 10.37 Range of effects of a hole in a horizontal diaphragm.

A possible approach for this horizontal structure is as follows:

1. Design for the diaphragm shear in the deck on the usual basis, using the net width at each location for the shear stress investigation.
2. Design the individual subdiaphragms for their independent actions, as shown in Figure 10.37.
3. Investigate for the chord forces in the whole diaphragm and the subdiaphragms to ensure that the steel framing can develop the forces.

As with the roof trusses, the horizontal structure must work first for the gravity loads. If lateral forces are minor (as they actually are for this example), the load combination with lateral forces may not be critical for design, except to ensure that detailing of the construction provides the necessary anchorages and the continuity of framing that performs chord, collector, tie, and drag strut functions.

10.5 BUILDING FIVE

As shown in Figure 10.38, Building Five consists of an open-sided, canopy roof structure. Vertical support for the roof is provided by eight columns. Each column is developed as a built-up steel member formed of three steel plates welded together with fillet welds. The columns are joined to the gabled roof edge members that achieve the 40-ft span and the 10-ft corner cantilever at the side of the building.

The individual T-shaped edge frame unit is shown in Figure 10.39. There are eight such units that form the four framed bents. These bents

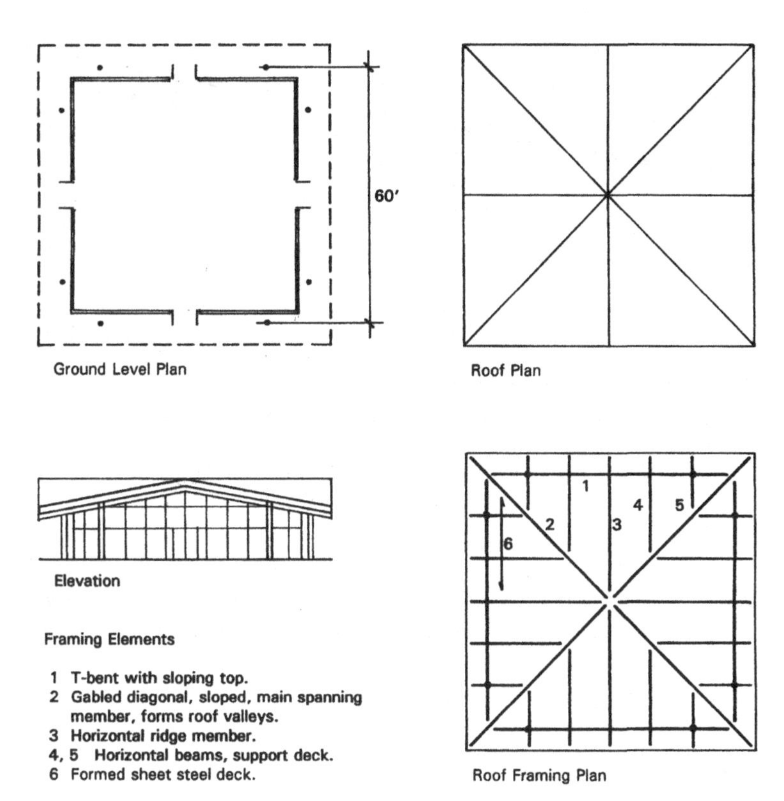

Figure 10.38 Building Five: General form.

provide both vertical support for the rest of the roof construction and lateral bracing for the building. Essential to the latter function is the development of the joint between the bent column and the gabled member as a moment-resisting joint. These bents are actually very similar in nature to those shown for the gabled roof example for Building One in Section 10.1. (See Figure 10.11.)

The interior portion of the roof is developed with the framing shown in the layout in Figure 10.38*d*. The gabled diagonals (No. 2 in the plan) span approximately 42 ft from the corner to the center and rise 10 ft. These are simple span elements that carry concentrated loads consisting of the end reactions from members 4 and 5. Their investigation is simple, including that for the horizontal thrust at the corner. The latter is resisted by tension in the edge framing members.

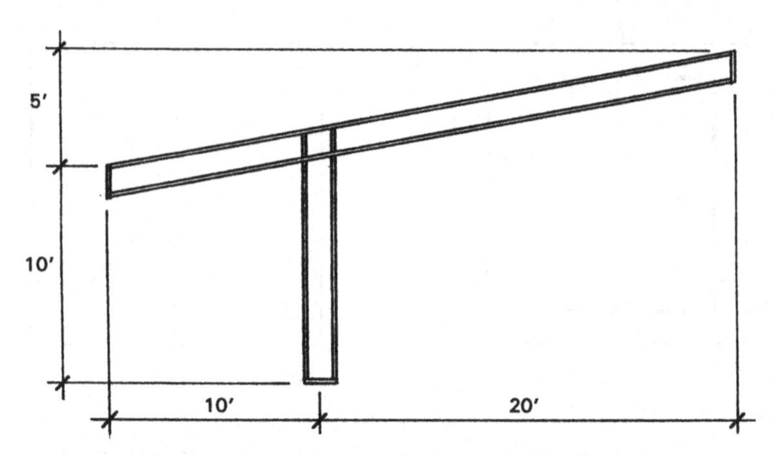

Figure 10.39 Form of the individual rigid frame bent unit.

Members 3, 4, and 5 are all simple beams with a generally uniform load, modified to some degree by the triangular load areas. There are some special connections to be designed for this structure, but they are not uncommon and can be reasonably easily achieved with special elements made from bent and welded steel plates.

10.6 BUILDING SIX

This section presents some possibilities for developing the roof structure and the exterior walls for a medium-size sports arena, one big enough for a swim stadium or a basketball court (see Figure 10.40). Options for the structure are strongly related to the desired building form.

General Design Requirements

Functional planning requirements derive from the specific activities to be housed and from the seating, internal traffic, overhead clearance, and exit and entrance arrangements. In spite of all of the requirements, there is usually some room for consideration of a range of alternatives for the general building plan and overall form. Choice of the truss system shown here, for example, relates to a commitment to a square plan and a flat roof profile. Other choices for the structure may permit more flexibility in the building form or also limit it. Selection of a domed roof, for example, would pretty much require a round plan.

In addition to the long-span structure in this case, there is also a major problem in developing the 42-ft-high curtain wall. Braced laterally only at the top and bottom, this is a 42-ft span structure sustaining wind pressure as a loading. With even modest wind conditions, producing pressures in the 20-psf range, this requires some major vertical mullion structure to span the 42 ft.

In this example, the fascia of the roof trusses, the soffit of the overhang, and the curtain wall are all developed with products ordinarily available for curtain wall construction. The vertical span of the tall wall is developed by a series of custom-designed trusses that brace the major vertical mullions. These vertical mullions are themselves custom-developed to relate to the standard units of the wall system, something that might be done by the supplier of the curtain wall system.

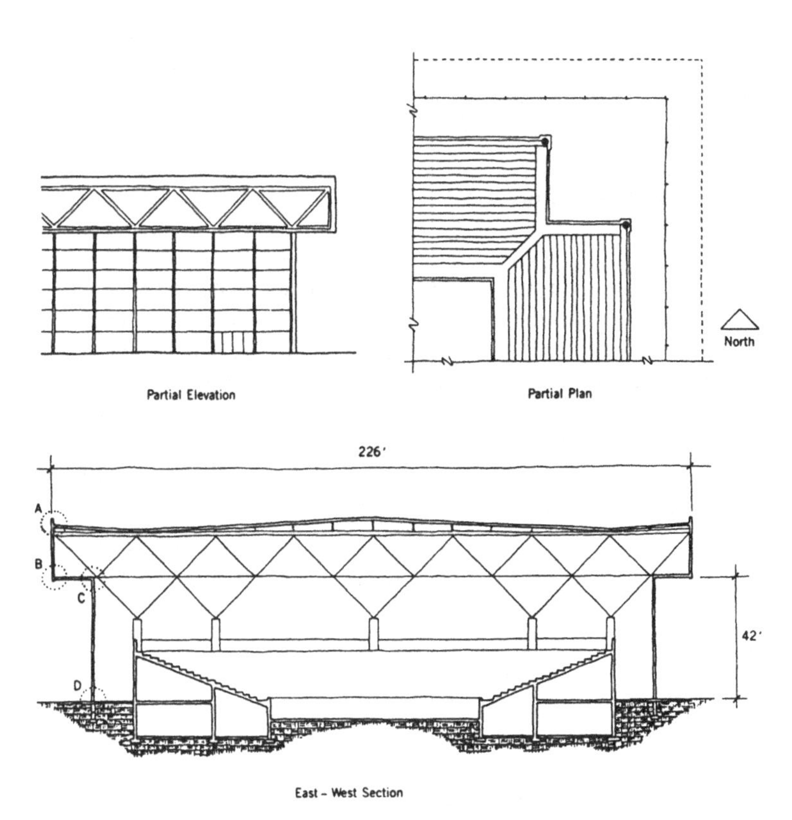

Figure 10.40 Building Six: General form.

As an exposed structure, the truss system is a major visual element of the building interior. The truss members define a pattern that is orderly and pervading. There will, however, most likely be many additional items overhead—within, and possibly beneath, the trusses. These may include the following:

Leaders and other elements of the roof drainage system
Ducts and registers for the HVAC system
A general lighting system
Signs, scoreboards, and so on
Elements of an audio system
Catwalks for access to the various equipment

To preserve some design order, these should be related to the truss geometry and detailing, if possible. However, some amount of independence is to be expected, especially with items installed or modified after the building is completed.

Structural Alternatives

The general form of the construction for Building Six is shown in Figure 10.41. Discussion for this building will be limited to the development of the roof structure and the tall exterior glazed wall. Some proposed details for the exterior wall are shown in Figure 10.42. The details indicate the use of a standard, proprietary window wall system for the wall surface development, with a custom-designed steel support structure occurring behind the wall and inside the building. The steel support structure for the wall is thus weather-protected by the glazed wall, which is ordinarily developed as a weather-resistive construction system.

The 168-ft clear span used here is definitely in the class of long-span structures, but not so great as to severely limit the options. A flat-spanning beam system is definitely out, but a one-way or two-way truss system is a feasible choice for a flat span. Most other structural options generally involve some form other than a flat profile; this includes domes, arches, shells, folded plates, suspended cables, cable-stayed systems, and pneumatic systems.

The structure shown in Figure 10.40 uses a two-way spanning steel truss system of the form described as an *offset grid*. The basic planning of this type of structure requires the development of a module relating to

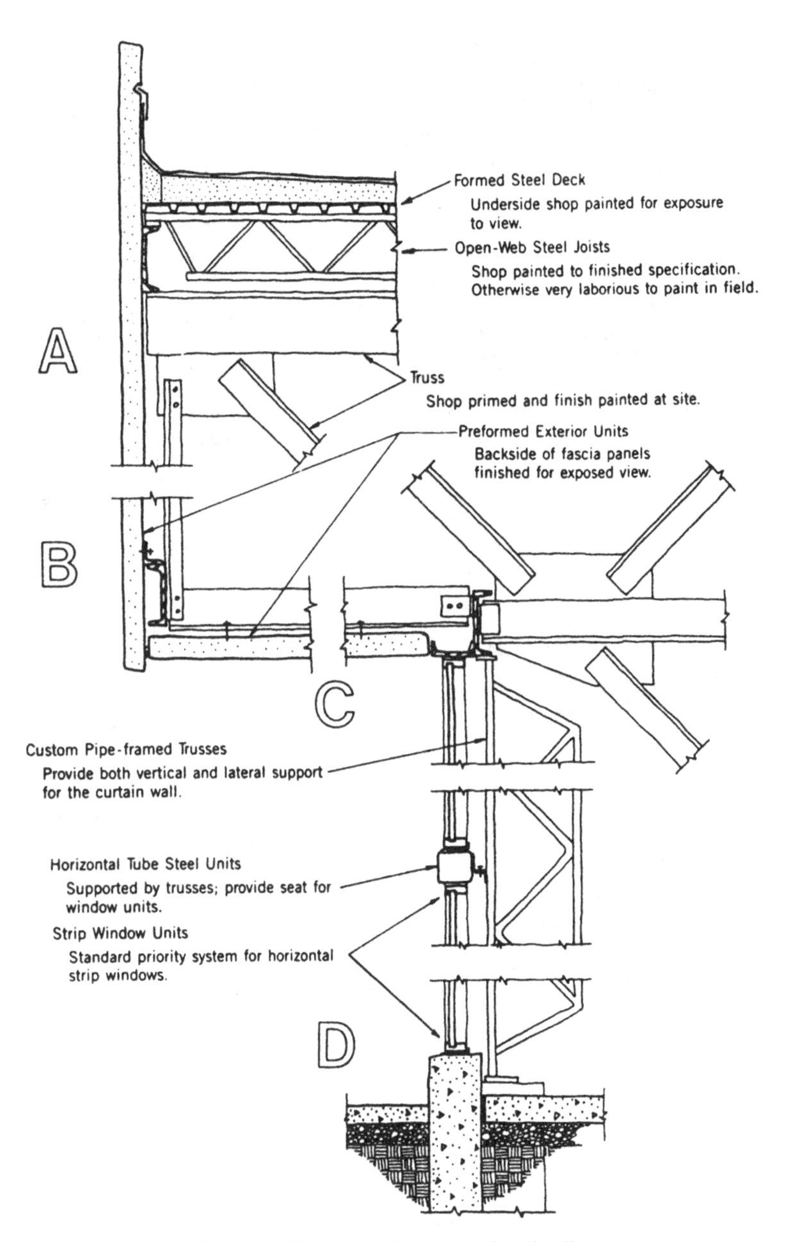

Figure 10.41 General construction details.

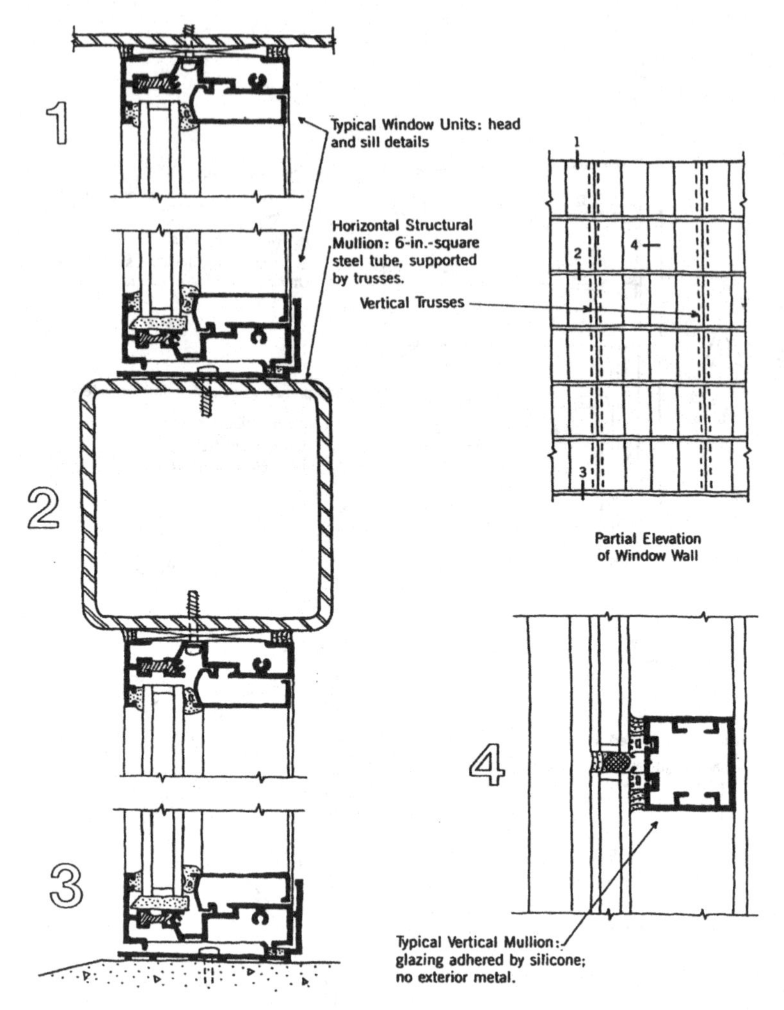

Figure 10.42 Wall construction details.

the frequency of nodal points (joints) in the truss system. Supports for the truss must be provided at nodal points, and any concentrated loads should preferably be applied at nodal points. Although the nodal point module relates basically to the formation of the truss system, its ramifications in terms of supports and loads typically extend it to other aspects of the building planning. At the extreme, this extension of the module may be used throughout the building—as has largely been done in this example.

The basic module here is 3.5 ft, or 42 in. Multiples and fractions of this basic dimension (X) are used throughout the building, in two and three dimensions. The truss nodal module is actually 8X or 28 ft. The height of the exterior wall, from ground to the underside of the truss, is 12X or 42 ft. And so on.

This is not exactly an ordinary building, although the need for such a building for various purposes has created many examples. There is an inevitable necessity for some innovation here, unless an exact duplicate of some previous example is used, which is not the usual case. There is nothing particularly unique about the construction shown here, but it cannot really be called common or standard.

Even in unique buildings, however, it is common to use as many standard products as possible. Thus, the roof structure on top of the truss and the curtain wall system for the exterior walls use off-the-shelf products (see Figures 10.41 and 10.42). The supporting columns and general seating structure may also be of conventional construction.

The Window Wall Support Structure

The window wall system consists of a metal frame that is assembled from modular units with glazing inserted in the framed openings. The frame is designed to support the glazing and is itself self-supporting to a degree. When used for ordinary multistory buildings, it is usually capable of spanning the clear height of a single story (12 to 15 ft or so), although some slightly stronger vertical mullions than shown here might be used. It is thus basically a complete structure for a single-story wall.

In this situation, however, the 42-ft height is about three times the height of a normal single story. Thus, some additional support structure is indicated for both the vertical gravity weight of the wall and for lateral forces (probably from wind, because the wall is quite light in weight). The details in Figures 10.41 and 10.42 show the use of a two-component system consisting of trusses that span the 42-ft height and are spaced 14 ft on center and horizontal steel tubes that span the 14 ft between trusses and support 7 ft of vertical height of the wall.

It is possible, because of the relatively light weight of the wall, that the vertical loads are actually carried by the closely spaced vertical mullions. In this case, the primary function of the horizontal steel tubes is to span between trusses and resist the wind forces on 14 ft of the wall. The tube shown here is certainly adequate for this task, although stiffness is probably more critical than bending stress in this situation. It is not really good for the wall construction to flip-flop during either seismic movements or fast changes in wind direction.

Although not shown in the drawings, the form of the truss would probably be that of a so-called *delta truss,* with two chords opposing a single one, thus creating a triangular form in cross section (like the Greek capital letter delta). A principal advantage of this form of truss is its relatively high lateral stability, permitting it to be used without the usual required cross-bridging.

The Truss System

The truss system shown here might be produced from various available proprietary systems, produced by different manufacturers. In general, the designer should pursue the availability of these when beginning the design of such a structure. A completely custom-designed structure of this kind requires enormous investments of design time for basic development and planning of the structural form, development of nodal joint construction, and reliable investigation of structural behavior of the highly indeterminate structure. Developments of assemblage and erection processes are themselves a major design problem.

If the project deserves this effort, the time for the design work is available, and the budget can absorb the cost, the end result might justify the expenditures. But if what is really desired and needed can be obtained with available products and systems, a lot of time and money can most likely be saved.

Aside from the considerations of the plan form, planning module, and development of the supports, there is the basic issue of the particular form and general nature of the truss structure. The square plan and general biaxial plan symmetry seem to indicate the logic for a two-way spanning system here. This is indeed what is shown in Figure 10.40. The particular system form here is called an *offset grid,* which describes the relationship between the layouts of the top and bottom chords of the truss system. The squares of the top chords are offset from those of the bottom chords so that the top chord nodules (joints) lie over the center of the bottom squares. As a result, there are no vertical web members and generally no vertical planar sets in the system.

At this point, refer to the discussion of two-way trusses in Section 7.8.

The remaining discussion of Building Six will make several references to issues and illustrations presented in that section.

The general form of the building as shown in Figure 10.40 can be developed with a considerable variety of truss systems for the roof. This section discusses several options for the truss system, all of which can achieve the form as proposed for this building. Basic economics would surely figure heavily in a decision regarding the final solution, although many other factors also would be considered.

Development of the Roof Infill System

All of the schemes for the roof truss system shown here involve the providing of a grid of truss chords at 28-ft centers in both directions at the general level of the roof surface (top of the trusses). This provides a roof support system, but it does not develop a roof as such. Something else must be done to achieve the infill surface-developing construction, with its top surface consisting of some form of water-resistive roofing.

Although some imaginative two-way spanning system might be developed for this structure, it is also possible to use various simpler systems merely to fill in the 28-ft-span voids. Figure 10.43 shows a very simple system consisting of simple-span steel open-web joists and formed sheet steel decking. For the simplest construction, the joists would use only the truss chords in one direction for support, thus creating a somewhat unsymmetrical loading for the two-way system. Except for the related connection details, however, this would most likely not affect the basic development of the larger truss system.

The steel deck could be exposed on its underside, along with the open-web joists. However, for extra thermal insulation and possibly for sound control, it may be desirable to provide a ceiling surface at the bottom of the open-web joists.

Roof drainage is also a problem that relates to the form of the two-way truss system. If the open-web joists are placed as shown in Figure 10.43 (the usual installation), it may be necessary to slope the two-way truss system top chords to achieve roof drainage. This may indicate the logic of using some additional elements between the two-way trusses and the roof infill structure, permitting the two-way system to be flat on top—surely a simplifying condition for the truss system.

This basic infill system is assumed to be reasonable for all of the following proposed alternatives for the two-way system.

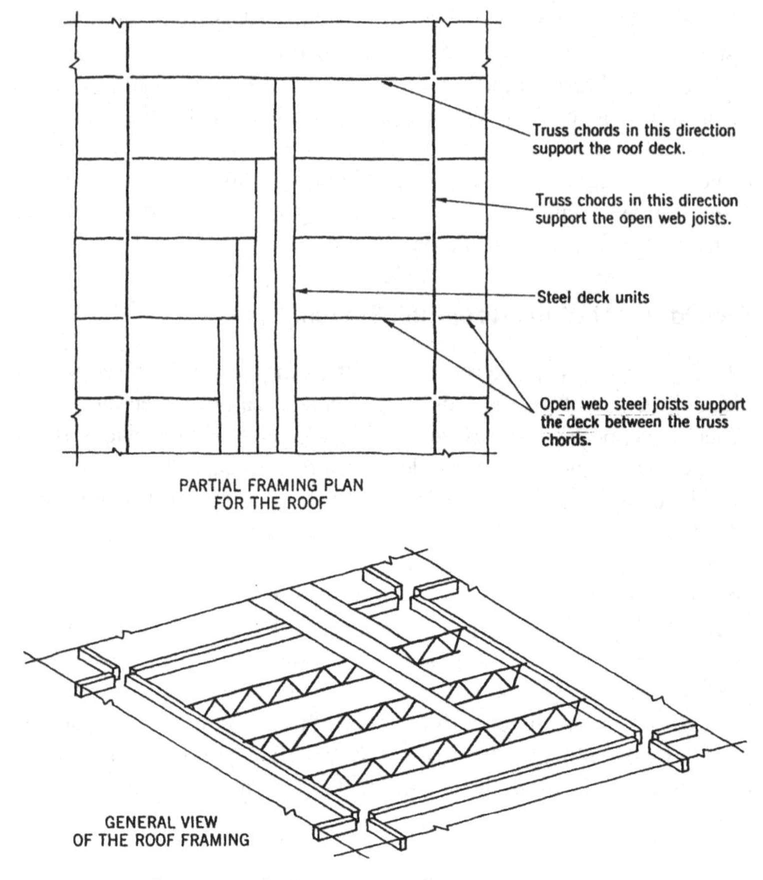

Figure 10.43. Development of the roof infill system.

Alternative One: The Offset Grid System

Figures 10.44 and 10.45 show the form of an offset grid system for the truss roof for Building Six. The plan shows the placement of the grid system resulting in the locations of supports beneath top chord joints. This is the general form indicated in the drawings in Figures 10.44 and 10.45.

The drawing in Figure 10.45 indicates the use of three columns on each side of the structure, providing a total of 12 supports for the truss system. The tops of the columns are dropped below the spanning truss to permit

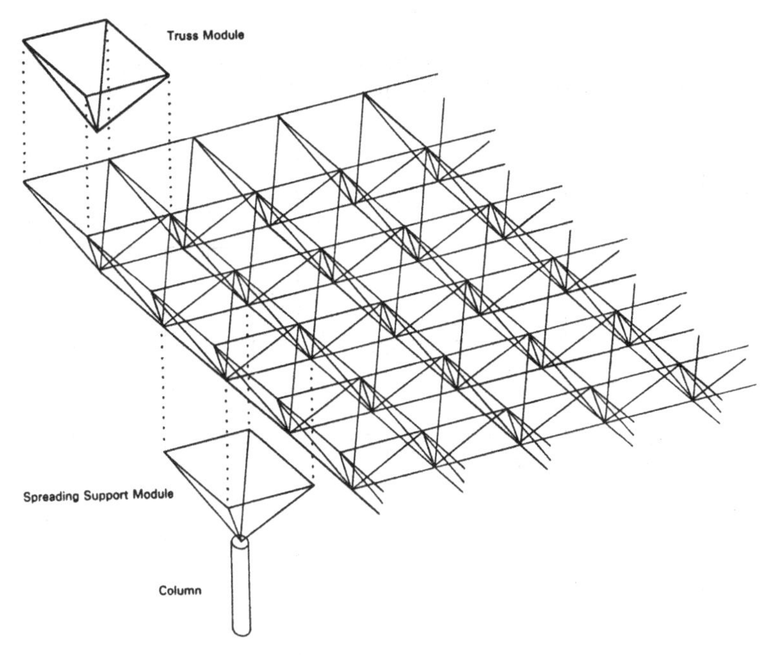

Figure 10.44 General form of the offset grid system.

the use of a pyramidal module of four struts between the top of the column and the bottom chords of the truss system. This considerably reduces the maximum shear required by the truss interior members, with the entire gravity load being shared by 12 times 4 equals 48 struts. If the total vertical factored design gravity load is approximately 140 psf (0.14 kips/ft^2), the load in a single diagonal column strut is thus approximately:

$$P_u = \frac{(226)^2(0.14)}{48} = 149 \text{ kips}$$

These struts are approximately 28 ft long and are inclined at 45 degrees. Thus, the design factored compression force for the strut is

$$C = \frac{1.414 P_u}{\phi} = \frac{1.414(149)}{0.85} = 248 \text{ kips}$$

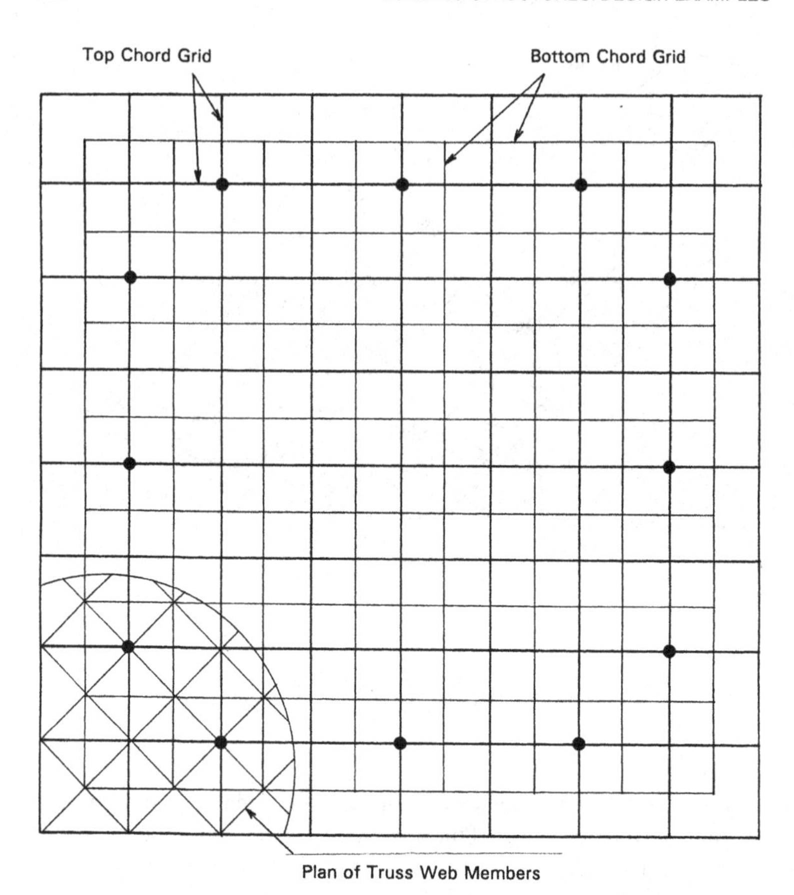

Figure 10.45 Plan of the offset grid system.

Because each strut picks up a truss node with four interior diagonals, the maximum compression force in each diagonal is 248/4 = 62 kips. Using steel pipe, some options for these members are as follows:

For the struts, from Table 4.4: 10-in. std. pipe, 8-in. XXS pipe
For the truss diagonals: 6-in. std. pipe, 5-in. XXS pipe

The closely spaced edge columns plus the struts constitute almost a continuous edge support for the truss system, with only a minor edge cantilever. The spanning task is thus essentially that of a simple beam span in two directions. For an approximation, we may consider the span in each direction to carry half the load. Thus, taking half the clear span width of 168 ft as a middle strip, the total load for design of the "simple beam" is

$$W = (\text{span width})(\text{span length})(0.14 \text{ kips/ft}^2)(1/2)$$
$$= (84)(168)(0.14)(1/2) = 988 \text{ kips}$$

and the simple beam moment at midspan is

$$M = \frac{WL}{8} = \frac{988(168)}{8} = 20{,}748 \text{ kip-ft}$$

Sharing this with three top chord members in the middle strip, and assuming a center-to-center chord depth to be approximately 19 ft, the force in a single chord is

$$C = \frac{20{,}748}{(0.85)(3)(19)} = 428 \text{ kips}$$

If the compression chord is unbraced for its 28-ft length, options are as follows:

(No options for pipe from Table 4.4, although the AISC Manual has larger pipes.)
From Table 4.3, with $F_y = 50$ ksi: W 12 × 79, W 10 × 112

For the bottom chords, the critical problem will be development of joints at nodes or splices. It is unlikely that chord members will be more than a single module (28 ft) long, so each joint must be fully developed with welds or bolts.

These are not record sizes for large steel structures, but it does seem to indicate some strong efforts for reduction of the design loads (lightest possible general roof construction). It also probably means some variation of sizes will be used in the truss with some minimal members used for low-stress situations.

As discussed elsewhere in this book, this is a highly indeterminate structure, although its symmetry and the availability of computer-aided procedures for investigation make its design accessible to most professional structural designers.

Alternative Two: The Two-Way Vertical Planar Truss System

A second possibility for the truss form is shown in Figure 10.46 and consists of perpendicular, intersecting sets of vertical planar trusses. In this system, the top chord grid squares are directly above the bottom chord grid squares.

Figure 10.47 shows a plan layout for this system in which the vertical truss planes are offset from the columns. The principal structural reason for this is to permit the use of the spread unit at the column, similar to the one used in the example for Alternative One. This unit does not relate to the basic truss system form as it did in the preceding example, but it could take various shapes to fulfill its task. As shown here, it could literally be quite similar to the one for the offset grid structure.

Approximation of the chord forces for this two-way spanning system could be made in a manner similar to that for the preceding example. An advantage here is the possibility for use of interior vertical members to reduce the lateral support problem for the chords. This is possible with the addition of some members in the offset grid system, but not so neatly

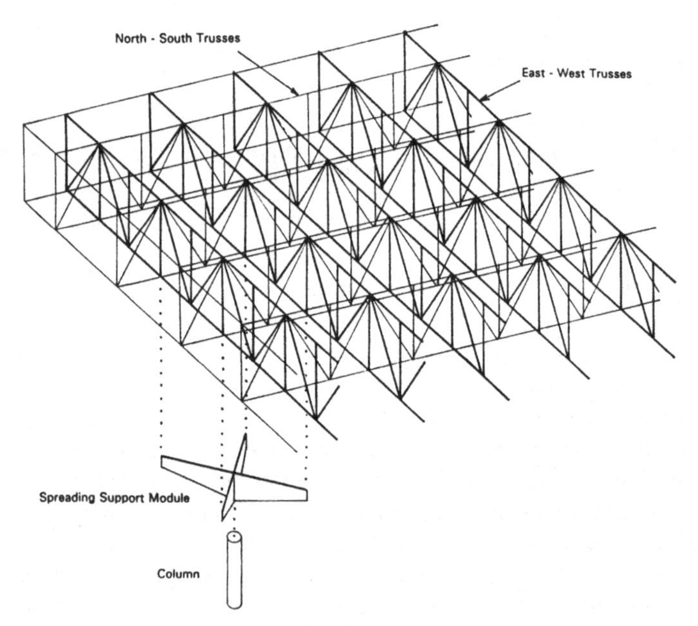

Figure 10.46 General form of the two-way truss system with vertical planar trusses.

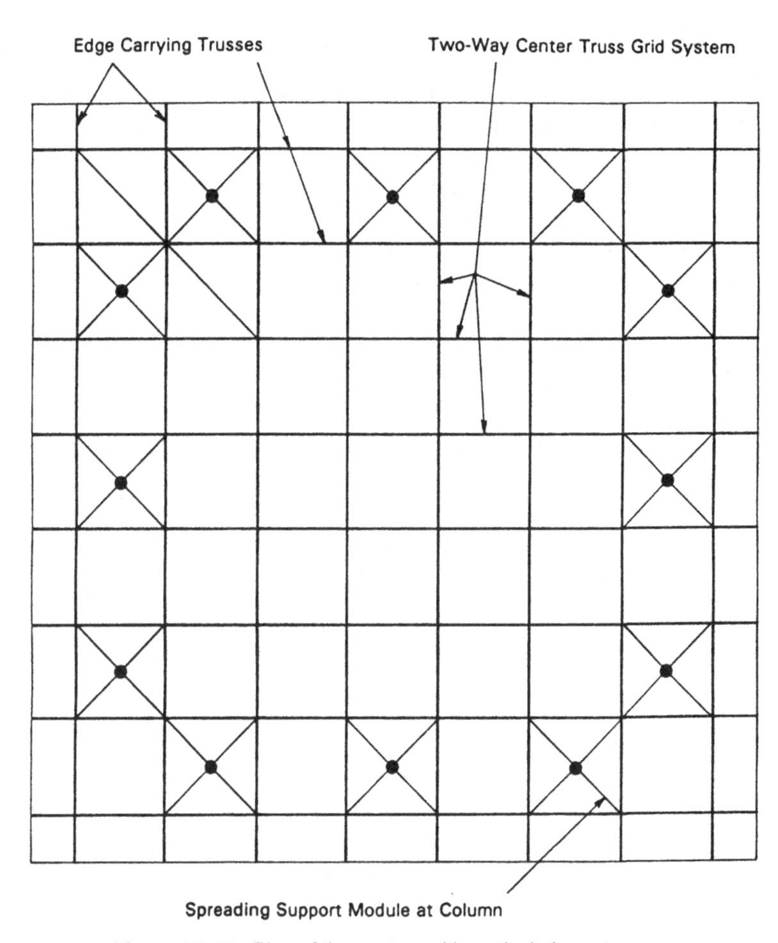

Figure 10.47 Plan of the system with vertical planar trusses.

achieved. Using the same force approximations as determined for the chords in the preceding example, but with unsupported lengths of only 14 ft, some considerably smaller members will be obtained.

For both of the two-way systems, a major consideration is the planning for the erection. Major considerations are what size and shape of unit can be assembled in the shop and transported to the site and what temporary support must be provided. Working this out may well relate to the design of the truss jointing details.

Alternative Three: The One-Way System

Figure 10.48 shows the form of a system that uses a set of one-way spanning, planar trusses to achieve a system with a general appearance very similar to that of the system for Alternative Two. In fact, the general truss form is identical; the difference consisting of the manner in which trusses are individually formed and joints are achieved.

In this example, the span is achieved by the set of trusses spanning in one direction, whereas the cross-trussing is used only for spanning be-

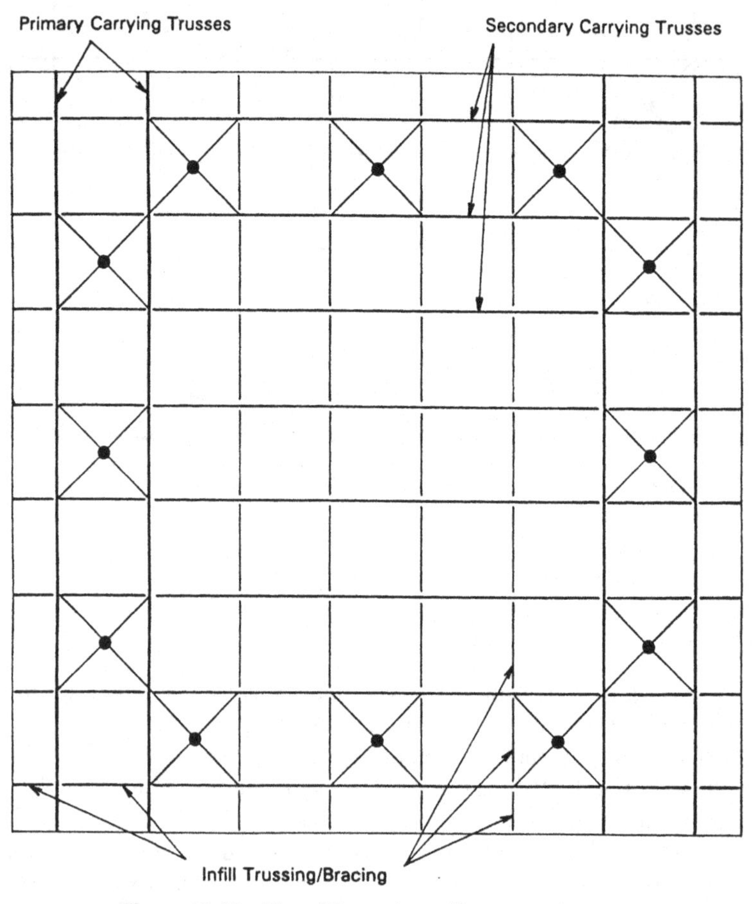

Figure 10.48 Plan of the system with one-way trusses.

tween the carrying trusses and providing lateral bracing for the system. The cross-trussing also cantilevers to develop the facia and soffit on two sides of the building.

This system lends itself to relatively simple design procedures, because the main trusses are simple, planar, determinate trusses. Although we will not do so, we could fully illustrate their investigation and design with the simple procedures developed in this book. This is maybe not a compelling reason for choosing this scheme, but it is food for thought when one considers the complexity of investigations of highly indeterminate systems.

The 28-ft on center carrying trusses will be slightly heavier than the trusses in Alternative Two, because their share of the load is slightly more in the one-way structure. However, this is compensated for by the minor structural demands for the cross-trussing system. The designer would have to make a philosophical decision about how far to go to make the structure appear to be symmetrical in the otherwise biaxially symmetrical building. In reality, however, if no effort is made, most nonprofessionals will probably never notice the lack of symmetry, unless someone points it out. This is most likely true of most of the subtlety we design into our buildings; it is lost on all but our fellow professionals.

A principal potential advantage for this scheme is its simplified assemblage and erection. Single carrying trusses may be erected in one piece with very little temporary support necessary. Once two trusses that straddle a column are in place, the development of the cross-trussing can begin, serving both temporary and permanent bracing functions. This may be the single critical deciding factor for favoring this scheme.

Appendix A

PROPERTIES OF STRUCTURAL SECTIONS

This appendix deals with various geometric properties of planar (two-dimensional) areas. The areas referred to are the cross-sectional areas of structural members. These geometric properties are used for the analysis of stresses and deformations in the design of the structural members.

A.1 CENTROIDS

The *center of gravity* of a solid is the point at which all of its weight can be considered to be concentrated. Because a planar area has no weight, it has no center of gravity. The point in a planar area that corresponds to the center of gravity of a very thin plate of the same area, and shape is called the *centroid* of the area. The centroid is a useful reference for various geometric properties of planar areas.

For example, when a beam is subjected to a bending moment, the materials in the beam above a certain plane in the beam are in compression and the materials below the plane are in tension. This plane is the *neutral stress plane,* also called the neutral surface or the zero stress plane (see Section 3.2). For a cross section of the beam, the intersection of the neutral stress plane is a line that passes through the centroid of the section and is called the *neutral axis* of the section. The neutral axis is very important for investigation of bending stresses in beams.

The centroid for symmetrical shapes is located on the axis of symmetry for the shape. If the shape is bisymmetrical—that is, it has two axes of symmetry—the centroid is at the intersection of these axes. Consider the rectangular area shown in Figure A.1*a;* obviously, its centroid is at its geometric center and is quite easily determined.

(*Note:* Tables A.3 through A.7 and Figure A.11, referred to in the discussion that follows, are located at the end of this appendix.)

For more complex forms, such as those of rolled steel members, the centroid will also be on any axis of symmetry. And as for the simple rectangle, if there are two axes of symmetry, the centroid is readily located.

For simple geometric shapes, such as those shown in Figure A.1, the location of the centroid is easily established. However, for more complex shapes, the centroid and other properties may have to be determined by computations. One method for achieving this is by use of the *statical moment,* defined as the product of an area times its distance from some reference axis. Use of this method is demonstrated in the following examples.

Example 1. Figure A.2 is a beam cross section that is unsymmetrical with respect to a horizontal axis (such as X-X in the figure). The area is symmetrical about its vertical centroidal axis, but the true location of the centroid requires the locating of the horizontal centroidal axis. Find the location of the centroid.

Solution: Using the statical moment method, first divide the area into units for which the area and location of the centroid are readily deter-

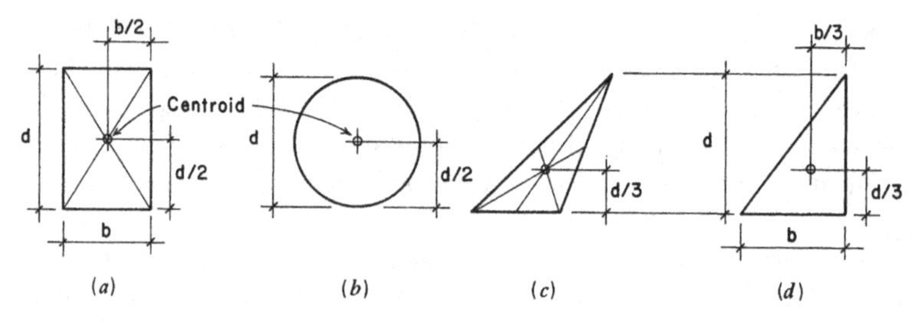

Figure A.1 Centroids of various planar shapes.

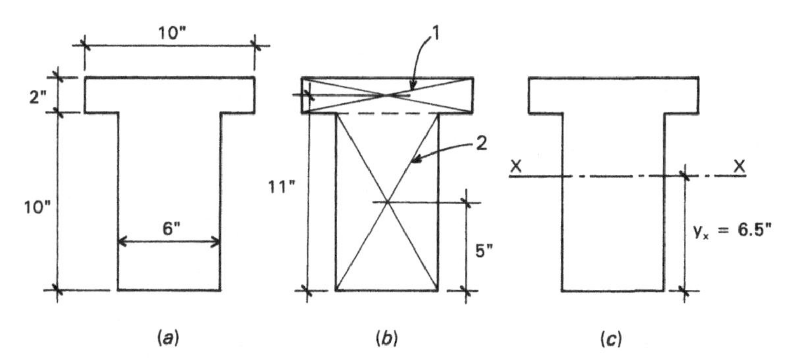

Figure A.2 Reference figure for Example 1.

mined. The division chosen here is shown in Figure A.2*b* with the two parts labeled 1 and 2.

The second step is to choose a reference axis about which to sum statical moments and from which the location of the centroid is readily measured. A convenient reference axis for this shape is one at either the top or bottom of the shape. With the bottom chosen, the distances from the centroids of the parts to this reference axis are shown in Figure A.2*b*.

We then determine the unit areas and their statical moments. This work is summarized in Table A.1, which shows the total area to be 80 in.^{2} and the total statical moment to be 520 in.^{3} Dividing this moment by the total area produces the value of 6.5 in., which is the distance from the reference axis to the centroid of the whole shape, as shown in Figure A.2*c*.

Problems A.1.A–F. Find the location of the centroid for the cross-sectional areas shown in Figure A.3. Use the reference axes indicated and compute the distances from the axes to the centroid, designated as c_x and c_y, as shown in Figure A.3*b*.

Table A.1 Summary of Computations for Centroid: Example 1

Part	Area (in.^{2})	y (in.)	$A \times y$ (in.^{3})
1	$2 \times 10 = 20$	11	220
2	$6 \times 10 = 60$	5	300
Σ	80		520

$$y_x = \frac{520}{80} = 6.5 \text{ in.}$$

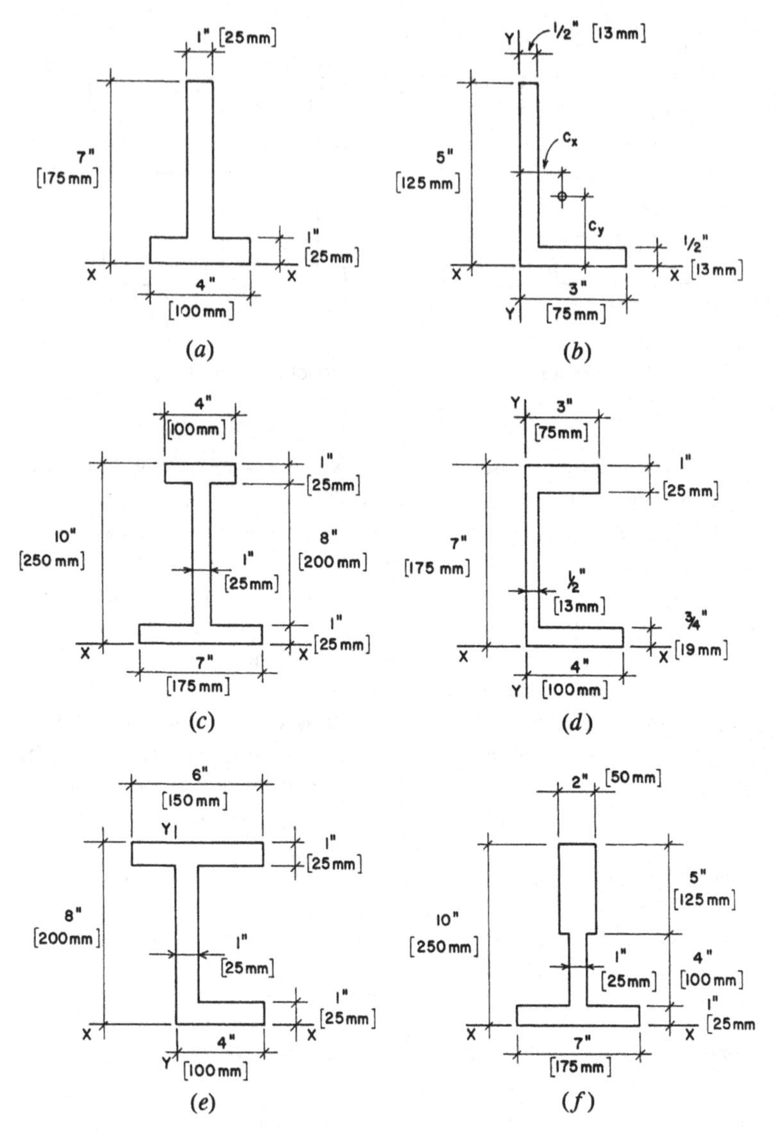

Figure A.3 Reference for Problem A.1.

A.2 MOMENT OF INERTIA

Consider the area enclosed by the irregular line in Figure A.4*a.* In this area, designated *A,* a small unit area *a* is indicated at *z* distance from the axis marked X-X. If this unit area is multiplied by the square of its distance from the reference axis, the result is the quantity az^2. If all of the units of the area are thus identified and the sum of these products is made, the result is defined as the *second moment* or the *moment of inertia* of the area, designated as *I.* Thus:

$$\Sigma az^2 = I, \text{ or specifically } I_{X\text{-}X}$$

which is the moment of inertia of the area about the X-X axis.

The moment of inertia is a somewhat abstract item, less able to be visualized than area, weight, or center of gravity. It is, nevertheless, a real geometric property that becomes an essential factor for investigating stresses and deformations due to bending. Of particular interest is the moment of inertia about a centroidal axis and, most significantly, about a principal axis for the shape. Figures A.4*b, c, e,* and *f* indicate such axes for various shapes. Tables A.3 through A.7 show the properties of moment of inertia about the principal axes of the shapes in the tables.

Moment of Inertia of Geometric Figures

Values for moment of inertia can often be obtained from tabulations of structural properties. Occasionally it is necessary to compute values for a given shape. This may be a simple shape, such as a square, rectangle, circle, or triangle. For such shapes, simple formulas are derived to express the value for the moment of inertia.

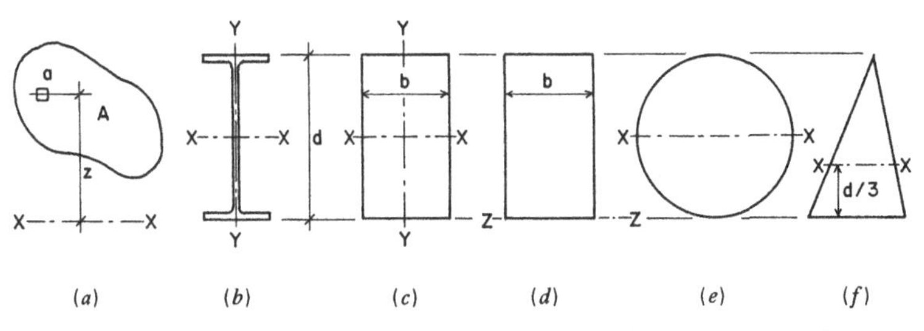

Figure A.4 Consideration of reference axis for the moment of inertia of various shapes of cross sections.

Rectangle. Consider the rectangle shown in Figure A.4*c*. Its width is b and its depth is d. The two principal axes are X-X and Y-Y, both passing through the centroid of the area. For this case, the moment of inertia with respect to the centroidal axis X-X is

$$I_{X\text{-}X} = \frac{bd^3}{12}$$

and the moment of inertia with respect to the Y-Y axis is

$$I_{Y\text{-}Y} = \frac{db^3}{12}$$

Example 2. Find the value of the moment of inertia for a 6-in.-×-12-in. wood beam about an axis through its centroid and parallel to the narrow dimension.

Solution: As listed in standard references for wood products, the actual dimensions of the section are 5.5 × 11.5 in. Then:

$$I = \frac{bd^3}{12} = \frac{5.5(11.5)^3}{12} = 697.1 \text{ in.}^4$$

which is in agreement with the value for $I_{\text{X-X}}$ in references.

Circle. Figure A.4*e* shows a circular area with diameter d and axis X-X passing through its center. For the circular area the moment of inertia is

$$I = \frac{\pi d^4}{64}$$

Example 3. Compute the moment of inertia of a circular cross section, 10 in. in diameter, about its centroidal axis.

Solution: The moment of inertia is

$$I = \frac{\pi d^4}{64} = \frac{3.1416(10)^4}{64} = 490.9 \text{ in.}^4$$

Triangle. The triangle in Figure A.4*f* has a height h and a base width b. The moment of inertia about a centroidal axis parallel to the base is

$$I = \frac{bh^3}{36}$$

Example 4. If the base of the triangle in Figure A.4*f* is 12 in. wide and the height from the base is 10 in., find the value for the centroidal moment of inertia parallel to the base.

Solution: Using the given values in the formula, we solve as

$$I = \frac{bh^3}{36} = \frac{12(10)^3}{36} = 333.3 \text{ in.}^4$$

Open and Hollow Shapes. Values of moment of inertia for shapes that are open or hollow may sometimes be computed by a method of subtraction. The following examples demonstrate this process. Note that this is possible only for shapes that are symmetrical.

Example 5. Compute the moment of inertia for the hollow box section shown in Figure A.5*a* about a centroidal axis parallel to the narrow side.

Solution: Find first the moment of inertia of the shape defined by the outer limits of the box.

$$I = \frac{bd^3}{12} = \frac{6(10)^3}{12} = 500 \text{ in.}^4$$

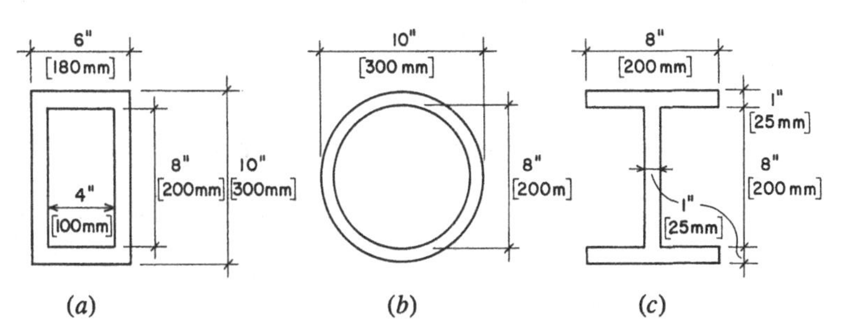

Figure A.5 Reference figures for Examples 5, 6, and 7.

Then find the moment of inertia for the shape defined by the void area.

$$I = \frac{4(8)^3}{12} = 170.7 \text{ in.}^4$$

The value for the hollow section is the difference, thus:

$$I = 500 - 170.7 = 329.3 \text{ in.}^4$$

Example 6. Compute the moment of inertia about the centroidal axis for the pipe section shown in Figure A.5*b*. The thickness of the shell is 1 in.

Solution: As in the preceding example, the two values may be found and subtracted. Or a single computation may be made as follows:

$$I = \frac{\pi}{64}(d_o^4 - d_i^4) = \frac{3.1416}{64}(10^4 - 8^4) = 491 - 201 = 290 \text{ in.}^4$$

Example 7. Referring to Figure A.5*c*, compute the moment of inertia of the I-shape section about the centroidal axis parallel to the flanges.

Solution: This is essentially similar to the computation for Example 5. The two voids may be combined into a single one that is 7 in. wide. Thus:

$$I = \frac{8(10)^3}{12} - \frac{7(8)^3}{12} = 667 - 299 = 368 \text{ in.}^4$$

Note that this method can only be used when the centroids of the outer shape and the void coincide. For example, it cannot be used to find the moment of inertia for the I-shape about its vertical centroidal axis. For this computation, the method discussed in the next section must be used.

A.3 TRANSFERRING MOMENTS OF INERTIA

Determination of the moment of inertia of unsymmetrical and complex shapes cannot be done by the simple processes illustrated in the preceding examples. An additional step that must be used involves the transfer of moment of inertia about a remote axis. The formula for achieving this transfer is as follows:

$$I = I_o + Az^2$$

where I = moment of inertia of the cross section about the required reference axis

I_o = moment of inertia of the cross section about its own centroidal axis, parallel to the reference axis

A = area of the cross section

z = distance between the two parallel axes

These relationships are illustrated in Figure A.6, where X-X is the centroidal axis of the area and Y-Y is the reference axis for the transferred moment of inertia.

Application of this principle is illustrated in the following examples.

Example 8. Find the moment of inertia of the T-shaped area in Figure A.7 about its horizontal (X-X) centroidal axis. (*Note:* The location of the centroid for this section was solved as Example 1 in Section A.1.)

Solution: A necessary first step in these problems is to locate the position of the centroidal axis if the shape is not symmetrical. In this case, the T shape is symmetrical about its vertical axis, but not about the horizontal axis. Locating the position of the horizontal axis was the problem solved in Example 1 in Section A.1.

The next step is to break the complex shape down into parts for which centroids, areas, and centroidal moments of inertia are readily found. As was done in Example 1, the shape here is divided between the rectangular flange part and the rectangular web part.

The reference axis to be used here is the horizontal centroidal axis. Table A.2 summarizes the process of determining the factors for the par-

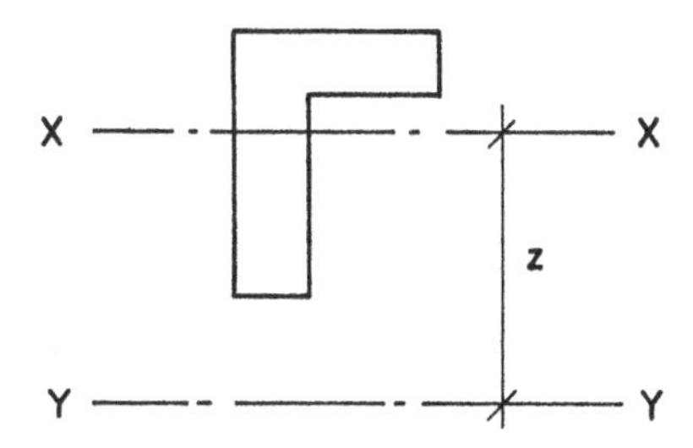

Figure A.6 Transfer of moment of inertia to a parallel axis.

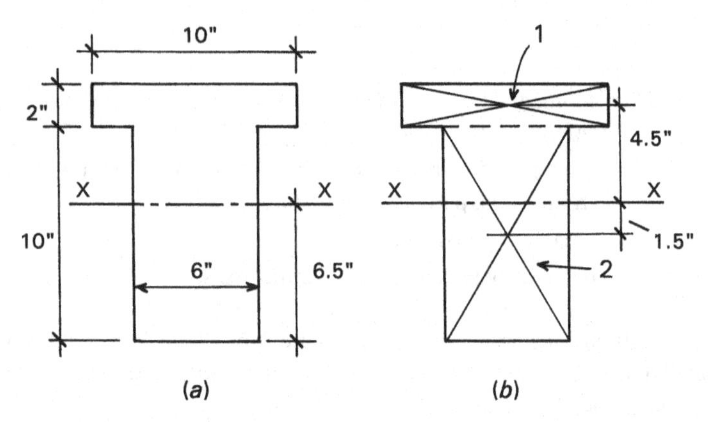

Figure A.7 Reference for Example 8.

allel axis transfer process. The required value for I about the horizontal centroidal axis is determined to be 1046.7 $in.^4$

A common situation in which this problem must be solved is in the case of structural members that are built up from distinct parts. One such section is that shown in Figure A.8, where a box-shaped cross section is composed by attaching two plates and two rolled channel sections. Although this composite section is actually symmetrical about both its principal axes, and the locations of these axes are apparent, the values for moment of inertia about both axes must be determined by the parallel axis transfer process. The following example demonstrates the process.

Example 9. Compute the moment of inertia about the centroidal X-X axis of the built-up section shown in Figure A.8.

Solution: For this situation, the two channels are positioned so that their centroids coincide with the reference axis. Thus, the value of I_o for the channels is also their actual moment of inertia about the required reference axis, and their contribution to the required value here is simply two

Table A.2 Summary of Computations for Moment of Inertia: Example 8

Part	Area ($in.^2$)	y (in.)	I_o ($in.^4$)	$A \times y^2$ ($in.^4$)	I_x ($in.^4$)
1	20	4.5	$10(2)^3/12 = 6.7$	$20(4.5)^2 = 405$	411.7
2	60	1.5	$6(10)^3/12 = 500$	$60(1.5)^2 = 135$	635
Σ					1046.7

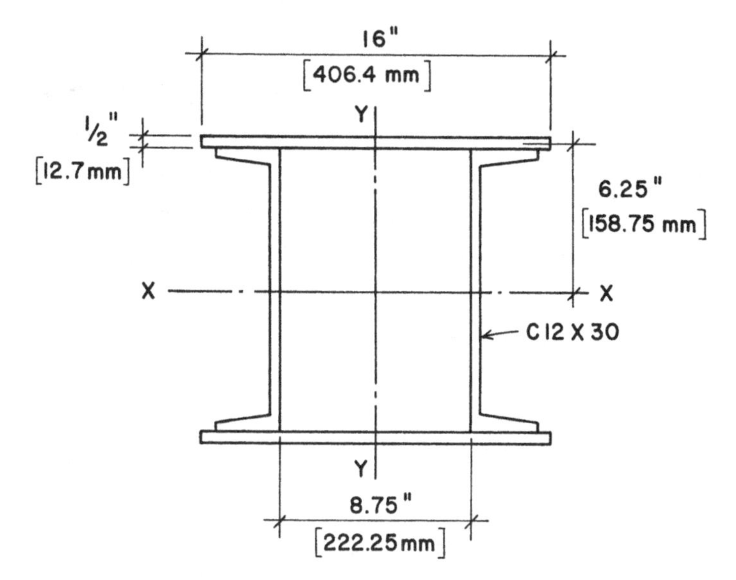

Figure A.8 Reference for Example 9.

times their listed value for moment of inertia about their X-X axis, as given in Table A.4: 2(162) = 324 in.^{4}

The plates have simple rectangular cross sections, and the centroidal moment of inertia of one plate is thus determined as:

$$I_o = \frac{bd^3}{12} = \frac{16 \times (0.5)^3}{12} = 0.1667 \text{ in.}^4$$

The distance between the centroid of the plate and the reference X-X axis is 6.25 in. And the area of one plate is 8 in.^{2} The moment of inertia for one plate about the reference axis is thus:

$$I_o + Az^2 = 0.1667 + (8)(6.25)^2 = 312.7 \text{ in.}^4$$

and the value for the two plates is twice this, or 625.4 in.^{4}

Adding the contributions of the parts, the answer is 324 + 625.4 = 949.4 in.^{4}

Problems A.3.A–F. Compute the moments of inertia about the indicated centroidal axes for the cross-sectional shapes in Figure A.9.

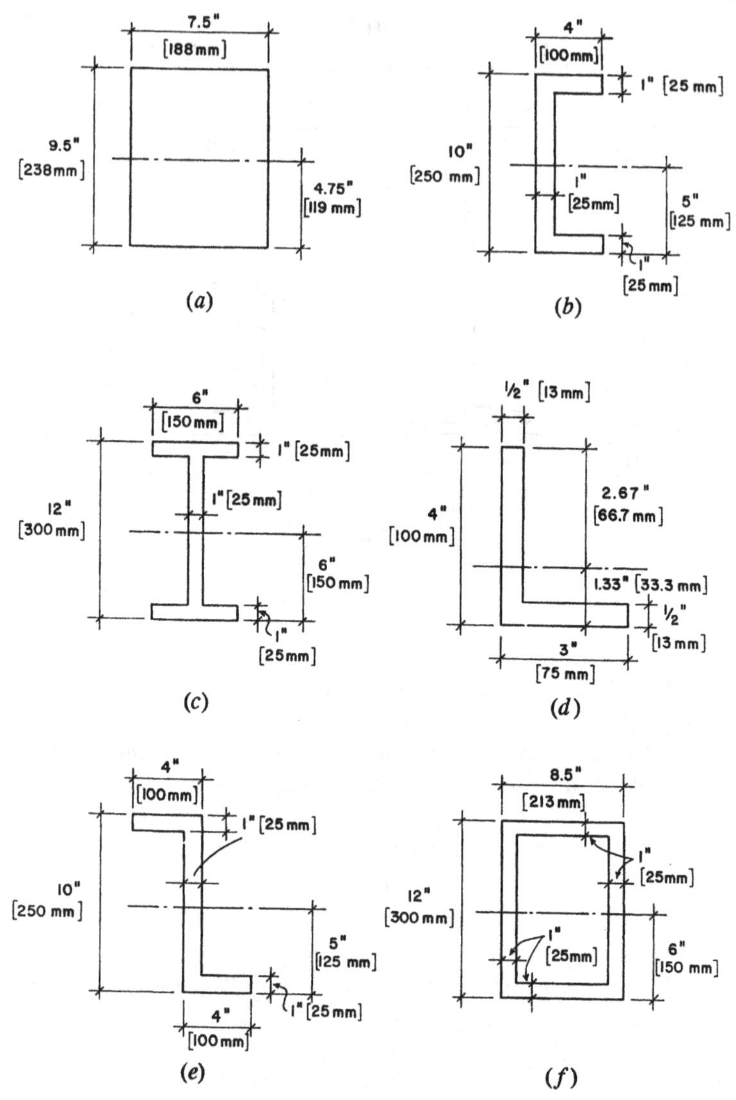

Figure A.9 Reference for Problems A.3.A–F.

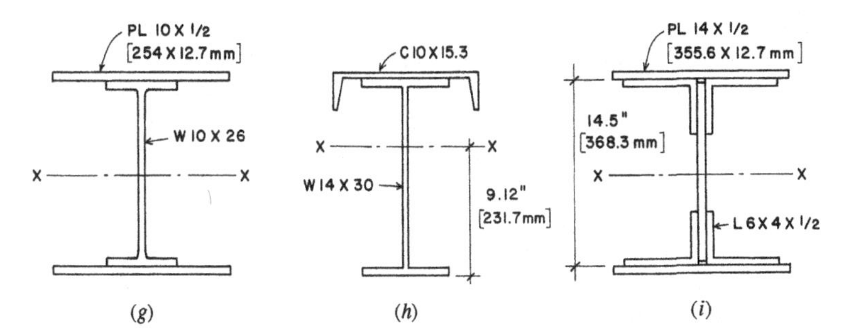

Figure A.10 Reference for Problems A.3.G–I.

Problems A.3.G–I. Compute the moments of inertia with respect to the centroidal X-X axes for the built-up sections in Figure A.10. Make use of any appropriate data from the tables of properties for steel shapes.

A.4 MISCELLANEOUS PROPERTIES

Elastic Section Modulus

As noted in Section 3.2, the term I/c in the formula for flexural stress is called the *elastic section modulus,* designated S. Using the section modulus permits a minor shortcut in the computations for flexural stress or the determination of the bending moment capacity of members. However, the real value of this property is in its measure of relative bending strength of members. As a geometric property, it is a direct index of bending strength for a given member cross section. Members of various cross sections may thus be rank-ordered in terms of their bending strength strictly on the basis of their S values. (See Table B.1 in Appendix B.) Because of its usefulness, the value of S is listed together with other significant properties in the tabulations for steel and wood members.

For members of standard form (structural lumber and rolled steel shapes), the value of S may be obtained from tables similar to those presented at the end of this chapter. For complex shapes not of standard form, the value of S must be computed, which is readily done once the centroidal axes are located and moments of inertia about the centroidal axes are determined.

Example 10. Verify the tabulated value for the section modulus of a 6×12 wood beam about the centroidal axis parallel to its narrow side.

Solution: The actual dimensions of this member are 5.5 × 11.5 in. And the value for the moment of inertia is 697.068 in.^{4} Then:

$$S = \frac{I}{c} = \frac{697.068}{5.75} = 121.229 \text{ in.}^3$$

which agrees with the value in reference sources.

Radius of Gyration

For design of slender compression members, an important geometric property is the *radius of gyration,* defined as:

$$r = \sqrt{\frac{I}{A}}$$

Just as with moment of inertia and section modulus values, the radius of gyration has an orientation to a specific axis in the planar cross section of a member. Thus, if the I used in the formula for r is that with respect to the X-X centroidal axis, then that is the reference for the specific value of r.

A value of r with particular significance is that designated as the *least radius of gyration.* Because this value will be related to the least value of I for the cross section, and because I is an index of the bending stiffness of the member, then the least value for r will indicate the weakest response of the member to bending. This relates specifically to the resistance of slender compression members to buckling. Buckling is essentially a sideways bending response, and its most likely occurrence will be on the axis identified by the least value of I or r. Use of these relationships is discussed for columns in Chapter 4.

Plastic Section Modulus

The plastic section modulus, designated Z, is used in a similar manner to the elastic stress section modulus S. The plastic modulus is used to determine the fully plastic stress moment capacity of a steel beam. Thus:

$$M_p = F_y \times Z$$

The use of the plastic section modulus is discussed in Section 3.3.

A.5 TABLES OF PROPERTIES OF SECTIONS

Figure A.11 presents formulas for obtaining geometric properties of various simple plane sections. Some of these may be used for single-piece structural members or for the building up of complex members.

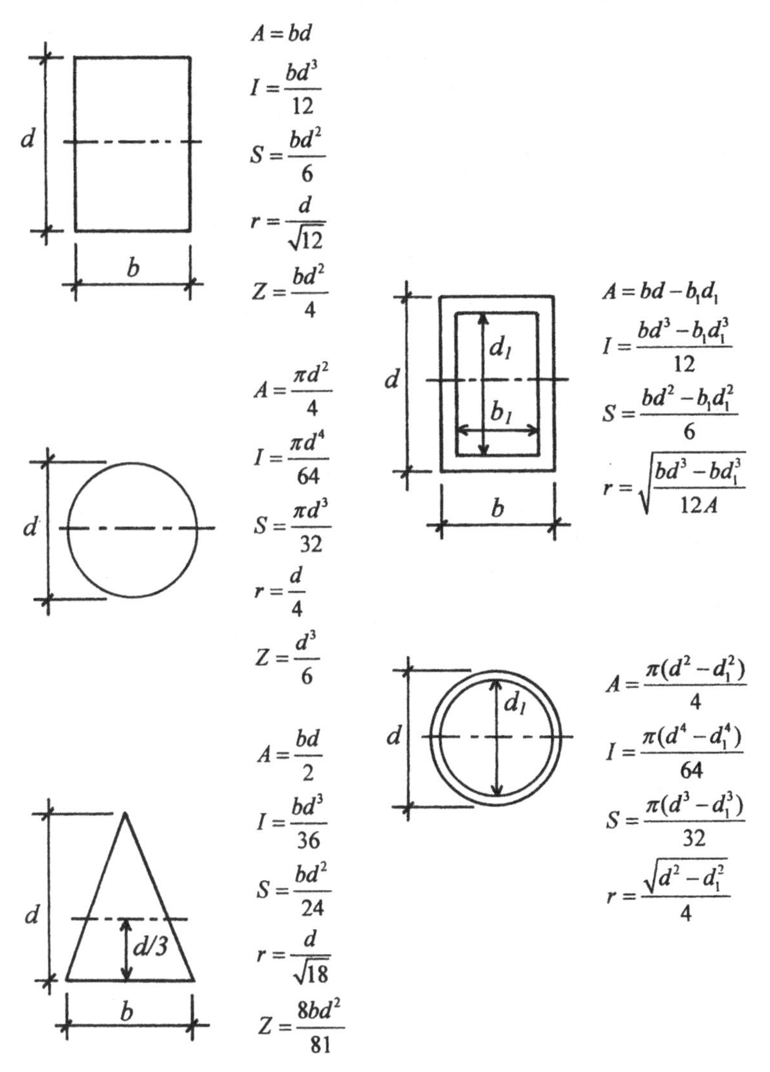

Figure A.11 Properties of various geometric shapes.

Tables A.3 through A.7 present the properties of various plane sections. These are sections identified as those of standard industry-produced sections of steel. Standardization means that the shapes and dimensions of the sections are fixed, and each specific section is identified in some way.

Structural members may be employed for various purposes, and thus they may be oriented differently for some structural uses. Of note for any plane section are the *principal axes* of the section. These are the two, mutually perpendicular, centroidal axes for which the values will be greatest and least, respectively, for the section; thus, the axes are identified as the major and minor axes. If sections have an axis of symmetry, it will always be a principal axis—either major or minor.

For sections with two perpendicular axes of symmetry (rectangle, H, I, etc.), one axis will be the major axis and the other the minor axis. In the tables of properties, the listed values for I, S, and r are all identified as to a specific axis, and the reference axes are identified in a figure for the table.

Other values given in the tables are for significant dimensions, total cross-sectional area, and the weight of a 1-ft-long piece of the member. The weight of steel members is given for W and channel shapes as part of their designation; thus, a W 8×67 member weighs 67 lb/ft. For steel angles and pipes, the weight is given in the table, as determined from the density of steel at 490 lb/ft^3.

The designation of some members indicates their true dimensions. Thus, a 10-in. channel and a 6-in. angle have true dimensions of 10 and 6 in. For W shapes and pipe, the designated dimensions are *nominal,* and the true dimensions must be obtained from the tables.

TABLE A.3 Properties of W Shapes

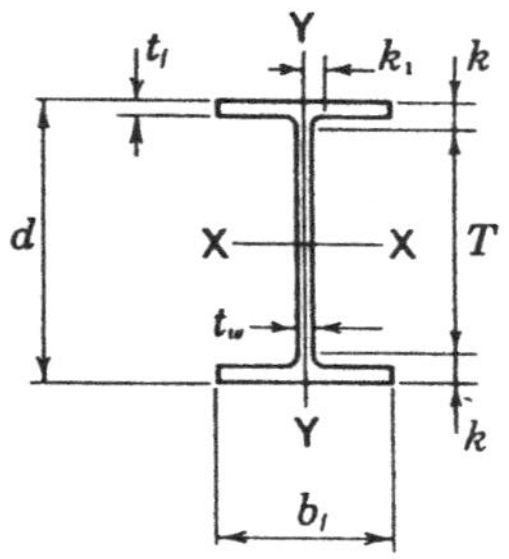

Shape	Area	Depth	Web Thickness	Flange Width	Flange Thickness		Elastic Properties Axis X-X			Elastic Properties Axis Y-Y			Plastic Modulus
	A in.^{2}	d in.	t_w in.	b_f in.	t_f in.	k in.	I in.^{4}	S in.^{3}	r in.	I in.^{4}	S in.^{3}	r in.	Z_x in.^{3}
W 30 × 116	34.2	30.01	0.565	10.495	0.850	1.625	4930	329	12.0	164	31.3	2.19	378
× 108	31.7	29.83	0.545	10.475	0.760	1.562	4470	299	11.9	146	27.9	2.15	346
× 99	29.1	29.65	0.520	10.450	0.670	1.437	3990	269	11.7	128	24.5	2.10	312
W 27 × 94	27.7	26.92	0.490	9.990	0.745	1.437	3270	243	10.9	124	24.8	2.12	278
× 84	24.8	26.71	0.460	9.960	0.640	1.375	2850	213	10.7	106	21.2	2.07	244
W 24 × 84	24.7	24.10	0.470	9.020	0.770	1.562	2370	196	9.79	94.4	20.9	1.95	224
× 76	22.4	23.92	0.440	8.990	0.680	1.437	2100	176	9.69	82.5	18.4	1.92	200
× 68	20.1	23.73	0.415	8.965	0.585	1.375	1830	154	9.55	70.4	15.7	1.87	177

(continued)

TABLE A.3 (Continued)

Shape	Area A in.^{2}	Depth d in.	Web Thickness t_w in.	Flange Width b_f in.	Flange Thickness t_f in.	k in.	Elastic Properties Axis X-X I in.^{4}	Axis X-X S in.^{3}	Axis X-X r in.	Axis Y-Y I in.^{4}	Axis Y-Y S in.^{3}	Axis Y-Y r in.	Plastic Modulus Z_x in.^{3}
W 21 × 83	24.3	21.43	0.515	8.355	0.835	1.562	1830	171	8.67	81.4	19.5	1.83	196
× 73	21.5	21.24	0.455	8.295	0.740	1.500	1600	151	8.64	70.6	17.0	1.81	172
× 57	16.7	21.06	0.405	6.555	0.650	1.375	1170	111	8.36	30.6	9.35	1.35	129
× 50	14.7	20.83	0.380	6.530	0.535	1.312	984	94.5	8.18	24.9	7.64	1.30	110
W 18 × 86	25.3	18.39	0.480	11.090	0.770	1.437	1530	166	7.77	175	31.6	2.63	186
× 76	22.3	18.21	0.425	11.035	0.680	1.375	1330	146	7.73	152	27.6	2.61	163
× 60	17.6	18.24	0.415	7.555	0.695	1.375	984	108	7.47	50.1	13.3	1.69	123
× 55	16.2	18.11	0.390	7.530	0.630	1.312	890	98.3	7.41	44.9	11.9	1.67	112
× 50	14.7	17.99	0.355	7.495	0.570	1.250	800	88.9	7.38	40.1	10.7	1.65	101
× 46	13.5	18.06	0.360	6.060	0.605	1.250	712	78.8	7.25	22.5	7.43	1.29	90.7
× 40	11.8	17.90	0.315	6.015	0.525	1.187	612	68.4	7.21	19.1	6.35	1.27	78.4
W 16 × 50	14.7	16.26	0.380	7.070	0.630	1.312	659	81.0	6.68	37.2	10.5	1.59	92.0
× 45	13.3	16.13	0.345	7.035	0.565	1.250	586	72.7	6.65	32.8	9.34	1.57	82.3
× 40	11.8	16.01	0.305	6.995	0.505	1.187	518	64.7	6.63	28.9	8.25	1.57	72.9
× 36	10.6	15.86	0.295	6.985	0.430	1.125	448	56.5	6.51	24.5	7.00	1.52	64.0
W 14 × 211	62.0	15.72	0.980	15.800	1.560	2.250	2660	338	6.55	1030	130	4.07	390
× 176	51.8	15.22	0.830	15.650	1.310	2.000	2140	281	6.43	838	107	4.02	320
× 132	38.8	14.66	0.645	14.725	1.030	1.687	1530	209	6.28	548	74.5	3.76	234
× 120	35.3	14.48	0.590	14.670	0.940	1.625	1380	190	6.24	495	67.5	3.74	212
× 74	21.8	14.17	0.450	10.070	0.785	1.562	796	112	6.04	134	26.6	2.48	126

× 68	20.0	14.04	0.415	10.035	0.720	1.500	723	103	6.01	121	24.2	2.46	115
× 48	14.1	13.79	0.340	8.030	0.595	1.375	485	70.3	5.85	51.4	12.8	1.91	78.4
× 43	12.6	13.66	0.305	7.995	0.530	1.312	428	62.7	5.82	45.2	11.3	1.89	69.6
× 34	10.0	13.98	0.285	6.745	0.455	1.000	340	48.6	5.83	23.3	6.91	1.53	54.6
× 30	8.85	13.84	0.270	6.730	0.385	0.937	291	42.0	5.73	19.6	5.82	1.49	47.3
W 12 × 136	39.9	13.41	0.790	12.400	1.250	1.937	1240	186	5.58	398	64.2	3.16	214
× 120	35.3	13.12	0.710	12.320	1.105	1.812	1070	163	5.51	345	56.0	3.13	186
× 72	21.1	12.25	0.430	12.040	0.670	1.375	597	97.4	5.31	195	32.4	3.04	108
× 65	19.1	12.12	0.390	12.000	0.605	1.312	533	87.9	5.28	174	29.1	3.02	96.8
× 53	15.6	12.06	0.345	9.995	0.575	1.250	425	70.6	5.23	95.8	19.2	2.48	77.9
× 45	13.2	12.06	0.335	8.045	0.575	1.250	350	58.1	5.15	50.0	12.4	1.94	64.7
× 40	11.8	11.94	0.295	8.005	0.515	1.250	310	51.9	5.13	44.1	11.0	1.93	57.5
× 30	8.79	12.34	0.260	6.520	0.440	0.937	238	38.6	5.21	20.3	6.24	1.52	43.1
× 26	7.65	12.22	0.230	6.490	0.380	0.875	204	33.4	5.17	17.3	5.34	1.51	37.2
W 10 × 88	25.9	10.84	0.605	10.265	0.990	1.625	534	98.5	4.54	179	34.8	2.63	113
× 77	22.6	10.60	0.530	10.190	0.870	1.500	455	85.9	4.49	154	30.1	2.60	97.6
× 49	14.4	9.98	0.340	10.000	0.560	1.312	272	54.6	4.35	93.4	18.7	2.54	60.4
× 39	11.5	9.92	0.315	7.985	0.530	1.125	209	42.1	4.27	45.0	11.3	1.98	46.8
× 33	9.71	9.73	0.290	7.960	0.435	1.062	170	35.0	4.19	36.6	9.20	1.94	38.8
× 19	5.62	10.24	0.250	4.020	0.395	0.812	96.3	18.8	4.14	4.29	2.14	0.874	21.6
× 17	4.99	10.11	0.240	4.010	0.330	0.750	81.9	16.2	4.05	3.56	1.78	0.844	18.7

Source: Adapted from data in the *Manual of Steel Construction,* with permission of the publishers, American Institute of Steel Construction. This table is a sample from an extensive set of tables in the reference document.

TABLE A.4 Properties of American Standard Channels

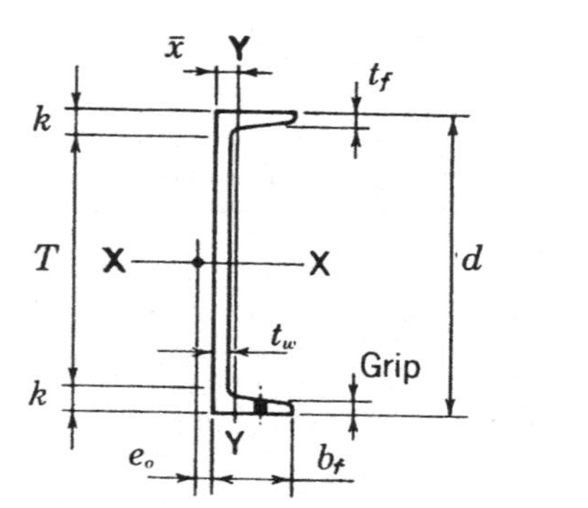

	Area	Depth	Web Thickness	Flange Width	Flange Thickness		Elastic Properties Axis X-X			Elastic Properties Axis Y-Y				
Shape	A (in.^{2})	d (in.)	t_w (in.)	b_f (in.)	t_f (in.)	k (in.)	I (in.^{4})	S (in.^{3})	r (in.)	I (in.^{4})	S (in.^{3})	r (in.)	x^a (in.)	e_o^b (in.)
C 15 × 50	14.7	15.0	0.716	3.716	0.650	1.44	404	53.8	5.24	11.0	3.78	0.867	0.798	0.583
× 40	11.8	15.0	0.520	3.520	0.650	1.44	349	46.5	5.44	9.23	3.37	0.886	0.777	0.767
× 33.9	9.96	15.0	0.400	3.400	0.650	1.44	315	42.0	5.62	8.13	3.11	0.904	0.787	0.896
C 12 × 30	8.82	12.0	0.510	3.170	0.501	1.13	162	27.0	4.29	5.14	2.06	0.763	0.674	0.618
× 25	7.35	12.0	0.387	3.047	0.501	1.13	144	24.1	4.43	4.47	1.88	0.780	0.674	0.746
× 20.7	6.09	12.0	0.282	2.942	0.501	1.13	129	21.5	4.61	3.88	1.73	0.799	0.698	0.870

C 10 × 30	8.82	10.0	0.673	3.033	0.436	1.00	103	20.7	3.42	3.94	1.65	0.669	0.649	0.369
× 25	7.35	10.0	0.526	2.886	0.436	1.00	91.2	18.2	3.52	3.36	1.48	0.676	0.617	0.494
× 20	5.88	10.0	0.379	2.739	0.436	1.00	78.9	15.8	3.66	2.81	1.32	0.692	0.606	0.637
× 15.3	4.49	10.0	0.240	2.600	0.436	1.00	67.4	13.5	3.87	2.28	1.16	0.713	0.634	0.796
C 9 × 20	5.88	9.0	0.448	2.648	0.413	0.94	60.9	13.5	3.22	2.42	1.17	0.642	0.583	0.515
× 15	4.41	9.0	0.285	2.485	0.413	0.94	51.0	11.3	3.40	1.93	1.01	0.661	0.586	0.682
× 13.4	3.94	9.0	0.233	2.433	0.413	0.94	47.9	10.6	3.48	1.76	0.962	0.669	0.601	0.743
C 8 × 18.75	5.51	8.0	0.487	2.527	0.390	0.94	44.0	11.0	2.82	1.98	1.01	0.599	0.565	0.431
× 13.75	4.04	8.0	0.303	2.343	0.390	0.94	36.1	9.03	2.99	1.53	0.854	0.615	0.553	0.604
× 11.5	3.38	8.0	0.220	2.260	0.390	0.94	32.6	8.14	3.11	1.32	0.781	0.625	0.571	0.697
C 7 × 14.75	4.33	7.0	0.419	2.299	0.366	0.88	27.2	7.78	2.51	1.38	0.779	0.564	0.532	0.441
× 12.25	3.60	7.0	0.314	2.194	0.366	0.88	24.2	6.93	2.60	1.17	0.703	0.571	0.525	0.538
× 9.8	2.87	7.0	0.210	2.090	0.366	0.88	21.3	6.08	2.72	0.968	0.625	0.581	0.540	0.647
C 6 × 13	3.83	6.0	0.437	2.157	0.343	0.81	17.4	5.80	2.13	1.05	0.642	0.525	0.514	0.380
× 10.5	3.09	6.0	0.314	2.034	0.343	0.81	15.2	5.06	2.22	0.866	0.564	0.529	0.499	0.486
× 8.2	2.40	6.0	0.200	1.920	0.343	0.81	13.1	4.38	2.34	0.693	0.492	0.537	0.511	0.599

[a]Distance to centroid of section.

[b]Distance to shear center of section.

Source: Adapted from data in the *Manual of Steel Construction*, with permission of the publishers, American Institute of Steel Construction. This table is a sample from an extensive set of tables in the reference document.

TABLE A.5 Properties of Single-Angle Shapes

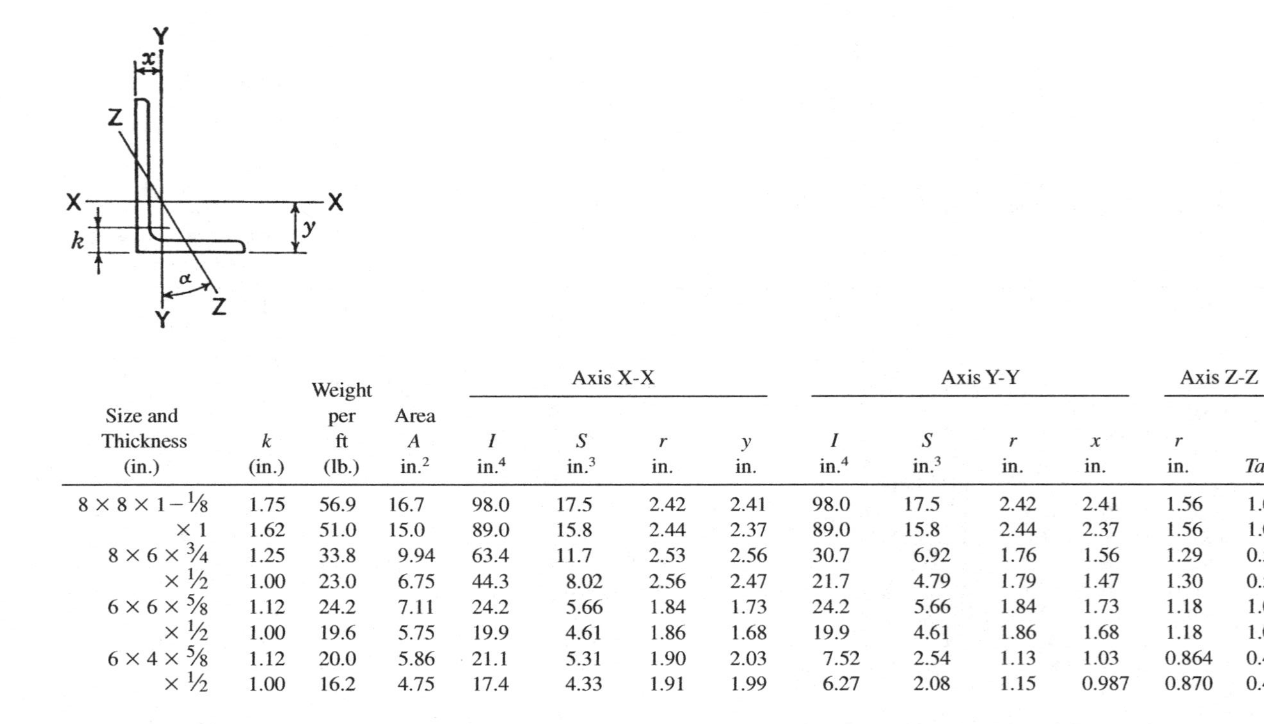

Size and Thickness (in.)	k (in.)	Weight per ft (lb.)	Area A in.^{2}	Axis X-X I in.^{4}	Axis X-X S in.^{3}	Axis X-X r in.	Axis X-X y in.	Axis Y-Y I in.^{4}	Axis Y-Y S in.^{3}	Axis Y-Y r in.	Axis Y-Y x in.	Axis Z-Z r in.	Axis Z-Z Tan α
8 × 8 × 1−1/8	1.75	56.9	16.7	98.0	17.5	2.42	2.41	98.0	17.5	2.42	2.41	1.56	1.000
× 1	1.62	51.0	15.0	89.0	15.8	2.44	2.37	89.0	15.8	2.44	2.37	1.56	1.000
8 × 6 × 3/4	1.25	33.8	9.94	63.4	11.7	2.53	2.56	30.7	6.92	1.76	1.56	1.29	0.551
× 1/2	1.00	23.0	6.75	44.3	8.02	2.56	2.47	21.7	4.79	1.79	1.47	1.30	0.558
6 × 6 × 5/8	1.12	24.2	7.11	24.2	5.66	1.84	1.73	24.2	5.66	1.84	1.73	1.18	1.000
× 1/2	1.00	19.6	5.75	19.9	4.61	1.86	1.68	19.9	4.61	1.86	1.68	1.18	1.000
6 × 4 × 5/8	1.12	20.0	5.86	21.1	5.31	1.90	2.03	7.52	2.54	1.13	1.03	0.864	0.435
× 1/2	1.00	16.2	4.75	17.4	4.33	1.91	1.99	6.27	2.08	1.15	0.987	0.870	0.440

× 3/8	0.87	12.3	3.61	13.5	3.32	1.93	1.94	4.90	1.60	1.17	0.941	0.877	0.446
5 × 3½ × ½	1.00	13.6	4.00	9.99	2.99	1.58	1.66	4.05	1.56	1.01	0.906	0.755	0.479
× 3/8	0.87	10.4	3.05	7.78	2.29	1.60	1.61	3.18	1.21	1.02	0.861	0.762	0.486
5 × 3 × ½	1.00	12.8	3.75	9.45	2.91	1.59	1.75	2.58	1.15	0.829	0.750	0.648	0.357
× 3/8	0.87	9.8	2.86	7.37	2.24	1.61	1.70	2.04	0.888	0.845	0.704	0.654	0.364
4 × 4 × ½	0.87	12.8	3.75	5.56	1.97	1.22	1.18	5.56	1.97	1.22	1.18	0.782	1.000
× 3/8	0.75	9.8	2.86	4.36	1.52	1.23	1.14	4.36	1.52	1.23	1.14	0.788	1.000
4 × 3 × ½	0.94	11.1	3.25	5.05	1.89	1.25	1.33	2.42	1.12	0.864	0.827	0.639	0.543
× 3/8	0.81	8.5	2.48	3.96	1.46	1.26	1.28	1.92	0.866	0.879	0.782	0.644	0.551
× 5/16	0.75	7.2	2.09	3.38	1.23	1.27	1.26	1.65	0.734	0.887	0.759	0.647	0.554
3½ × 3½ × 3/8	0.75	8.5	2.48	2.87	1.15	1.07	1.01	2.87	1.15	1.07	1.01	0.687	1.000
× 5/16	0.69	7.2	2.09	2.45	0.976	1.08	0.990	2.45	0.976	1.08	0.990	0.690	1.000
× 5/16	0.75	6.1	1.78	2.19	0.927	1.11	1.14	0.939	0.504	0.727	0.637	0.540	0.501
3 × 3 × 3/8	0.69	7.2	2.11	1.76	0.833	0.913	0.888	1.76	0.833	0.913	0.888	0.587	1.000
× 5/16	0.62	6.1	1.78	1.51	0.707	0.922	0.865	1.51	0.707	0.922	0.865	0.589	1.000
3 × 2½ × 3/8	0.75	6.6	1.92	1.66	0.810	0.928	0.956	1.04	0.581	0.736	0.706	0.522	0.676
× 5/16	0.69	5.6	1.62	1.42	0.688	0.937	0.933	0.898	0.494	0.744	0.683	0.525	0.680
3 × 2 × 3/8	0.69	5.9	1.73	1.53	0.781	0.940	1.04	0.543	0.371	0.559	0.539	0.430	0.428
× 5/16	0.62	5.0	1.46	1.32	0.664	0.948	1.02	0.470	0.317	0.567	0.516	0.432	0.435
2½ × 2½ × 3/8	0.69	5.9	1.73	0.984	0.566	0.753	0.762	0.984	0.566	0.753	0.762	0.487	1.000
× 5/16	0.62	5.0	1.46	0.849	0.482	0.761	0.740	0.849	0.482	0.761	0.740	0.489	1.000
2½ × 2 × 3/8	0.69	5.3	1.55	0.912	0.547	0.768	0.831	0.514	0.363	0.577	0.581	0.420	0.614
× 5/16	0.62	4.5	1.31	0.788	0.466	0.776	0.809	0.446	0.310	0.584	0.559	0.422	0.620

Source: Adapted from data in the *Manual of Steel Construction*, with permission of the publishers, American Institute of Steel Construction. This table is a sample from an extensive set of tables in the reference document.

TABLE A.6 Properties of Double-Angle Shapes with Long Legs Back-to-Back

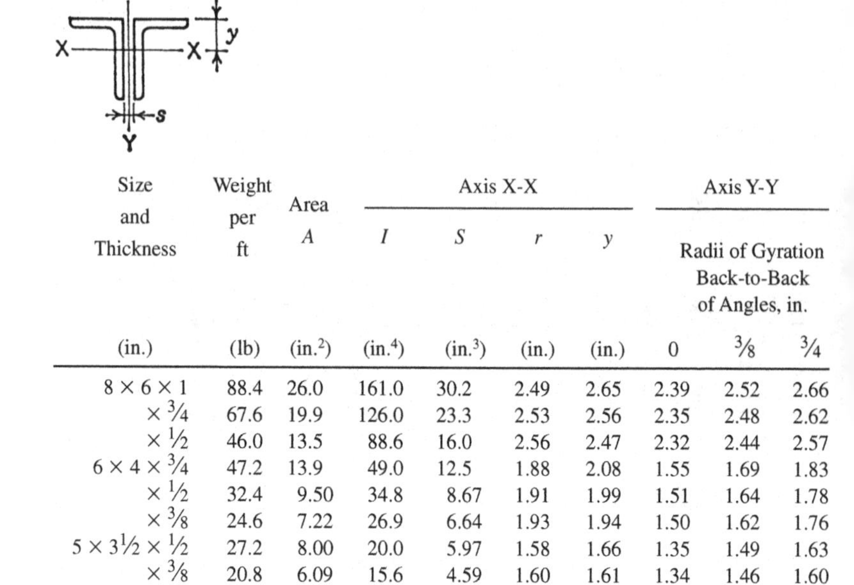

Size and Thickness	Weight per ft	Area A	Axis X-X				Axis Y-Y: Radii of Gyration Back-to-Back of Angles, in.		
			I	S	r	y			
(in.)	(lb)	(in.^{2})	(in.^{4})	(in.^{3})	(in.)	(in.)	0	3/8	3/4
8 × 6 × 1	88.4	26.0	161.0	30.2	2.49	2.65	2.39	2.52	2.66
× 3/4	67.6	19.9	126.0	23.3	2.53	2.56	2.35	2.48	2.62
× 1/2	46.0	13.5	88.6	16.0	2.56	2.47	2.32	2.44	2.57
6 × 4 × 3/4	47.2	13.9	49.0	12.5	1.88	2.08	1.55	1.69	1.83
× 1/2	32.4	9.50	34.8	8.67	1.91	1.99	1.51	1.64	1.78
× 3/8	24.6	7.22	26.9	6.64	1.93	1.94	1.50	1.62	1.76
5 × 3½ × 1/2	27.2	8.00	20.0	5.97	1.58	1.66	1.35	1.49	1.63
× 3/8	20.8	6.09	15.6	4.59	1.60	1.61	1.34	1.46	1.60
5 × 3 × 1/2	25.6	7.50	18.9	5.82	1.59	1.75	1.12	1.25	1.40
× 3/8	19.6	5.72	14.7	4.47	1.61	1.70	1.10	1.23	1.37
× 5/16	16.4	4.80	12.5	3.77	1.61	1.68	1.09	1.22	1.36
4 × 3 × 1/2	22.2	6.50	10.1	3.78	1.25	1.33	1.20	1.33	1.48
× 3/8	17.0	4.97	7.93	2.92	1.26	1.28	1.18	1.31	1.45
× 5/16	14.4	4.18	6.76	2.47	1.27	1.26	1.17	1.30	1.44
3½ × 2½ × 3/8	14.4	4.22	5.12	2.19	1.10	1.16	0.976	1.11	1.26
× 5/16	12.2	3.55	4.38	1.85	1.11	1.14	0.966	1.10	1.25
× 1/4	9.8	2.88	3.60	1.51	1.12	1.11	0.958	1.09	1.23
3 × 2 × 3/8	11.8	3.47	3.06	1.56	0.940	1.04	0.777	0.917	1.07
× 5/16	10.0	2.93	2.63	1.33	0.948	1.02	0.767	0.903	1.06
× 1/4	8.2	2.38	2.17	1.08	0.957	0.993	0.757	0.891	1.04
2½ × 2 × 3/8	10.6	3.09	1.82	1.09	0.768	0.831	0.819	0.961	1.12
× 5/16	9.0	2.62	1.58	0.932	0.776	0.809	0.809	0.948	1.10
× 1/4	7.2	2.13	1.31	0.763	0.784	0.787	0.799	0.935	1.09

Source: Adapted from data in the *Manual of Steel Construction*, with permission of the publishers, American Institute of Steel Construction. This table is a sample from an extensive set of tables in the reference document.

TABLE A.7 Properties of Standard Weight Steel Pipe

Dimensions				Weight	Properties			
Nominal Diameter (in.)	Outside Diameter (in.)	Inside Diameter (in.)	Wall Thickness (in.)	per ft (lb)	A (in.^{2})	I (in.^{4})	S (in.^{3})	r (in.)
3	3.500	3.068	0.216	7.58	2.23	3.02	1.72	1.16
3½	4.000	3.548	0.226	9.11	2.68	4.79	2.39	1.34
4	4.500	4.026	0.237	10.79	3.17	7.23	3.21	1.51
5	5.563	5.047	0.258	14.62	4.30	15.2	5.45	1.88
6	6.625	6.065	0.280	18.97	5.58	28.1	8.50	2.25
8	8.625	7.981	0.322	28.55	8.40	72.5	16.8	2.94
10	10.750	10.020	0.365	40.48	11.9	161	29.9	3.67
12	12.750	12.000	0.375	49.56	14.6	279	43.8	4.38

Source: Adapted from data in the *Manual of Steel Construction*, with permission of the publishers, American Institute of Steel Construction. This table is a sample from an extensive set of tables in the reference document.

Appendix B

BEAM DESIGN AIDS

This section presents materials that support the design of steel beams.

Table B.1. Allowable Stress Design Selection for Shapes Used as Beams. This table is a shortened version of one from the AISC Manual. The primary use of the table is for rapid selection of shapes for a maximum bending moment. Shapes are listed in the table in descending order of the magnitudes of their section modulus values for the primary bending axis: S_x for the *x-x* axis. Also included in the table is the safe service load bending moment corresponding to the S_x value and an allowable stress based on a yield stress of 36 ksi. Shapes are grouped in the table, with each group headed by a shape appearing in bold type; this is the *least-weight member*—there being no other beam as strong with less weight. The table also yields values for the actual beam depth and for the lateral unsupported length limits of L_c and L_u; the latter two dimensions are the limits that apply to changes in allowable bending related to critical lengths for buckling.

TABLE B.1 Allowable Stress Design Selection for Shapes Used as Beams

S_x	Shape	Depth d	F_y'	$F_y = 36$ ksi		
				L_c	L_u	M_R
In.^{3}		In.	Ksi	Ft.	Ft.	Kip-ft.
1110	**W 36x300**	$36\frac{3}{4}$	—	17.6	35.3	2220
1030	**W 36x280**	$36\frac{1}{2}$	—	17.5	33.1	2060
953	**W 36x260**	$36\frac{1}{4}$	—	17.5	30.5	1910
895	**W 36x245**	$36\frac{1}{8}$	—	17.4	28.6	1790
837	**W 36x230**	$35\frac{7}{8}$	—	17.4	26.8	1670
829	W 33x241	$34\frac{1}{8}$	—	16.7	30.1	1660
757	**W 33x221**	$33\frac{7}{8}$	—	16.7	27.6	1510
719	**W 36x210**	$36\frac{3}{4}$	—	12.9	20.9	1440
684	**W 33x201**	$33\frac{5}{8}$	—	16.6	24.9	1370
664	**W 36x194**	$36\frac{1}{2}$	—	12.8	19.4	1330
663	W 30x211	31	—	15.9	29.7	1330
623	**W 36x182**	$36\frac{3}{8}$	—	12.7	18.2	1250
598	W 30x191	$30\frac{5}{8}$	—	15.9	26.9	1200
580	**W 36x170**	$36\frac{1}{8}$	—	12.7	17.0	1160
542	**W 36x160**	36	—	12.7	15.7	1080
539	W 30x173	$30\frac{1}{2}$	—	15.8	24.2	1080
504	**W 36x150**	$35\frac{7}{8}$	—	12.6	14.6	1010
502	W 27x178	$27\frac{3}{4}$	—	14.9	27.9	1000
487	W 33x152	$33\frac{1}{2}$	—	12.2	16.9	974
455	W 27x161	$27\frac{5}{8}$	—	14.8	25.4	910
448	**W 33x141**	$33\frac{1}{4}$	—	12.2	15.4	896
439	**W 36x135**	$35\frac{1}{2}$	—	12.3	13.0	878
414	W 24x162	25	—	13.7	29.3	828
411	W 27x146	$27\frac{3}{8}$	—	14.7	23.0	822
406	**W 33x130**	$33\frac{1}{8}$	—	12.1	13.8	812
380	W 30x132	$30\frac{1}{4}$	—	11.1	16.1	760
371	W 24x146	$24\frac{3}{4}$	—	13.6	26.3	742
359	**W 33x118**	$32\frac{7}{8}$	—	12.0	12.6	718
355	W 30x124	$30\frac{1}{8}$	—	11.1	15.0	710
329	**W 30x116**	30	—	11.1	13.8	658
329	W 24x131	$24\frac{1}{2}$	—	13.6	23.4	658
329	W 21x147	22	—	13.2	30.3	658
299	**W 30x108**	$29\frac{7}{8}$	—	11.1	12.3	598
299	W 27x114	$27\frac{1}{4}$	—	10.6	15.9	598
295	W 21x132	$21\frac{7}{8}$	—	13.1	27.2	590
291	W 24x117	$24\frac{1}{4}$	—	13.5	20.8	582
273	W 21x122	$21\frac{5}{8}$	—	13.1	25.4	546
269	**W 30x 99**	$29\frac{5}{8}$	—	10.9	11.4	538
267	W 27x102	$27\frac{1}{8}$	—	10.6	14.2	534
258	W 24x104	24	58.5	13.5	18.4	516
249	W 21x111	$21\frac{1}{2}$	—	13.0	23.3	498
243	**W 27x 94**	$26\frac{7}{8}$	—	10.5	12.8	486
231	W 18x119	19	—	11.9	29.1	462
227	W 21x101	$21\frac{3}{8}$	—	13.0	21.3	454
222	**W 24x 94**	$24\frac{1}{4}$	—	9.6	15.1	444
213	**W 27x 84**	$26\frac{3}{4}$	—	10.5	11.0	426
204	W 18x106	$18\frac{3}{4}$	—	11.8	26.0	408
196	**W 24x 84**	$24\frac{1}{8}$	—	9.5	13.3	392
192	W 21x 93	$21\frac{5}{8}$	—	8.9	16.8	384
190	W 14x120	$14\frac{1}{2}$	—	15.5	44.1	380
188	W 18x 97	$18\frac{5}{8}$	—	11.8	24.1	376
176	**W 24x 76**	$23\frac{7}{8}$	—	9.5	11.8	352
175	W 16x100	17	—	11.0	28.1	350
173	W 14x109	$14\frac{3}{8}$	58.6	15.4	40.6	346
171	W 21x 83	$21\frac{3}{8}$	—	8.8	15.1	342
166	W 18x 86	$18\frac{3}{8}$	—	11.7	21.5	332
157	W14x 99	$14\frac{1}{8}$	48.5	15.4	37.0	314
155	W 16x 89	$16\frac{3}{4}$	—	10.9	25.0	310
154	**W 24x 68**	$23\frac{3}{4}$	—	9.5	10.2	308
151	W 21x 73	$21\frac{1}{4}$	—	8.8	13.4	302
146	W 18x 76	$18\frac{1}{4}$	64.2	11.6	19.1	292
143	W 14x 90	14	40.4	15.3	34.0	286
140	**W 21x 68**	$21\frac{1}{8}$	—	8.7	12.4	280
134	W 16x 77	$16\frac{1}{2}$	—	10.9	21.9	268
131	**W 24x 62**	$23\frac{3}{4}$	—	7.4	8.1	262
127	**W 21x 62**	21	—	8.7	11.2	254
127	W 18x 71	$18\frac{1}{2}$	—	8.1	15.5	254
123	W 14x 82	$14\frac{1}{4}$	—	10.7	28.1	246
118	W 12x 87	$12\frac{1}{2}$	—	12.8	36.2	236
117	W 18x 65	$18\frac{3}{8}$	—	8.0	14.4	234
117	W 16x 67	$16\frac{3}{8}$	—	10.8	19.3	234
114	**W 24x 55**	$23\frac{5}{8}$	—	7.0	7.5	228
112	W 14x 74	$14\frac{1}{8}$	—	10.6	25.9	224
111	W 21x 57	21	—	6.9	9.4	222
108	W 18x 60	$18\frac{1}{4}$	—	8.0	13.3	216
107	W 12x 79	$12\frac{3}{8}$	62.6	12.8	33.3	214
103	W 14x 68	14	—	10.6	23.9	206
98.3	**W 18x 55**	$18\frac{1}{8}$	—	7.9	12.1	197
97.4	W 12x 72	$12\frac{1}{4}$	52.3	12.7	30.5	195

S_x	Shape	Depth d	F_y'	F_y = 36 ksi		
				L_c	L_u	M_R
In.^{3}		In.	Ksi	Ft.	Ft.	Kip-ft.
94.5	**W 21x50**	**$20\frac{7}{8}$**	—	**6.9**	**7.8**	**189**
92.2	W 16x57	$16\frac{3}{8}$	—	7.5	14.3	184
92.2	W 14x61	$13\frac{7}{8}$	—	10.6	21.5	184
88.9	**W 18x50**	**18**	—	**7.9**	**11.0**	**178**
87.9	W 12x65	$12\frac{1}{8}$	43.0	12.7	27.7	176
81.6	**W 21x44**	**$20\frac{5}{8}$**	—	**6.6**	**7.0**	**163**
81.0	W 16x50	$16\frac{1}{4}$	—	7.5	12.7	162
78.8	W 18x46	18	—	6.4	9.4	158
78.0	W 12x58	$12\frac{1}{4}$	—	10.6	24.4	156
77.8	W 14x53	$13\frac{7}{8}$	—	8.5	17.7	156
72.7	W 16x45	$16\frac{1}{8}$	—	7.4	11.4	145
70.6	W 12x53	12	55.9	10.6	22.0	141
70.3	W 14x48	$13\frac{3}{4}$	—	8.5	16.0	141
68.4	**W 18x40**	**$17\frac{7}{8}$**	—	**6.3**	**8.2**	**137**
66.7	W 10x60	$10\frac{1}{4}$	—	10.6	31.1	133
64.7	**W 16x40**	**16**	—	**7.4**	**10.2**	**129**
64.7	W 12x50	$12\frac{1}{4}$	—	8.5	19.6	129
62.7	W 14x43	$13\frac{5}{8}$	—	8.4	14.4	125
60.0	W 10x54	$10\frac{1}{8}$	63.5	10.6	28.2	120
58.1	W 12x45	12	—	8.5	17.7	116
57.6	**W 18x35**	**$17\frac{3}{4}$**	—	**6.3**	**6.7**	**115**
56.5	W 16x36	$15\frac{7}{8}$	64.0	7.4	8.8	113
54.6	W 14x38	$14\frac{1}{8}$	—	7.1	11.5	109
54.6	W 10x49	10	53.0	10.6	26.0	109
51.9	W 12x40	12	—	8.4	16.0	104
49.1	W 10x45	$10\frac{1}{8}$	—	8.5	22.8	98
48.6	**W 14x34**	**14**	—	**7.1**	**10.2**	**97**
47.2	**W 16x31**	**$15\frac{7}{8}$**	—	**5.8**	**7.1**	**94**
45.6	W 12x35	$12\frac{1}{2}$	—	6.9	12.6	91
42.1	W 10x39	$9\frac{7}{8}$	—	8.4	19.8	84
42.0	**W 14x30**	**$13\frac{7}{8}$**	**55.3**	**7.1**	**8.7**	**84**
38.6	**W 12x30**	**$12\frac{3}{8}$**	—	**6.9**	**10.8**	**77**
38.4	**W 16x26**	**$15\frac{3}{4}$**	—	**5.6**	**6.0**	**77**
35.3	**W 14x26**	**$13\frac{7}{8}$**	—	**5.3**	**7.0**	**71**
35.0	W 10x33	$9\frac{3}{4}$	50.5	8.4	16.5	70
33.4	**W 12x26**	**$12\frac{1}{4}$**	**57.9**	**6.9**	**9.4**	**67**
32.4	W 10x30	$10\frac{1}{2}$	—	6.1	13.1	65
31.2	W 8x35	$8\frac{1}{8}$	64.4	8.5	22.6	62

S_x	Shape	Depth d	F_y'	F_y = 36 ksi		
				L_c	L_u	M_R
In.^{3}		In.	Ksi	Ft.	Ft.	Kip-ft.
29.0	**W 14x22**	**$13\frac{3}{4}$**	—	**5.3**	**5.6**	**58**
27.9	W 10x26	$10\frac{3}{8}$	—	6.1	11.4	56
27.5	W 8x31	8	50.0	8.4	20.1	55
25.4	**W 12x22**	**$12\frac{1}{4}$**	—	**4.3**	**6.4**	**51**
24.3	W 8x28	8	—	6.9	17.5	49
23.2	**W 10x22**	**$10\frac{1}{8}$**	—	**6.1**	**9.4**	**46**
21.3	**W 12x19**	**$12\frac{1}{8}$**	—	**4.2**	**5.3**	**43**
21.1	**M 14x18**	**14**	—	**3.6**	**4.0**	**42**
20.9	W 8x24	$7\frac{7}{8}$	64.1	6.9	15.2	42
18.8	W 10x19	$10\frac{1}{4}$	—	4.2	7.2	38
18.2	W 8x21	$8\frac{1}{4}$	—	5.6	11.8	36
17.1	**W 12x16**	**12**	—	**4.1**	**4.3**	**34**
16.7	W 6x25	$6\frac{3}{8}$	—	6.4	20.0	33
16.2	W 10x17	$10\frac{1}{8}$	—	4.2	6.1	32
15.2	W 8x18	$8\frac{1}{8}$	—	5.5	9.9	30
14.9	**W 12x14**	**$11\frac{7}{8}$**	**54.3**	**3.5**	**4.2**	**30**
13.8	W 10x15	10	—	4.2	5.0	28
13.4	W 6x20	$6\frac{1}{4}$	62.1	6.4	16.4	27
13.0	M 6x20	6	—	6.3	17.4	26
12.0	**M 12x11.8**	**12**	—	**2.7**	**3.0**	**24**
11.8	W 8x15	$8\frac{1}{8}$	—	4.2	7.2	24
10.9	W 10x12	$9\frac{7}{8}$	47.5	3.9	4.3	22
10.2	W 6x16	$6\frac{1}{4}$	—	4.3	12.0	20
10.2	W 5x19	$5\frac{1}{8}$	—	5.3	19.5	20
9.91	W 8x13	8	—	4.2	5.9	20
9.72	W 6x15	6	31.8	6.3	12.0	19
9.63	M 5x18.9	5	—	5.3	19.3	19
8.51	W 5x16	5	—	5.3	16.7	17
7.81	**W 8x10**	**$7\frac{7}{8}$**	**45.8**	**4.2**	**4.7**	**16**
7.76	**M 10x 9**	**10**	—	**2.6**	**2.7**	**16**
7.31	W 6x12	6	—	4.2	8.6	15
5.56	**W 6x 9**	**$5\frac{7}{8}$**	**50.3**	**4.2**	**6.7**	**11**
5.46	W 4x13	$4\frac{1}{8}$	—	4.3	15.6	11
5.24	M 4x13	4	—	4.2	16.9	10
4.62	**M 8x 6.5**	**8**	—	**2.4**	**2.5**	**9**
2.40	**M 6x 4.4**	**6**	—	**1.9**	**2.4**	**5**

Source: Adapted from the *Manual of Steel Construction,* with permission of the publishers, American Institute of Steel Construction.

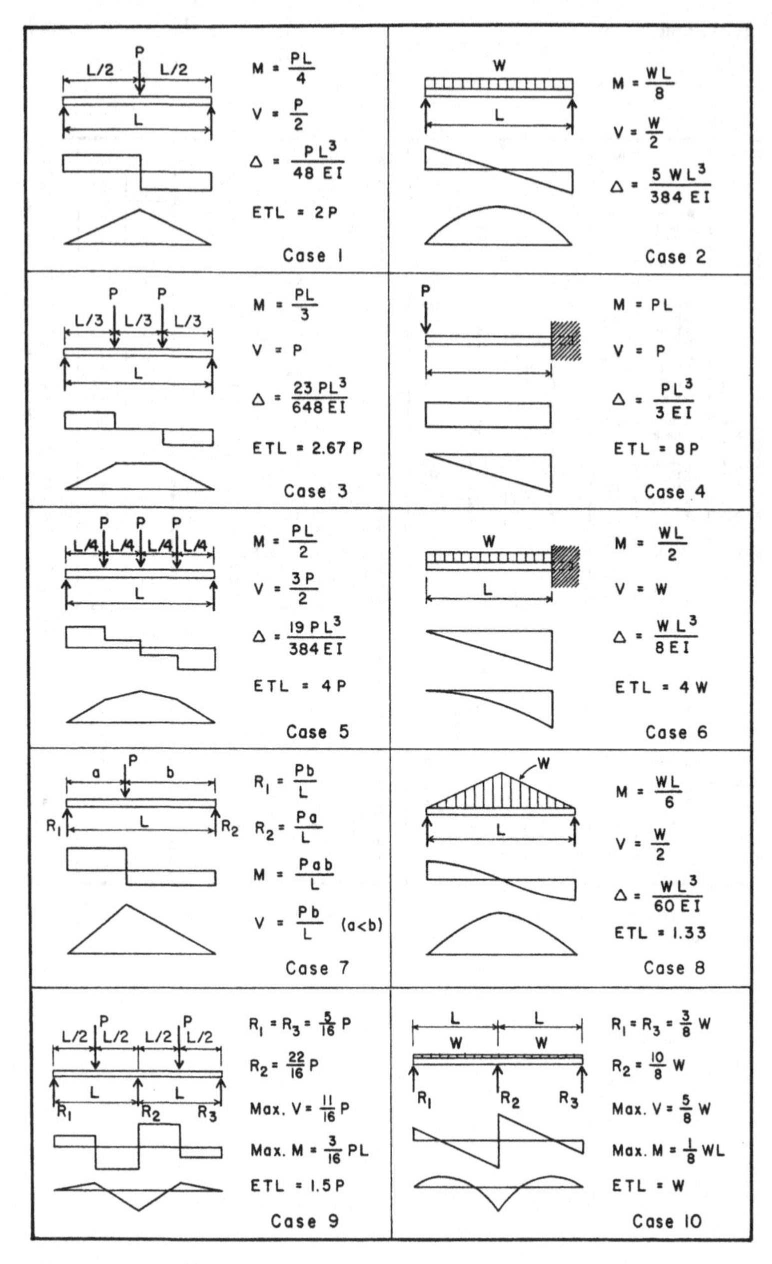

Figure B.1 Values for typical beam loadings.

This table is provided for quick comparisons of design capacities for steel bending members as determined by the stress method using unfactored service loads. The following example demonstrates the use of the table.

Example. Design for flexure a simple beam 14 ft in length and having a total uniformly distributed dead load of 13.2 kips and a total uniformly distributed live load of 26.4 kips.

Solution: The total service load is 13.2 + 26.4 = 39.6 kips, and the maximum bending moment for the simple beam is

$$M = \frac{WL}{8} = \frac{39.6(14)}{8} = 69.3 \text{ kip-ft}$$

From Table B.1, the lightest shape is a W 14 × 26 (or a W 16 × 26).

Figure B.1. Values for Typical Beam Loadings. This figure presents ten loading and support configurations for beams, with formulas for determination of values for reaction forces, maximum shear, maximum moment, and maximum deflection. Use of the *ETL* values given in the figure is explained in Section 3.8.

Appendix C

STUDY AIDS

The materials in this section are provided for readers to test their general understanding of the book presentations. After reading an individual chapter, you should use the materials here as a review.

TERMS

Using the text of the chapter indicated, together with the index, review the definitions and significance of the following terms.

Introduction

AISC
AISC Manual
ASCE
ASD
ASTM

Cold-forming
Hot-rolling
LRFD
Miscellaneous metals
SDI
Service load
SJI
Structural steel

Chapter 1

Allowable stress
Deformation limits
Ductility
Field assemblage
Modulus of elasticity
Plastic range
Rolled shapes
Shop assemblage
Stability
Stiffness
Strain hardening
Ultimate limit
Yield point

Chapter 2

Combined load
Continuous beam
Cut section
Deformed shape
Factored load
Free-body diagram
Moment-resistive joint (connection)
Resistance factor
Rigid frame

Safety
Strength reduction factor
Structural investigation

Chapter 3

Bearing
Deck
Deflection
Elastic buckling
Elastic section modulus (S)
Equivalent uniform load (EUL)
Formed sheet steel deck
Framing system, layout, plan
Inelastic buckling
Joist girder
Lateral buckling
Lateral unsupported length
Lightest (least-weight) section
Open-web steel joist
Plastic hinge
Plastic moment
Plastic section modulus (Z)
Safe load
Shear center
Steel joist (truss form)
Superimposed load
Torsional buckling
Web crippling
Web stiffener
Wide-flange beam

Chapter 4

Bending factor
Column interaction

Double angle
Effective column buckling length
P-delta effect
Radius of gyration
Slenderness ratio (*L/r*)
Strut

Chapter 5

Bent
Captive column
Eccentric bracing
Rigid-frame bent
Sidesway
Trussed bent

Chapter 6

Composite structural element
Flitched beam
Net section

Chapter 7

Chord member
Maxwell diagram
Method of joints
Truss panel
Two-way spanning structures

Chapter 8

Boxing
Butt joint
Double shear
Edge distance
Effective area (in tension)

Effective length (of weld)
Fastener
Fillet weld
Framed beam connection
Gage (distance for angles)
Groove weld
Lap joint
Penetration (of weld)
Pitch
Plug weld
Single shear
Slot weld
Tearing
Tee joint (weld)
Throat (of weld)
Unfinished bolt (A307)
Upset end

Chapter 9

Building code
Dead load
Lateral load
Live load
Live load reduction
Periphery, load
Tributary area (of load)

Appendix A

Centroid
Moment of inertia
Radius of gyration
Section modulus

QUESTIONS

Answers follow the last question.

Chapters 1–2

1. Why is the yield point generally of more concern than the ultimate strength of the steel for most structural steel elements?
2. Why is the depth indicated in the designation for a W shape (i.e., 12 in. in W 12 × 36) referred to as a *nominal dimension?*
3. Steel is generally considered to be vulnerable in exposed conditions. What are the primary concerns in this situation?

Chapter 3

1. What single property of a beam cross section is most predictive of bending strength?
2. What is significant about the properties for a rolled W shape designated as L_p and L_r?
3. For shapes of A36 steel used as beams, what makes it possible to say that most beams with the same depth will have the same deflection at their limiting loads on a given span?
4. What is the primary hot-rolled steel product that is used to form light-gage structural elements?
5. When concrete fill is placed on top of a formed sheet steel deck, what are the possible ways for the concrete and steel to interact in response to load on the deck?
6. What is the most common means for dealing with torsion on a steel beam?
7. What single property of the cross section is most critical to the resistance of web crippling in a W-shape beam?

Chapter 4

1. For evaluation of simple axial compression capacity, what are the significant properties of the cross section of a steel column?
2. What situation usually makes it necessary to consider the effects of buckling on both axes of a W-shape steel column?
3. How do end support conditions affect column buckling?
4. What is the *P*-delta effect?

Chapter 5

1. What interaction is required between rigid-frame members?
2. When structural frames and structural walls interact, why do the walls tend to take most of the load?
3. Why is wind load not always a critical concern for design of individual structural members?
4. Regarding deformation of a rigid frame, what is significant about the relative stiffness of the frame members?

Chapter 7

1. Why should trusses be loaded only at their panel points?
2. What are the necessary conditions for equilibrium of a coplanar, concurrent force system for algebraic analysis?
3. What is the basic geometric principle of structural analysis that permits the use of the Maxwell diagram for the investigation of internal forces in a truss?

Chapter 8

1. When high-strength steel bolts are used for a connection between steel members, what basic action develops the initial load resistance in the joint?
2. Tearing in a bolted connection is resisted by what combination of stress developments?
3. Other than spacing and edge distances, what basic dimension limits the number of bolts that can be used in a framed beam connection?
4. Why shouldn't supporting steel beams have the same depth as the beams they support?
5. What is significant about the throat dimension in a fillet weld?
6. Boxing welds in a joint layout gains what structural advantage?

Chapter 9

1. Of what does the design dead load primarily consist?
2. What primary factor affects the percentage of live load reduction?
3. In what unit is the load periphery for a column expressed?

4. Why is the achievement of optimal structural efficiency not always a dominant concern for the general cost of building construction?

Appendix A

1. Although there is a single centroid for an area, why is there not a single value for the centroidal moment of inertia for most areas? For what geometric form is there a single value for the centroidal moment of inertia?
2. For a column subjected to buckling, what is the significance of the least radius of gyration of the member cross section?

ANSWERS TO QUESTIONS

Chapters 1–2

1. A steel structure's maximum practical strength limit is usually determined by excessive deformation in ductile (yield) response of the material.
2. This dimension is the approximate one for a group of shapes; true depth varies within the group.
3. For safety, fire is most critical. Loss of material by rusting is the other major concern.

Chapter 3

1. Section modulus, elastic or plastic.
2. They establish limits with regard to lateral unsupported conditions.
3. Stress and strain are proportional to each other and to the load magnitude. For a given depth, if maximum stress is a constant (under service loads), then strain will be a constant for all beams of that depth.
4. Sheet steel.
5. (a) With the concrete as structurally inert fill on a steel structure. (b) With the steel deck serving only as a form for a structural concrete slab. (c) With the steel and concrete interacting as a composite structure.
6. Bracing of the beam to prevent its rotation, and thus prevent the development of torsional moment.
7. Thickness of the beam web.

Chapter 4

1. Area and radius of gyration.
2. When effective buckling length (*KL*) is different for the two axes.
3. They may alter the effective buckling length of the column, thus changing the column's resistance to buckling.
4. The *P*-delta effect is a bending moment (*P* times column deformation) that produces more deflection (i.e., additional delta), which results in greater *P*-delta, and so on.

Chapter 5

1. Transfer of bending through connections.
2. They are typically much stiffer than the frames in resisting lateral forces.
3. The different load factors for load combinations with wind may produce a less critical design load.
4. If some members are exceptionally stiff or flexible, the character of deformation of the frame may be affected.

Chapter 7

1. To avoid shear and bending in truss members.
2. The sum of the forces is zero—usually established by algebraic summations of the vertical and horizontal force components.
3. Closing of a force polygon establishes equilibrium for the forces at a truss joint. Close all of the joint force systems and find all the internal forces in the truss members.

Chapter 8

1. Friction between the two connected parts, induced by the tight clamping (squeezing) action of the highly tightened bolts.
2. Shear and tension.
3. Depth of the beam.
4. Because both flanges of the supported beam must be cut back to achieve the usual framing connection (see Figure 8.9*f*). Doing so results in major shear strength loss for the supported beam.
5. It defines the critical weld cross section for shear stress, which in turn establishes the weld strength.

6. Increased resistance to tearing caused by twisting actions on the welded joint.

Chapter 9

1. Weight of the building construction.
2. Total area of the loaded surface being carried by the member being considered.
3. Square feet of supported surface area.
4. Structural efficiency, while an important basic concept for engineering design, may be less important economically than the effect of the structure on the cost of the rest of the construction.

Appendix A

1. Except for a circle, an infinite number of different reference axes exist.
2. It indicates the weakest axis of bending resistance and the direction in which buckling is most likely to occur.

Appendix D

ANSWERS TO PROBLEMS IN CHAPTERS

Chapter 2

2.4.A. R = 10 kips up and 110 kip-ft counterclockwise
2.4.B. R = 5 kips up and 24 kip-ft counterclockwise
2.4.C. R = 6 kips to the left and 72 kip-ft counterclockwise
2.4.D. Left R = 4.5 kips up, right R = 4.5 kips down and 12 kips to the right
2.4.E. Left R = 4.5 kips down and 6 kips to left, right R = 4.5 kips up and 6 kips to left

Chapter 3

3.2.A. 13.5%
3.2.B. 13.2%
3.3.A. (1) 1038 kip-ft, (2) 648 kip-ft, (3) 527 kip-ft
3.3.B. (1) 192 kip-ft, (2) 170 kip-ft, (3) 106 kip-ft
3.4.A. W 14 × 26
3.4.B. W 21 × 44
3.4.C. W 10 × 26
3.4.D. W 24 × 55

3.4.E. W 12 × 26
3.4.F. W 14 × 34
3.4.G. W 10 × 19
3.4.H. W 16 × 36
3.4.I. W 16 × 36
3.4.J. W 16 × 26
3.5.A. (a) W 30 × 90, (b) W 30 × 108, (c) W 27 × 114
3.5.B. (a) W 24 × 55, (b) W 24 × 55, (c) W 24 × 62
3.5.C. (a) W 24 × 62, (b) W 24 × 76, (c) W 24 × 76
3.5.D. (a) W 24 × 62, (b) W 21 × 68. (c) W 24 × 76
3.6.A. 220 kips
3.6.B. 78.5 kips
3.6.C. 54.8 kips
3.7.A. (a) 0.800 in., (b) 0.9 in.
3.7.B. (a) 0.692 in., (b) 0.7 in.
3.7.C. (a) 0.829 in., (b) 0.8 in.
3.7.D. (a) 0.881 in., (b) 0.9 in.
3.8.A. (a) W 16 × 57, (b) W 10 × 88
3.8.B. (a) W 16 × 26, (b) W 10 ×45
3.8.C. (a) W 21 × 50, (b) W 18 × 86
3.8.D. (a) W 30 × 90, (b) W 21 × 122
3.8.E. (a) W 12 × 16 (b) W 10 × 19
3.8.F. (a) W 30 × 108, (b) W 18 × 158
3.8.G. (a) W 24 × 76 (b) W 21 × 83
3.8.H. (a) W 14 × 34, (b) W 14 × 34
3.9.A. 26K7
3.9.B. 30K7
3.9.C. (a) 24K4, (b) 20K7
3.9.D. (a) 20K3, (b) 16K6
3.10.A. WR20
3.10.B. IR22
3.10.C. WR18
3.10.D. WR22
3.10.E. IR22 or WR22
3.10.F. WR20
3.11.A. 102 kips [455 kN]
3.11.B. For crippling, P_u = 121.3 kips [539 kN], just barely adequate so no stiffeners required
3.14.A. PL 1.75 × 10 × 14
3.14.B. PL 1.25 × 6 × 11

Chapter 4

4.3.A. 361 kips [1606 kN]
4.3.B. 921 kips [4097 kN]

4.3.C. 401 kips [1957 kN]
4.3.D. 717 kips [3189 kN]
4.4.A. W 8 × 31
4.4.B. W 12 × 53
4.4.C. W 10 × 68
4.4.D. W 14 × 145
4.4.E. 4 in. standard pipe
4.4.F. 5 in. standard pipe
4.4.G. 6 in. standard pipe
4.4.H. 8 in. standard pipe
4.4.I. 98 kips [436 kN]
4.4.J. 47.5 kips [211 kN]
4.4.K. HHS 6 × 6 × 3/16
4.4.L. HHS 10 × 10 × 3/16
4.4.M. 104 kips [463 kN]
4.4.N. 172 kips [765 kN]
4.4.O. 4 × 3 × 5/16
4.4.P. 8 × 6 × 3/4
4.5.A. W12 × 45
4.5.B. Complies
4.5.C. W12 × 96
4.5.D. Complies
4.5.E. W14 × 82
4.5.F. W 14 × 120
4.7.A. 23 by 26 by 2½ in. plate
4.7.B. 16 by 19 by 1–5/8 in. plate

Chapter 6

6.2.A. 19,700 lb [87.6 kN]
6.2.B. ½ in. [13 mm]
6.3.A. 56.9 kips [253 kN]; elongation is critical
6.3.B. 116.2 kips [517 kN]; elongation is critical
6.3.C. At support: $V = 5$ kips, $H = 12.5$ kips; tension in cable is 13.46 kips
6.3.D. At the left support: $V = 6.67$ kips, $H = 6.67$ kips; at the right support: $V = 3.33$ kips, $H = 6.67$ kips; maximum tension in cable is 9.43 kips
6.3.E. Combined action = 0.79, bar is OK
6.3.F. Combined action is 0.77, bar is OK

Chapter 7

7.4.A. Sample values: $CI = 2000$C, $IJ = 812.5$T, $JG = 1250$T
7.4.B. Sample values: $BI = 2828$C, $IJ = 1000$C, $IH = 2000$T
7.5.A. Same as 7.4.A.
7.5.B. Same as 7.4.B

Chapter 8

8.3.A. 6 bolts, outer plates ½ in., middle plate 11⁄16 in.
8.3.B. 6 bolts, outer plates ½ in., middle plate 11⁄16 in.
8.7.A. Rounded up, 11 in. and 5 in.
8.7.B. 4.75 in. per side

Appendix A

A.1.A. c_y = 2.6 in. [66 mm]
A.1.B. c_y = 1.75 in. [43.9 mm], c_x = 0.75 in. [18.9 mm]
A.1.C. c_y = 4.2895 in. [107.24 mm]
A.1.D. c_y = 3.4185 in. [85.185 mm], c_x 1.293 in. [32.2 mm]
A.1.E. c_y = 4.4375 in. [110.9 mm], c_x = 1.0625 in. [26.6 mm]
A.1.F. c_y = 4.3095 in. [107.7 mm]
A.3.A. I = 535.86 in.^{4} [223 × 10^6 mm^4]
A.3.B. I = 205.33 in.^{4} [80.21 × 10^6 mm^4]
A.3.C. I = 447.33 in.^{4} [198 × 10^6 mm^4]
A.3.D. I = 5.0485 in.^{4} [2.036 × 10^6 mm^4]
A.3.E. I = 205.33 in.^{4} [80.21 × 10^6 mm^4]
A.3.F. I = 682.33 in.^{4} [267 × 10^6 mm^4]
A.3.G. I = 438 in.^{4} [182 × 10^6 mm^4]
A.3.H. I = 420.1 in.^{4} [175 × 10^6 mm^4]
A.3.I. I = 1672.49 in.^{4} [696 × 10^6 mm^4]

REFERENCES

1. *Minimum Design Loads for Buildings and Other Structures,* SEI/ASCE 7–02, American Society of Civil Engineers, Reston, VA, 2003.
2. *Uniform Building Code, Volume 2: Structural Engineering Provisions,* International Conference of Building Officials, Whittier, CA, 1997.
3. *Manual of Steel Construction, Load and Resistance Factor Design,* 3d ed., American Institute of Steel Construction, Chicago, IL, 2001.
4. *Standard Specifications, Load Tables, and Weight Tables for Steel Joists and Joist Girders,* Steel Joist Institute, Myrtle Beach, SC, 2002.
5. *Steel Deck Institute Design Manual for Composite Decks, Form Decks, and Roof Decks,* Steel Deck Institute, St. Louis, MO, 2000.
6. *Simplified Building Design for Wind and Earthquake Forces,* 3d ed., James Ambrose and Dimitry Vergun, Wiley, Hoboken, NJ, 1995.
7. *Structural Steel Design, LRFD Method,* 3d ed., Jack McCormac and James Nelson, Pearson Education, Upper Saddle River, NJ, 2003.
8. Uang, C. M., S. W. Wattar, and K. M. Leet, "Proposed Revision of the Equivalent Axial Load Method for LRFD Steel and Composite Beam—Column Design," *Engineering Journal,* AISC, Fall, 1990.

INDEX

A36 steel, 13
Accuracy of computations, 9
AISC (American Institute of Steel Construction), 5
AISC Manual, 2, 5
Allowable deflection, 112
Allowable stress design (ASD), 3, 50, 439
American Institute of Steel Construction (AISC), 5
American Society of Civil Engineers (ASCE), 6
American Socity for Testing and Materials (ASTM), 6
Angles, 20
  double, 178, 436
  gage for, 281
  properties of cross sections, 434, 436
  single, 20, 434
  in tension, 222
Approximate investigation, 68
Arc welding, 298
ASD (allowable stress design), 3, 50, 439
Assemblage of steel structures, 17, 31

Bar joist, 129
Bars and plates, 21
Base plate for column, 192
Beam bearing plate, 156
Beams:
  bearing plates, 156
  bending in, 76

buckling of, 85, 151
concentrated load effects in, 57, 144
connections, 292
continuous, 58
crippling of web, 316
deflection of, 108
design aids (ASD), 439
design for buckling, 89
design factor for, 73
design procedure, general, 76
end support conditions, 55
fireproofing for, 30
fixed end in, 55
flexure in, 76
flitched, 215
framed connections, 292
with internal pins, 338
lateral support for, 85, 151
load-span values for, 118
multiple span, 58, 337
plastic behavior of, 78
restrained, 55
safe load tables for, 118
shear in, 103
shear center for, 148
simple, 55
torsion in, 85, 148
typical loadings, 442
web crippling in, 144
web shear in, 103
web stiffeners for, 146
web tearing in, 278
Bearing plates:
for beams, 156
for columns, 192
Bearing pressure, 156, 192
Bending in:
columns, 180
truss chords, 260
Bending factors for columns, 172, 173, 183, 186
Bents and frames, 196
Biaxial bending in columns, 186
Biaxial bracing for columns, 169
Block shear, 276
Bolted connections, 272
bearing in, 274
bending in, 275
block shear in, 276
capacities, 279
design of, 284
framed, for beams, 292
framing with, 290
layout of, 280, 289
net section in, 274
pitch and edge distance for, 280, 289
shear in, 273
tearing in, 276
for trusses, 295
Bolts:
angle gage for, 281
capacity, 279
edge distance for, 280
high strength, 278
pitch, 280
types, 278
unfinished, 278
Bounce of floors, 39, 364
Box system for lateral load, 339
Braced frame, 207
Buckling:
of beams, 85, 89, 151
of columns, 163
inelastic, 86
torsional, 86
Building codes, 315
Built-up sections, 162

C shapes, 19, 432
Cable structures, 224
Cantilever frame, 36, 61
Captive frame, 206
Casting, 16
Ceiling structure, 37
Centroid, 413
Channels, 19, 432

Chevron bracing, 209
Choice of building construction, 312
Choosing design methods, 73
CMU (concrete masonry unit), 332, 379
Cold-formed steel products, 2, 22
Columns:
  base plates for, 192
  bending factor for, 172, 173, 183, 186
  bending in, 180, 183
  biaxial bending in, 186
  biaxial bracing for, 169
  buckling, 163
  built-up, 162
  connections, 188
  critical stress for, 166
  design of, 170, 367
  eccentrically loaded, 184
  effective length for, 163
  framing, 188
  interaction, 182
  K-factor, 163
  load determination for, 365
  multistory, 367
  P-delta effect, 183
  pipe shapes, 176, 437
  safe loads, 164
  sections, 161
  shapes, 163
  slenderness, 163
  splices for, 191
  structural tubing shapes, 178
  strut, 178
  W shapes, 170
Combined loads, 318, 326
Compact section, 84
Composite elements142, 214
Computations, structural, 9
Concentrated load, effect in beams, 57, 144
Concrete, strength in bearing, 156, 192
Connections, 25, 269
  basic considerations, 269
  beam, 292
  bolted, 272
  column, 188
  control joint, 307, 338
  field, 32, 271
  framed, for beams, 292
  moment-resisting, 61, 197
  shop, 32, 271
  special concerns, 271
  structural functions, 270
  tearing in, 276
  tension, 282
  for trusses, 295
  types of, 270
  welded, 298
Continuity of:
  beams, 58
  columns, 367
Continuous action of beams, 58
Conversion of units, 9
Core bracing systems, 358
Corrosion, 29
Cost factors for steel structures, 32, 329
Crippling of beam webs, 316
Cut section, 48

Dead load, 315
Decks:
  floor, 137
  formed sheet steel, 140
  roof, 139
Deck-beam-girder system, 35, 335, 359
Deflection:
  allowable, 112
  approximation graph for steel beams,
  of beams, 108, 364
  of column-beam bents,
  computations for, 112
  limits, 112
Deformation limits, 29, 112
Delta truss, 402

Design method, 3
  choice of, 73
Design references, 5, 313
Designations for steel elements:
  bolts, 278
  formed sheet steel decks, 142
  joist girders, 135
  open web joists, 130
  rolled shapes, 22
Double angles, 436
  as struts, 178
  as truss members, 296, 348
Double shear, 274
Dual bracing, 204
Ductility of steel, 13, 78
Dynamic behavior, 39

Earthquake load, 318
Eccentrically braced frames, 210
Eccentrically loaded columns, 184
Electric arc welding, 298
Equivalent loading:
  for approximation of deflection, 128
  axial, for column with bending, 183
Equivalent tabular loading, 128
Extrusion, 16

Fabricated steel products, 23
Factored load, 326
Field connection, 32
Fillet weld, 299
Fireproofing, 30, 318
Flitched beam, 215
Floor framing, 359
Floors:
  framing for, 359
  as horizontal diaphragms, 392
  loads on, 321
Forging, 16
Formed sheet steel, 2, 140
Forming processes, 16
Framed beam connections, 292
Frames and bents, 41, 60, 196
Framing:
  for beams, 292
  with bolts, 290
  for columns, 188
  for floors, 359, 384
  for roofs, 334
  with welds, 307
Free-body diagram, 47

Gabled roof:
  with rigid frame bent, 349, 394
  with truss, 346, 389
Gage lines for angles, 281
Geometric shapes. Properties, 427
Girders:
  in deck-beam systems, 362
  joist, 135, 352
Grades of steel, 13

High-strength bolts, 278
Hinge, plastic, 81
Hole:
  in floors and roofs, 40
  in horizontal diaphragm, 393
Horizontal bracing, 211, 393
Hot-rolled products, 2

I-beams, 19
Indeterminate structures, approximate investigation of, 68
Inelastic:
  behavior, 77
  buckling, 86
  stresses, 78
Interaction in columns, 182
Internal forces in trusses, 242
Investigation of:
  beams, 53
  columns, 53, 164
  rigid frames, 60
  structures, general, 45
  trusses, 242, 346, 389

Joist:
  bar, 130
  open web, 129, 334, 350, 368
Joist girder, 135, 352, 370

K-bracing, 208
K factor for column, 163
Knee bracing, 208

Lateral bracing:
  for beams, 85, 151
  box system, 339
  for columns, 168
  rigid frame bents, 197, 397
  trussed bents, 207, 358, 372
  for trusses, 238
  for wind and earthquake forces, 339, 358
Lateral loads, 322
Laterally unbraced length, 85
Least weight selection, 77
Light-gage steel products, 140
Live load, 317, 318
  element factor, 320
  reduction of, 321
Load and resistance factor design (LRFD), 3, 51
Load sharing in lateral bracing, 202
Loads, 314
  building code, 315
  combinations, 318, 326
  computation, 326, 364
  dead, 315
  duration, 318
  earthquake, 318, 325
  equivalent axial, 184
  equivalent tabular, 128
  factored, 326
  floor, 321
  lateral, 322
  live, 317, 318
  movable partitions, 322
  periphery, 326, 364
  ponding, 321
  roof, 319
  seismic, 318
  service, 72
  tributary, 327, 364
  wind, 317, 322, 341, 373, 390
LRFD (load and resistance factor design), 3, 51

Manual of Steel Construction (AISC Manual), 2
Manufactured:
  systems, 213
  trusses, 129
Maxwell diagram for truss, 244
Measurement, units of, 6
Method of joints, 242, 249
Methods of investigation and design, 3, 50
Miscellaneous metals, 2
Mixed frame and wall systems, 44, 201
Modulus of elasticity, 417
Modulus, section, 425
  elastic, 425
  plastic, 426
Moment:
  connection, 61, 197
  elastic limit, 79
  plastic, 80
Moment of inertia, 417
Movable partitions, 322
Multistory structures, 354
  wind forces on, 358, 372

Net section in tension, 222, 274
Neutral axis, 414
Nomenclature, 10
Notation, standard, 10

Offset grid truss form, 402
Open web steel joist, 129, 334, 350, 368
  safe load table for, 131

P-delta effect, 183
Partitions, movable, 322
Perimeter (peripheral) bracing, 358
Peripheral load, 326, 366
Pin connection in beam, 338
Pipe, steel, 437
Pipe column, 176, 437
Pitch, of bolts, 280
Planning of framing, 33, 328
Plastic:
  hinge, 81
  moment, 80
  range of stress and strain, 78
  section modulus, 81
Plates and bars, 21
Plug weld, 306
Ponding, on flat roof, 111
Properties for designing:
  angles, 434
  double angles, 436
  geometric shapes, 427
  pipe, 437
  W shapes, 429
Properties of sections: 413
  centroid, 413
  moment of inertia, 417
  neutral axis, 414
  radius of gyration, 426
  section modulus, 425
  statical moment, 414
Properties of steel, 12

Radius of gyration, 426
Reactions, 47
Reduction of live load, 321
Reference sources for design, 5, 313
Relative stiffness of bracing elements, 202
Resistance factor, 52, 73
Response values for beams, 56 to 60, 442
Restrained beams, 55
Rigid frame, 43, 60, 197, 397
Rolled shapes, 16
  as beams, 75
  in built-up sections, 162, 422
  as columns, 161
Roof framing, 334
Roof live load, 319
Roofs,
  drainage for, 39
  loads on, 319
Roof trusses, 346, 389
Rust, 29

S shapes, 19
Safe load tables:
  beams, 118
  bolts, 279
  columns:
    double angle struts, 181
    pipe, 177
    tubes, 179
    W shapes, 172
  fillet welds, 301
  open web joists, 131
  roof deck, 139
  struts, double angle, 181
  W shape beams, 120
Safety, 46
Section, cut, 48
Section modulus, 425
  elastic, 425
  plastic, 81, 426
Section properties, *see* Properties of sections
Service conditions versus limit states, 72
Service load, 72
Shapes, structural:
  cold-formed, 22
  designations for, 21
  rolled, 17
Shear:
  in beams, 103
  in bolts, 273
  center, 148
  double, 274

single, 274
stud, 214
Sheet steel, 2, 140
Shop assemblage, 32
Sidesway, 69
Simple beam, 55
Single shear, 274
Single-span bent, 69
Slenderness of columns, 163
Slot weld, 306
Sources of information, 5, 313
Splice, of column, 191
Stability, 28
S shapes, 19
Staggered bolts, 283
Standard channels, 19
Statical moment, 414
Steel:
A36, 13
bolted connections, 272
cold-formed, 2
corrosion resistance, 29
deck, 140
grade, 13
light gage, 2
properties of, 12
sheet, 2
stress-strain behavior, 13
uses of, 1
Steel Deck Institute, (SDI), 6
Steel Joist Institute (SJI), 6
Steel products, 16
Stiffeners for beam webs, 146
Strain hardening, 79
Strength method, 51
Stress-strain behavior of steel, 13, 26
Stress-strain diagram, 14
Structural:
design standards, 313
investigation and design, 50
planning, 33, 328
tees, 20
tubing, 178
Struts, 178
Symbols, standard, 10

Tearing in bolted connections, 276
Tees, structural, 20
Tension:
and bending, 231
cable, 224
connections, 282
effective area in, 222
elements, 220
on net section of members, 222, 274
Three-dimensional frames, 41, 199
Torsion, 148
Torsional buckling, 85, 152
Tributary area, for load, 327
Trussed bent bracing system, 207, 372
Trusses, 234
algebraic analysis, 249
bolted connections for, 295
bracing for, 238
coefficients for internal forces, 257
combined stress in truss chords, 260
delta, 402
design considerations for, 257, 260, 389
graphical analysis, 242
gusset plate, 296
internal forces in, 242
joints in, 295
joist girder, 135
as lateral bracing, 207, 358, 372
loads on, 239
manufactured, 129
Maxwell diagram for, 244
member design, 348, 392
offset grid, 402
open web joist, 129
roof, 346, 389
two-way, 262, 394
types, 237

Trusses (*Continued*)
use of, 234
weight, 241
welded connections for, 296
Truss systems, 43
Tubular steel members, 178
Two-way trusses, 262, 394

Ultimate strength, 79
method, 3
Unfinished bolt, 278
Units of measurement, 6
conversion of, 9
Upset end, on rod, 280
Use of steel for structures, 1, 26

V-bracing, 209

W shapes, 18, 429
Web crippling of beams, 144
Weight of building materials, 316
Welded connections, 298
design of, 303
electric arc, 298
framing, 307
for trusses, 296
Welding, 298
Welds:
boxing of, 302
butt, 299
fillet, 299
groove, 299
penetration of, 299
plug and slot, 306
size limits, 302
strength of, fillet, 301
stress in, 299
Wide flange (W) shapes, 18, 429
Wind loads, 317, 322, 341, 373, 390
Wire, 16
Working stress method, 3

X-brace, 208

Yield point, 13, 78

Printed in the United States
By Bookmasters